BIO-BASED POLYMERS AND COMPOSITES

Properties, Durability, and Applications

BIO-BASED POLYMERS AND COMPOSITES

Properties, Durability, and Applications

Edited by

Amadou Belal Gueye

Sabu Thomas, PhD

Nandakumar Kalarikkal, PhD

Modou Fall, PhD

APPLE ACADEMIC PRESS

First edition published 2025

Apple Academic Press Inc.
1265 Goldenrod Circle, NE,
Palm Bay, FL 32905 USA

760 Laurentian Drive, Unit 19,
Burlington, ON L7N 0A4, CANADA

CRC Press
2385 NW Executive Center Drive,
Suite 320, Boca Raton FL 33431

4 Park Square, Milton Park,
Abingdon, Oxon, OX14 4RN UK

Library and Archives Canada Cataloguing in Publication

Title: Bio-based polymers and composites : properties, durability, and applications / edited by Amadou Belal Gueye, Sabu Thomas, PhD, Nandakumar Kalarikkal, PhD, Modou Fall, PhD.
Names: Gueye, Amadou Belal, editor. | Thomas, Sabu, editor. | Kalarikkal, Nandakumar, editor. | Fall, Modou (Lecturer in chemistry), editor.
Description: First edition. | Includes bibliographical references and index.
Identifiers: Canadiana (print) 20240307062 | Canadiana (ebook) 20240307143 | ISBN 9781774915325 (hardcover) | ISBN 9781774915332 (softcover) | ISBN 9781032669243 (ebook)
Subjects: LCSH: Biopolymers. | LCSH: Nanocomposites (Materials) | LCSH: Nanostructured materials.
Classification: LCC QP801.B69 B56 2024 | DDC 572/.33—dc23

Library of Congress Cataloging-in-Publication Data

CIP data on file with US Library of Congress

ISBN: 978-1-77491-532-5 (hbk)
ISBN: 978-1-77491-533-2 (pbk)
ISBN: 978-1-03266-924-3 (ebk)

About the Editors

Amadou Belal Gueye
Research Scholar, School of Chemical Sciences, Mahatma Gandhi University, Kottayam, Kerala, India

Amadou Belal Gueye is a Research Scholar at the School of Chemical Sciences at Mahatma Gandhi University, Kottayam, Kerala, India. He received his bachelor's degree in Physics-chemistry and master's degree in Physical Chemistry Applied to Energy and Analysis from Cheikh Anta Diop University of Dakar, Senegal. He works, within the framework of his thesis, in the field of energy storage.

Sabu Thomas, PhD
Chairman, Trivandrum Engineering Science & Technology Research Park (TrEST Research Park), Kerala, India

Sabu Thomas, PhD, is currently the Chairman of Trivandrum Engineering Science & Technology Research Park (TrEST Research Park). He is also the former Vice Chancellor Mahatma Gandhi University, Kottayam, Kerala; Director School of Energy Materials; Director School of Nanoscience and Nanotechnology; Director International and Inter-University Centre for Nanoscience and Nanotechnology; former Director School of Chemical Sciences; founder School of Polymer Science and Technology, Mahatma Gandhi University, Kottayam, Kerala, India; retired full professor of Polymer Science and Engineering at the School of Chemical Sciences of Mahatma Gandhi University, Kottayam, Kerala, India. Prof. Thomas is an outstanding leader with sustained international acclaims for his work in nanoscience, polymer science and engineering, polymer nanocomposites, elastomers, polymer blends, interpenetrating polymer networks, polymer membranes, green composites, nanomedicine and green nanotechnology. Professor Thomas has received a number of national and international awards which include: fellowship of the Royal Society of Chemistry, London FRSC; Distinguished Professorship from Josef Stefan Institute, Slovenia; MRSI medal; Nano Tech Medal; CRSI medal; Distinguished Faculty Award; Dr. APJ Abdul Kalam Award for Scientific Excellence 2016; Mahatma Gandhi University Award for Outstanding Contribution, Nov. 2016; Lifetime Achievement Award of the Malaysian Polymer Group; Indian Nano Biologists Award 2017; and Sukumar Maithy Award for the Best Polymer Researcher in the country. Due to his outstanding contributions to the field of Nanoscience and Polymer Science and Engineering, Prof. Thomas has been conferred Honoris Causa (D.Sc) Doctorate by the University of South Brittany, Lorient, France and University of Lorraine, Nancy, France. Professor Thomas has published over 1000 peer reviewed research papers, reviews, and book chapters. He has co-edited more than 120 books published by Royal Society, Wiley, Wood head, Elsevier, CRC Press, Springer, and Nova etc. He is the inventor of six patents. The H-index of Prof. Thomas is 103 and has more than 50,000 citations. Prof. Thomas has delivered over 300 plenary/inaugural and invited lectures in national/international meetings in over 30 countries.

Nandakumar Kalarikkal, PhD
Director of School of Nanoscience and Nanotechnology, Mahatma Gandhi University, India.

Nandakumar Kalarikkal, PhD, is the Director of School of Nanoscience and Nanotechnology, Mahatma Gandhi University, India. Additionally, he is the Chair of the School of Pure and Applied Physics. He was also, formerly, the director of that school. Prof. Nandakumar is Honorary Director of International and Inter University Centre for Nanoscience and Nanotechnology, Mahatma Gandhi University, India. His research group is specialized in synthesis, characterization and applications of various nanomaterials. The different topics of current research works include nanomultiferroics, nanosemiconductors and nanophosphors, nanocomposites, nanomedicine, nanosensors, ion beam irradiation effects and phase transitions and laser ablation and nonlinear optics using Nd-YAG Laser. He is the recipient of research fellowships and associateships from prestigious government organizations such as the Department of Science and Technology and Council of Scientific and Industrial Research of Government of India. He has active collaborations with national and international scientific institutions in India, South Africa, Slovenia, Canada, France, Germany, Malaysia, Australia and US. He has co-edited more than 20 books published by Royal Society, Wiley, Wood head, Elsevier, CRC Press, Springer, and Nova etc. He has over 390 publications with more than 68,000 citations and his h-index is 43.

Prof. Modou Fall, PhD

Director, Doctoral School of Physics, Chemistry, Sciences of the Earth, of the Universe and Engineering, Cheikh Anta Diop University of Dakar, Senegal.

Prof. Modou Fall, PhD, is the Director of the Doctoral School of Physics, Chemistry, Sciences of the Earth, of the Universe and Engineering at Cheikh Anta Diop University of Dakar, Senegal. His main research interests include the synthesis, characterization and applications of conducting polymers, energy storage and conversion, corrosion studies and electroanalysis of hazardous compounds. He has supervised five PhD theses, and over 15 master's theses. Prof. Fall is the Secretary General of the Senegalese Committee for Chemistry, and is titular member of the IUPAC Division of Physical and Biophysical Chemistry.

Contents

Contributors .... *xi*

Abbreviations .... *xv*

Preface .... *xix*

1. **Biopolymer-Derived Superabsorbent for Environmental Sustainability: A Review 1**
   Sweta Sinha

2. **Applications of Green Nanocomposites in Polycyclic Aromatic Hydrocarbons (PAHs) Bioremediation: Current Status and Future Direction 21**
   Paramita Chakravarty, Hemen Deka, and Devasish Chowdhury

3. **Lignin- and Chitosan-Based Biosorbents for Waste Water Treatment 51**
   Jyothy G. Vijayan

4. **Advancement of Natural Polymers from Waste and Their Biotechnological Applications 77**
   Mohammad Amir, Naushin Bano, Aamina Shahab, Shareen Fatima Rizvi, and Roohi

5. **Pectin–Carbon Nanosphere: A Green Nanocomposite for the Sensing of Glucose 109**
   S. K. Suja, J. Shamili, and S. Geethanjali

6. **Structural and Functional Diversities of Lactic Acid Bacterial Polysaccharide 129**
   Bindhumol Ismail, Soumya M. P., Reeba Parameswaran, K. Madhavan Nampoothiri, A. Jayakumaran Nair, and A. Gangaprasad

7. **Comparison of Thermal Resistance by Insulation Board Made from Rice Husk and Bagasse Waste 163**
   Sowjanya Madireddi, Deva Moukthika K., Archana C., and Kavya Ch.

8. **Effect of Micro- and Nanofillers on Polarization, Depolarization Currents, and Dielectric Properties in Fiber-Reinforced Epoxy Composites 177**
   Suchitra M., Geethashree A., and Panchami S. V.

9. Ni/Co/Cu and TiO_2 Incorporate Ragi Husk-Derived Biosilica Nanocomposite as Photocatalyst and Adsorbent for the Degradation of Methyl Blue Dye 197
A. Ananthi, D. Geetha, and P. S. Ramesh

10. Effect of Zinc Oxide Nanoparticles on Polymerization Shrinkage and Mechanical Properties of ORMO-48 211
Lizymol P. P., Vibha C., Deepu D. R., and Sonalilaxman Waghmare

11. Structural and Optical Characterization of PVP-Stabilized Ag-Doped CuO Nanoflakes via Hydrothermal Method 225
P. S. Ramesh and D. Geetha

12. PVA-Stabilized Zn-Doped CoS/Graphene Nanocomposites as Active Electrode Materials for Supercapacitors and Excellent Nano-Photocatalists 243
P. Priya, D. Geetha, and P. S. Ramesh

13. Antimicrobial Activity of Poly Vinyl Pyrrolidine-Assisted Pure and Transition Metals (Fe/Zn/Ag)-Doped CuO Nanocomposites 253
D. Geetha, P. S. Ramesh, and K. Kannaki

14. Effect of Sulfur Sources on the Properties of PVA-Stabilized Indium Sulfide for Photo Degradation Application 265
Lilly Mary, D. Geetha, and P. S. Ramesh

15. PLA Stabilized/Graphene-Based Copper Sulfide-Hole Transport Material for Organic Photovoltaics 275
V. Menaka, D. Geetha, and P. S. Ramesh

16. Structural Properties of Ag-Doped PVP-Stabilized SnO_2 Engineered with TiO_2 for Gas Sensing Applications 287
V. Vidhya, P. Anbarasu, and D. Geetha

17. Struvite Fertilization in Horticulture as the Closed Phosphorus Cycle 299
Anna Jama-Rodzeńska, Józef Sowiński, Katarzyna Adamzcewska-Sowińska, and Andrzej Dziuba

Index 307

Contributors

Geethashree A.
Vidyavardhaka College of Engineering, Mysuru, Karnataka, India

Mohammad Amir
Protein Research Laboratory, Department of Bioengineering, Integral University, Lucknow, Uttar Pradesh, India

A. Ananthi
Department of Physics, Annamalai University, Tamil Nadu, India

P. Anbarasu
Department of Electronics, GAC, Trichy, Bharathidasan University, Tiruchirappalli, Tamil Nadu, India

Naushin Bano
Protein Research Laboratory, Department of Bioengineering, Integral University, Lucknow, Uttar Pradesh, India

Archana C.
CVR College of Engineering, Hyderabad, India

Vibha C.
Division of Dental Products, BMT Wing, Sree Chitra Tirunal Institute of Medical Sciences and Technology (An Institute of National Importance under D.S.T., Government of India), Thiruvananthapuram, Kerala, India

Kavya Ch.
CVR College of Engineering, Hyderabad, India

Paramita Chakravarty
Ecology and Environmental Remediation Laboratory, Department of Botany, Gauhati University, Guwahati, Assam, India

Devasish Chowdhury
Institute of Advanced Study in Science and Technology, Paschim Boragaon, Guwahati, Assam, India

Hemen Deka
Ecology and Environmental Remediation Laboratory, Department of Botany, Gauhati University, Guwahati, Assam, India

Deepu D. R.
Division of Dental Products, BMT Wing, Sree Chitra Tirunal Institute of Medical Sciences and Technology (An Institute of National Importance under D.S.T., Government of India), Thiruvananthapuram, Kerala, India

Andrzej Dziuba
Krevox Europejskie Centrum Ekologiczne Sp. z o.o., Żurawia, Warszawa, Poland

A. Gangaprasad
Department of Biotechnology, University of Kerala, Kariavattom, Trivandrum, Kerala, India

D. Geetha
Department of Physics, Annamalai University, Annamalai Nagar, Chidambaram, Tamil Nadu, India

S. Geethanjali
Department of Chemistry, Lady Doak College, Madurai, Tamil Nadu, India

Bindhumol Ismail
Department of Biotechnology, University of Kerala, Kariavattom, Trivandrum, Kerala, India

Deva Moukthika K.
CVR College of Engineering, Hyderabad, India

K. Kannaki
Government Womens College, Bargor, Tamil Nadu, India

Suchitra M.
Vidyavardhaka College of Engineering, Gokulam 3rd stage, Mysuru, Karnataka, India

Sowjanya Madireddi
CVR College of Engineering, Hyderabad, India

Lilly Mary
Department of Physics, Annamalai University, Chidambaram, Tamil Nadu, India

V. Menaka
Department of Physics, Annamalai University, Chidambaram, India

Soumya M. P.
Microbial Processes and Technology Division, CSIR-National Institute for Interdisciplinary Science and Technology (CSIR-NIIST), Trivandrum, Kerala, India
Academy of Scientific and Innovative Research (AcSIR), Ghaziabad, India

A. Jayakumaran Nair
Department of Biotechnology, University of Kerala, Kariavattom, Trivandrum, Kerala, India

K. Madhavan Nampoothiri
Microbial Processes and Technology Division, CSIR- National Institute for Interdisciplinary Science and Technology (CSIR-NIIST), Trivandrum, Kerala, India
Academy of Scientific and Innovative Research (AcSIR), Ghaziabad, India

Lizymol P. P.
Division of Dental Products, BMT Wing, Sree Chitra Tirunal Institute of Medical Sciences and Technology (An Institute of National Importance under D.S.T., Government of India), Thiruvananthapuram, Kerala, India

Reeba Parameswaran
Microbial Processes and Technology Division, CSIR-National Institute for Interdisciplinary Science and Technology (CSIR-NIIST), Trivandrum, Kerala, India

P. Priya
Department of Physics, Annamalai University, Chidambaram, Tamil Nadu, India

P. S. Ramesh
Thiru Kolanjiappar Government Arts College, Virudhachalam, Tamil Nadu, India

Shareen Fatima Rizvi
Protein Research Laboratory, Department of Bioengineering, Integral University, Lucknow, Uttar Pradesh, India

Anna Jama-Rodzeńska
Institute of Agroecology and Plant Production, Wroclaw University of Environmental and Life Sciences, Wroclaw, Poland

Roohi
Protein Research Laboratory, Department of Bioengineering, Integral University, Lucknow, Uttar Pradesh, India

Aamina Shahab
Protein Research Laboratory, Department of Bioengineering, Integral University, Lucknow, Uttar Pradesh, India

J. Shamili
Department of Chemistry, Lady Doak College, Madurai, Tamil Nadu, India

Sweta Sinha
Department of Chemistry, Amity University Jharkhand, Ranchi, Jharkhand, India

Katarzyna Adamzcewska-Sowińska
Department of Horticulture, Wroclaw University of Environmental and Life Sciences, Wroclaw, Poland

Józef Sowiński
Institute of Agroecology and Plant Production, Wroclaw University of Environmental and Life Sciences, Wroclaw, Poland

S. K. Suja
Department of Chemistry, Lady Doak College, Madurai, Tamil Nadu, India

Panchami S. V.
Vidyavardhaka College of Engineering, Mysuru, Karnataka, India

V. Vidhya
Department of Electronics, GAC, Trichy, Bharathidasan University, Tiruchirappalli, Tamil Nadu, India

Jyothy G. Vijayan
Department of Chemistry, M. S. Ramaiah University of Applied Sciences, Bengaluru, India

Sonalilaxman Waghmare
Division of Dental Products, BMT Wing, Sree Chitra Tirunal Institute of Medical Sciences and Technology (An Institute of National Importance under D.S.T., Government of India), Thiruvananthapuram, Kerala, India

Abbreviations

ADA	American Dental Association
AFM	atomic force microscopy
AN	ammonium nitrate
ANOVA	analysis of variance
ATH	alumina trihydrate
CNS	carbon nanosphere
CuO	copper oxide
CV	crystal violet
CV	cyclic voltammetry
DLS	dynamic light scattering
DP	deacidified peat
DTS	diametral tensile strength
DW	deionized water
EDS	energy dispersive spectrometer
EIS	electrochemical impedance spectroscopy
EPA	Environmental Protection Agency
EPS	equilibrium percentage swelling
EPS	exopolysaccharides
FESEM	field emission scanning electron microscope
FRP	fiber-reinforced polymer
FT-IR	Fourier transform infrared spectroscopy
FWHM	full width half maximum
GB	green band
GOx	glucose oxidase
GRAS	generally regarded as safe
HAP	hydroxy apatite
IPNs	interpenetrating polymer networks
ITZ	interfacial transition zone
KBDD	Kirby–Bauer disk diffusion
LAB	lactic acid bacteria
LOD	limit of detection
LOQ	limit of quantification
MALLS	multi-angle laser light scattering
MB	methylene blue

MBA	methylenebisacrylamide
MH	Mueller Hinton
MHA	Mueller Hinton Agar
MO	methyl orange
MOS	metal–oxide semiconductor
MS	mineral soil
NaOH	sodium hydroxide
NBE	near band edge
NC	nanocrystals
NCA	natural coarse aggregate
PAHs	polycyclic aromatic hydrocarbons
PC	polycarbonate
PDC	polarization and depolarization currents
PE	polyethylene
PEEK	poly(etheretherketone)
PEG	poly(ethylene glycol)
PHA	polyhydroxyalkanoates
PI	polyimide
PL	photoluminescence spectra
PLA	polylactic acid
PLGA	poly(lactic-co-glycolic acid)
PMRAC	polymer-modified recycled-aggregate-concrete
PP	polypropylene
PS	potassium sulphate
Pt	platinum wire
PVC	poly(vinyl chloride)
PVE	plasma volume expander
PVP	polyvinylpyrrolidone
P3(HB)	poly(3-hydroxybutyrate)
RB	rose bengal
RBA	rose bengal agar
RCA	recycled coarse aggregate
SBR	styrene butadiene rubber
SCE	saturated calomel electrode
SCFAs	short-chain fatty acids
SEM	scanning electron micrographs
STEM	spirit TEM machine with optional scanning mode
STR	struvite
SUP	superphosphate

TEM	transmission electron micrographs
TM	transition metal
TPE	thermoplastic elastomer
TPS	thermoplastic starch
TW	total weight of the mixture
UN	United Nations
VLC	visible light cure
VOCs	volatile organic compounds
XRD	X-ray diffraction
ZnO	zinc oxide nanoparticles
ZnONPs	zinc oxide nanostructure

Preface

Nature is the source of innumerable and priceless resources. Research on renewable resources for biocomposites and bio-nano composites is expanding, and significant advancement has been made. A polymer is synthesized by connecting small molecules into a single giant molecule through a chemical approach. Monomers are the tiny molecules that are employed to create a polymer. These polymers can be prepared using condensation or addition polymerization. Numerous polymers, including proteins and nucleic acids, are present in our body. Another organic polymer that is essential to plants' structural integrity is cellulose. The bulk of polymers are synthesized from monomers via the condensation polymer process, which results in the production of water as a byproduct.

The latest breakthroughs in biopolymers, green composites, and green nanocomposites are covered in this book. Nanoscale materials have many fascinating properties that are ideal for different uses. This book consolidates information on the synthesis, and applications of biopolymers, green composites, and green nanocomposites for various applications with detailed characterization, mechanical approaches, and theoretical considerations. Recent developments in advanced devices using nanostructured materials are also covered. This book provides fundamentals and advanced concepts to the readers for developing nanostructured materials based on biopolymers and their composites.

—**Amadou Belal Gueye**

CHAPTER 1

Biopolymer-Derived Superabsorbent for Environmental Sustainability: A Review

SWETA SINHA

Department of Chemistry, Amity University Jharkhand, Ranchi, Jharkhand, India

ABSTRACT

Increased environmental degradation and solid waste production has resulted in a higher demand for biodegradable materials based on biopolymers. Superabsorbents are three-dimensional cross-linked polymeric material, which allow huge amounts of liquid to absorb. Synthetic SABs are extensively used as an absorbent of aqueous solutions for various applications, mainly as adult incontinence, feminine hygiene products etc. The major components of synthetic SABs are acrylic/methacrylic acid. According to the report of the US Environmental Protection Agency, the acrylamide used in the production of synthetic superabsorbent is extremely carcinogenic when exposed to people. On the other hand, a large number of synthetic superabsorbent polymers are quite resistant to the environmental degradation process, and thus responsible for the accumulation of solid waste material. However, the presence of hazardous and undedgraded content in the synthetic SABs leads to the designing of biopolymer-based superabsorbents. These developed hybrid materials of SABs are a combination of renewable raw materials and synthetic polymers. Biopolymer-based superabsorbents have a lot of potential due to their excellent biocompatibility, biodegradability, and nonlethal qualities. This review paper discusses the general perspective, characterization, and synthesis strategies for the production of hybrid superabsorbent. In

Bio-Based Polymers and Composites: Properties, Durability, and Applications. Amadou Belal Gueye, Sabu Thomas, Nandakumar Kalarikkal, & Modou Fall (Eds.)

addition, the methodologies for illustrating these superabsorbent hydrogels, as well as their prospective uses, are discussed.

1.1 INTRODUCTION

Growing environmental challenges lead to the demand of green materials based on natural polymers. Natural polymers can be obtained from a variety of natural feedstocks such as vegetables (e.g., starch, zein, cellulose, and pectin), livestock, and dairy products (e.g., whey protein, chitosan, casein, and gelatin). Biopolymer modification has been reported in a variety of contexts, including the generation of new biopolymer-derived materials also called hybrid materials.[16]

Polymer modifications are an excellent opportunity to enhance bulk and surface features of the biopolymers. It also expands the range of applications and opens up new possibilities. There are several techniques applied to modify the surface and chemical characteristics of a polymeric material, such as grafting, crosslinking, drying, and mixing (Figure 1.1).

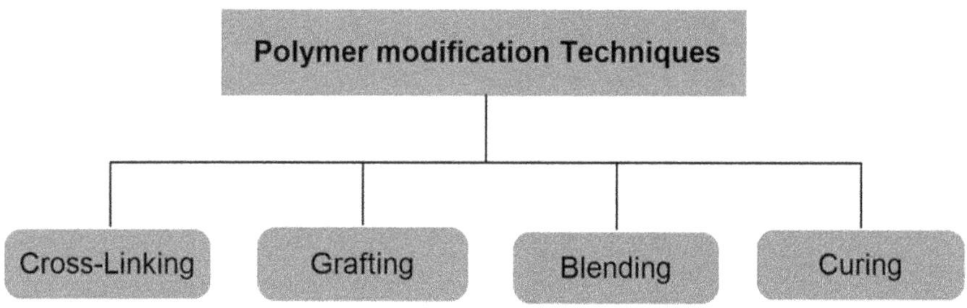

FIGURE 1.1 Schematic diagram represents various methods of polymer modification.

1.1.1 BIOPOLYMER-BASED SUPERABSORBENT: A CROSS-LINKED POLYMER

The maximum number of superabsorbent may be synthesized by cross-linking of biological molecules (proteins/polysaccharides) with crosslinking agents that covalently bonded with multi-functional monomeric units in a three-dimensional hydrophilic network. These hydrophilic functional groups attached with polymeric backbone provide capability to absorb and hold a huge amount of water, whereas cross-links between network chains provide resistance to dissolution and maintain the physical dimension structure of

SABs (Figure 1.2).[11–14] Urine, sweat, and blood are biological solutions that can be absorbed by SABs up to several times of their own weight in a short amount of time, but they do not dissolve in water.[7] The swelling behavior of superabsorbent is primarily influenced by the type of polymers used and the degree of crosslinking. Chemical crosslinking involves covalent interactions between one or more comonomers, whereas physical crosslinking involves ionic, hydrogen, or Van der Waals interactions between the polymeric chain.

Superabsorbents are used in healthcare products, as matrixes for drug delivery systems, agriculture and horticulture, wastewater treatment, and other applications.[9,10]

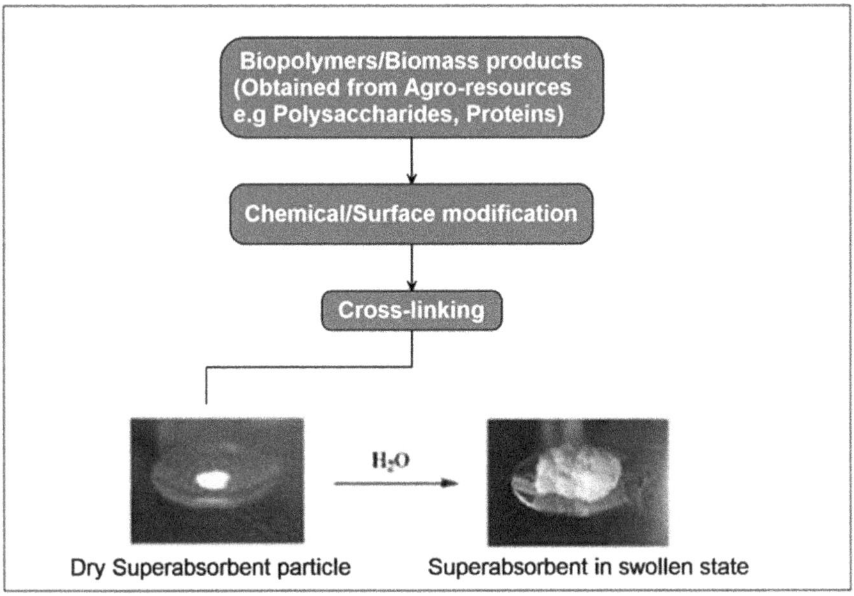

FIGURE 1.2 Illustration of typical superabsorbent in dry and swollen states.

1.2 CLASSIFICATION

Superabsorbent can be classified:

- Based on the sources of polymer:
  Natural and synthetic (petrochemical-based products)[17] (Figure 1.3),
- Based on the crosslinking techniques,
  Physical and chemical crosslinking.[18]

Cotton, wood pulp, and plant fibers are examples of natural absorbents. According to the Gale Encyclopedia of Science, natural absorbent materials like cotton and pulp estimated water retention potential ranging up to 20 times of their dry weight until the 1980s. The water-holding capacity improved after the introduction of synthetic SABs, leading to the rapid growth of many industries. Synthetic SABs are derived from petroleum products,[19] and many of them are synthesized by crosslinking unsaturated carboxylic acid monomers with acrylamide, such as acrylic or methacrylic acid. Several organizations, however, have raised objections about the presence of hazardous content in the synthetic SABs. The acrylamide used in the production of human sanitary items and diapers is extremely carcinogenic when exposed to people, according to the US Environmental Protection Agency[20] in a chemical profile for acrylamide. Neurotoxicity is another negative effect of acrylamide, which results in drowsiness, parathesias in the fingers, numbness, and ataxia.

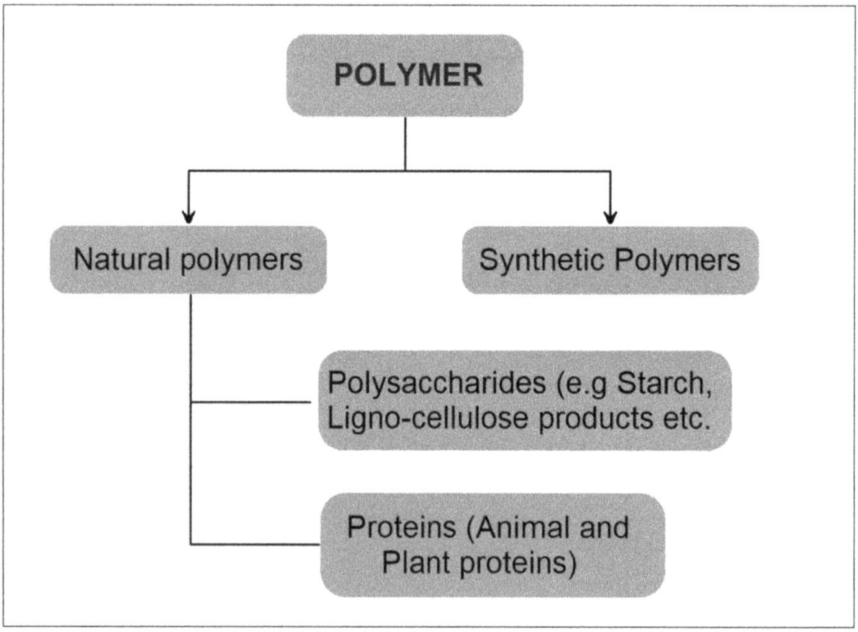

FIGURE 1.3 Occurrence-based classification of polymer.

Synthetic SAB material is petroleum-based chemical derivatives with a high cost due to the scarcity and high cost of petroleum. Synthetic SABs have raised a lot of interest despite their widespread use and production due to the advantage of being able to modify mechanical characteristics and produce a wider variety of materials with complex functions. However, in the development of these synthetic SABs, biodegradability, biocompatibility and cell-material interactions may be an issue. Natural polymers like polysaccharides and proteins have intrinsic macromolecular properties similar to the extracellular matrix, but they can also induce complex cell–material interactions. Biopolymer-based SABs are generally prepared by grafting a variety of synthetic monomers onto the backbone of a natural polymer. Table 1.1 summarizes the advantages, disadvantages, and recent examples of synthetic, natural, and semisynthetic SABs.

TABLE 1.1 Synthetic vs. Biopolymer-Based SABs

Types of SABs	Advantages	Disadvantages	Examples
Natural	Renewable, biocompatible, biodegradable, Decreased environmental impact, easily available, sustainable	Extraction methods needed for the starting product; modifications required to create SABs, shelf-life/sensitive storage conditions	Gelatin Chitosan Guar gum Starch cellulose
Synthetic	Abundant, multiple applications, high purity	Often not biocompatible, no cell adhesion, not biodegradable, not renewable, possible toxicity, lack of reaction control	Poly(acrylate) Poly(ethylene glycol)
Semi-synthetic	Advantages of both synthetic/natural, fine-tuned properties	Phase separation, biodegradable, Homogeneity	Starch-poly(acrylamide) Chitin-acrylate Chitosan-poly(acrylic acid) Cellulose/polyethylenimine

1.3 PREPARATION TECHNIQUES

Graft polymerization, crosslinking polymerization, water-soluble polymer network synthesis, and radiation crosslinking are just a few of the SAB preparation techniques available. A number of polymerization/crosslinking techniques have been reported for the synthesis of SABs.[21]

1.3.1 PHYSICAL CROSSLINKING

Physical crosslinking occurs when secondary forces such as electrostatic, H-bonding, or hydrophobic interactions hold polymeric networks together[22-25] (Figure 1.4). When environmental parameters such as pH, temperature, ionic pressure, tension, and the presence of solutes change, the physical interaction between the network of superabsorbent can be reversed in their association.[25]

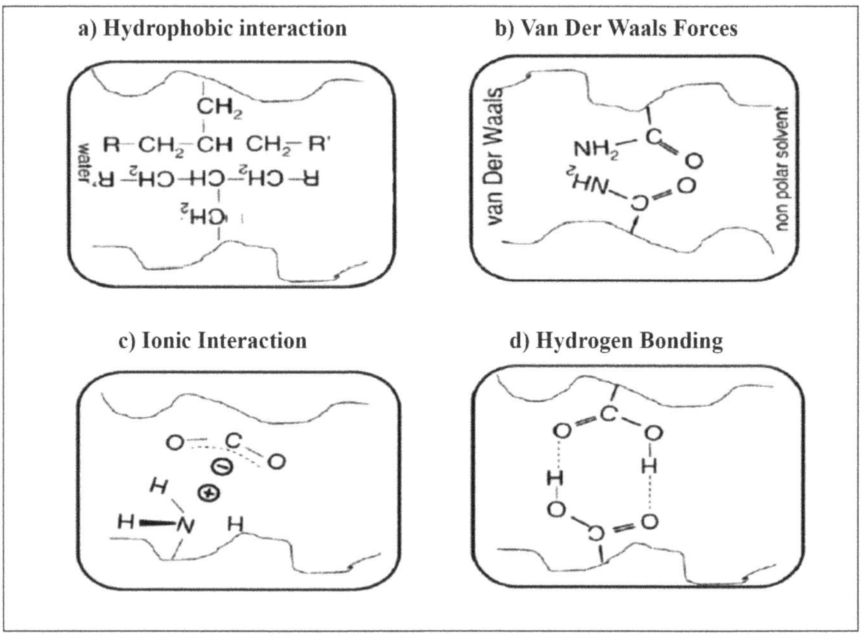

FIGURE 1.4 Common forces that influence the behavior of SABs.

1.3.2 CHEMICAL CROSSLINKING

Various methodologies for the formation of chemically crosslinked SABs exist, but the majority of them focus on free radical copolymerization. Crosslinking or copolymerization between one or more monomers and one multi-functional monomer is the first step. Two water soluble polymers can be crosslinked via a similar method called free radical polymerization. Polymerization reactions can be triggered by a chemical initiator (peroxide, ceric ammonium nitrate, and azo-compounds) or irradiation (UV, electron beams, -rays, X-rays, microwaves), and they can occur in bulk, solution, or suspension.[26–28]

The second method involves reacting a linear/branched polymer with a di/multi-functional polymer using a crosslinking agent (N, N′-methylenebisacrylamide (MBA)). Crosslinking agents usually contain di-functional or multi-functional groups that bind two high-molecular-weight polymeric chains together. Combining a bi-functional/crosslinking agent with polymeric chains with functional groups (–OH, –COOH, –NH2) on their backbones may also result in crosslinked hydrogels/SABs.

1.3.3 COPOLYMERIZATION CROSSLINKING

Small amounts of crosslinkers are needed to change the properties of superabsorbent polymers. Crosslinks between polymer chains are created by crosslinking agents. Each polymer chain's side growth makes a chemical bond with another polymer chain in a crosslinked polymer.[29]

A number of factors influence the effectiveness of crosslinking agents. Buchholz and Graham emphasize the crosslinker's solubility in monomer mixtures, steric hindrance, and mobility issues around dangling double bonds, and the crosslinker's susceptibility for intermolecular addition reactions like cyclo-polymerization.[30]

The following are some examples of copolymerization techniques:[31–33]

a. Solution polymerization/crosslinking

The ionic or neutral monomers are combined with the multifunctional crosslinking agent in solution copolymerization/crosslinking reactions. In the presence of solvent, the polymerization is initiated thermally, or by UV irradiation, or by a redox initiator mechanism.

The unreacted monomers, oligomers, crosslinking agent, initiator, soluble and extractable polymer, and other impurities must be washed out of the prepared SABs with distilled water (Figure 1.5).

b. Bulk polymerization

Bulk polymerization is the most basic method, including mainly monomer and monomer-soluble initiators. The high concentration of monomer causes a high rate of polymerization and degree of polymerization. However, as the conversion progresses, the viscosity of the process increases significantly, generating heat during polymerization. Controlling the reaction at low conversions can help avoid these issues. Bulk polymerization has the advantage of producing high molecular weight polymers with good purity.

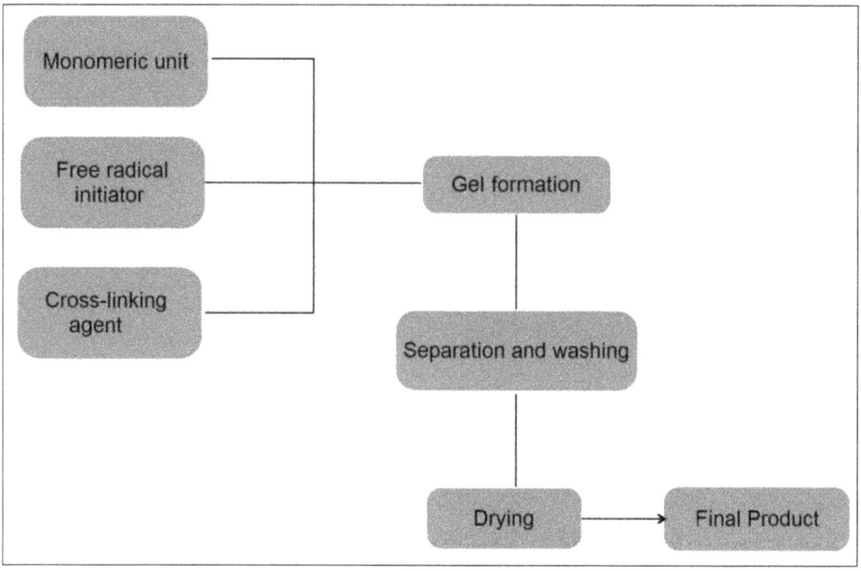

FIGURE 1.5 Block diagram representing solution polymerization.

c. Suspension polymerization or inverse suspension polymerization

Suspension polymerization is a method for the synthesis of spherical SAB microparticles in the 1 m to 1 mm size range. The monomer solution is dispersed in the nonsolvent, forming fine monomer droplets that are stabilized by the addition of a stabilizer in suspension polymerization. Radicals formed by the thermal decomposition of an initiator start the polymerization process. The unreacted residual monomers, crosslinking agent, and initiator are then washed out of the newly formed microparticles. Because of its ease of removal and management of the neurotoxic residual of acrylamide monomer during the synthesis of polyacrylamide-based SABs, the inverse suspension technique has recently become popular (Figure 1.6).

d. Irradiation-induced polymerization

The SABs of unsaturated compounds have been prepared using ionizing high-energy radiation such as gamma rays[31] and electron beams.[32] The formation of radicals on the polymer chains occurs when an aqueous solution of polymeric material is irradiated. In addition, radiolysis of water molecules produces hydroxyl radicals, which attack polymer chains and result in the formation of macroradicals. The formation of covalent bonds results from the recombination of macroradicals on various chains, resulting in the formation of a cross-linked structure. Polymers such as poly (vinyl alcohol), poly (ethylene glycol), and poly (propylene glycol) (acrylic acid) are examples of radiation-crosslinked polymers. The development of relatively pure and initiator-free SABs is a major benefit of radiation-initiated polymerization over chemical-initiated polymerization.

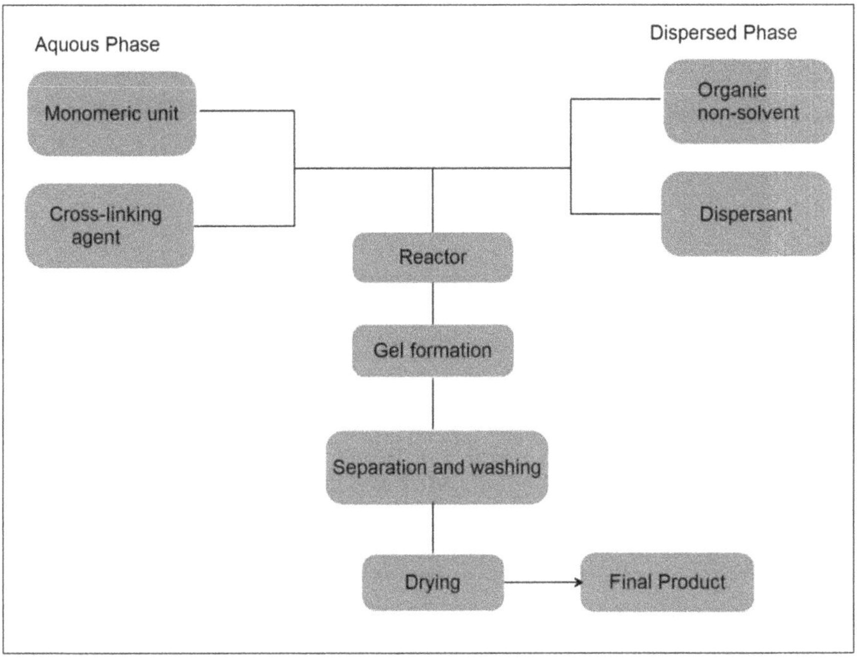

FIGURE 1.6 Block diagram representing inverse suspension polymerization.

1.4 CROSSLINKING IN SABS

The chemical composition of the polymeric material influences the swelling behavior of SABs, while crosslinking is normally independent of the polymer composition (i.e., degree of crosslinking). The degree of crosslinking increases as the amount of crosslinking agent introduced into the hydrogel matrix increases. The further crosslinking agent introduced into the matrix, the stronger the crosslinked hydrogel's structure is, and the lower the

SAB's swelling potential is relative to equivalent hydrogels prepared with a lower degree of crosslinking. Heavy crosslinking reduces the mobility of the polymer chain, lowering the swelling potential.

Crosslinking density of chemically synthesized SABs is regulated by the concentration of the crosslinking agent, temperature, reaction time, and other variables, according to Crini (2005).[34] Since the chemical structure of the SAB is influenced by the existence of the crosslinker and the degree of crosslinking, the crosslinking agent chosen has a major impact on various SAB properties, such as adsorption.

N,N′-methylenebisacrylamide (MBA, N-[(Prop-2-enoylamino) methyl] prop-2-enamid) is a bifunctional water-soluble crosslinking agent. During the polymerization process, a polyfunctional crosslinking agent (MBA) can generate a new macro-radical with four reactive sites, according to Singh et al. (2007).[35]

Both biopolymers (e.g., starch, casein) and the synthetic polymer (PAM) used during SAB synthesis can be linked to the crosslinker's reactive sites (MBA). Consequently, a three-dimensional network is created inexorably.

1.5 PROPERTIES OF AN IDEAL SABS

A superabsorbent polymer has the following properties:

- Extensive water absorption capacity under load,
- Desired rate of water absorption based on application requirement.
- Nontoxic, odorless, and colorless,
- Low cost,
- Re-swelling capacity,
- Biocompatibility
- Longevity and stability in a swollen climate
- Maximum durability and stability in a swelling environment as well as during storage.
- Maximum biodegradability without the generation of hazardous species after decomposition.
- Probability of finding soluble and residual content in the SABs is the lowest.
- Rewetting capabilities (if required): Depending on the applications (e.g., agricultural or hygienic), the hydrogel must be able to return or sustain the swallowed fluid.

1.6 SWELLING BEHAVIOR OF SUPERABSORBENT

The degree of swelling is the most emphasized and important factor to consider in the designing of a commercial SAB. The following mechanisms cause the SAB to swell very large:[36,37]

- Diffusion of aqueous solution through the SABs,
- Expansion of complex polymeric chain of SABs upon relaxation, and
- Relaxation of SABs polymeric network by the process of hydration.

A diffusive mechanism helps SABs to absorb liquid. Water begins to diffuse through a SAB particle because the activity of water in the interior of it is initially lower. To absorb the water molecules, the polymer chains must unavoidably migrate in the opposite direction of the water molecules as the volume of particle increases. Polymer chains with larger sizes as compared to water molecules have a slower diffusion rate.

The presence of crosslinks prevents the polymer from dissolving in water. These crosslinks, on the other hand, cause the polymer chains to comply, slowing the expansion of the polymer chains. Water stops flowing into the particle when the force resulting from the elongation of the polymer chains equals the force resulting from the difference in water concentration between the inside and outside of the SABs. This is the equilibrium state, where SABs consume as much water as they can, a condition known as equilibrium percentage swelling (EPS). In the equilibrium state, the expansion and contraction of the polymeric networks of SABs are balanced.[38,39]

1.7 APPLICATION OF SUPERABSORBENT

SABs can be used in a number of applications, including the following:

- **Pharmaceuticals:** Hydrogels are used in a variety of pharmaceutical formulations because of their high water absorbance capability, nontoxicity, biocompatibility, and delicate consistency, which makes them look like living tissues.[40] Hydrogels are useful for pharmaceutical applications because of their diffusional and swelling properties.[41] Proper material selection, synthesis process, and surface texture are essential characteristics to consider when selecting hydrogels for pharmaceutical formulation.[42]
- **Drug delivery systems:** Hydrogels have properties such as biocompatibility, biodegradability, hydrophilicity, and nontoxicity that make

them a good option for drug delivery applications. SABs have such a porous structure that can be adjusted by changing the cross-link density of the gel matrix, as well as the type and concentration of the porogen used; this feature allows for drug incorporation as well as controlled drug delivery.[44] In vitro drug release polymers include copolymers of acrylic acid and acrylamide, and biocompatible glucose acrylate-co-methacrylic acid polymer.[45,46] Smart hydrogels are being developed, such as stimuli-responsive hydrogels, that absorb and release active ingredients in response to various stimuli such as temperature, pH, electric, and magnetic fields. Polymers such as Gelatin, Chitosan, polyethylene oxide, ethylene-co-vinyl acetate, poly (2-hydroxyethyl methacrylate), poly (N-vinyl pyrrolidone), and poly (N-vinyl pyrrolidone) can be used to make stimuli-responsive hydrogels (N-isopropyl acrylamide).[47]

- **Bio-medication:** Hydrogels are useful in biomedical applications because of their hydrophilicity, biocompatibility, nontoxicity, and increased diffusivity. Bioremediation employs synthetic polymers such as polyethylene oxide and polyvinyl pyrrolidone, along with natural polymers such as proteins and various polysaccharides.[48]
- **Tissue engineering:** Due to their high water content, superabsorbents are usually well tolerated by cells, and they can bind to cell receptors in either a specific or nonspecific manner.[49] The hydrogels are soft, rubbery, and biocompatible, making them suitable for tissue engineering. Natural polymers include hyaluronic acid, alginate, collagen, gelatin, agar, and dextran; synthetic polymers include polyoxymethylene, polyethylene glycol, polyvinyl alcohol, polylactic acid, polyacrylamide, and polyhydroxylethyl methacrylate.[50,51]
- **Wound dressing:** Hydrogel-based dressings cool the wound site while adhering to the tissue, significantly lowering discomfort. Hydrogels are effective in the treatment of dry wounds due to their high water content.[52] Hydrogels can adapt to wounds in various body parts and absorb and retain wound exudate due to their tunable mechanical properties. The small mesh size of hydrogels prevents bacterial contamination while allowing for efficient biomolecule transport. Hydrogels' small mesh size prevents bacterial contamination while facilitating biomolecule transport.[53] Natural polymers include agrose, chitosan, pectin, collagen, dextran, pullulan, carrageenan, and fibrin, while synthetic polymers include polyethylene oxide, polyvinyl alcohol, polyvinyl pyrrolidone and various polyacrylates are used for this application.

- **Hygienic products:** SABs are commonly used to absorb water and aqueous solutions in baby diapers, adult incontinence products, and feminine sanitary products. Traditional absorbing materials such as cotton, fabric, and paper, which have disadvantages such as low performance, low sustainability, and leakage, are substituted by SABs in this application. The majority of the hydrogels used in this application were based on cellulose.[55,56]
- **Cosmetics:** Because of their high water content, which makes them moisturizing, SABs are utilized in skin-care products. Due to its hydrophilic base, which offers a matrix to the skin that is useful in transporting active chemicals into the skin, thermosensitive hydrogels are employed for localized, controlled, and continuous distribution of active compounds.[57] traditional hydrogels offer one advantage over bioadhesive hydrogels: they can stay on the application site longer, allowing for a higher local concentration of active agent in the surrounding tissues. This is useful for an active substance that has a surface activity but also affects the deeper layers of the skin. Various hydrogels were made with carrageenan, xanthan gum, guar gum, and carbomers (derivative of acrylic acid).[58] Polysaccharides (konjac mannan, xanthan gum, pullalan, and carrageenan) are mixed in a patent to generate a stable, flexible, translucent, and slightly sticky hydrogel for use as a skin care product.[59]
- **SABs:** SABs are used in skin-care formulations because of their high water content, which makes them moisturizing. Thermosensitive hydrogels are used for localized, regulated, and continuous delivery of active ingredients due to their hydrophilic foundation, which provides a matrix to the skin that is effective in delivering active ingredients into the skin.[57] Bioadhesive hydrogels have one advantage over traditional hydrogels: they allow for longer residence periods on the application site, allowing for a higher local concentration of active agent in the surrounding tissues for a longer period of time. This is advantageous for an active ingredient that is intended for superficial action but also acts on the skin's deeper layers. Carrageenan, xanthan gum, guar gum, and carbomers were used to produce various hydrogels (derivative of acrylic acid).[58] A patent shows how polysaccharides (konjac mannan, xanthan gum, pullalan, and carrageenan) are combined to create a stable, flexible, translucent, and slightly sticky hydrogel that will be used as a skin care cosmetic.[59]

- **Agriculture:** Because of their strong water retention qualities, SABs are utilized in agriculture as water-saving products and soil conditioners.[60] SABs promote porosity in clayey soils by allowing polymeric particles to expand and contract during moisture cycles. Plant growth was supported by the use of SABs, which helped to offset the negative effects of irrigation deficits and drought stress.[61] To achieve controlled or sustained release as well as water retention, fertilizers are combined with superabsorbent. Plant nutrition is improved, water evaporation losses are reduced, and irrigation frequency is reduced with this fertilizer and superabsorbent combination.[62] SABs are used to regulate the release of pests from SAB granules in both aquatic and terrestrial environments.[63] This application primarily employs polyacrylate and polyacrylamide-based superabsorbents. Agar and starch-based hydrogels act as soil conditioner and encourage plants growth and are used to control herbicide release and enhance soil health.[64]
- **Food industries:** The tunable features of micro-hydrogels are employed in processed foods for advanced rheological control and fat reduction. Simplesse, a fat substitute derived from microparticulate whey protein, is a commercially available microgel.[65] Food packaging uses edible polymer-based hydrogels.[66]
- **Water purification:** Water is the most essential substance for the existence of life on the earth. Water contamination is expanding rapidly, and industries are introducing lots of new dangerous chemicals into the water system, posing a serious environmental threat. Heavy metal ions and dyes are the major pollutants of waste water. Hydrogels are used to treat waste water and available in a range of sizes, including beads, films, and nanocomposites. Strong metal ions such as Cu^{2+}, Ni^{2+}, and Pb^{2+} are removed using hydrogel beads. The hydrogel beads are also used to adsorb dyes such as Congo red and Crystal violet. Radioactive waste is very often eliminated from water using hydrogels.[67]

1.8 CONCLUSION

A type of three-dimensional crosslinked polymeric material is known as SUBs that do not dissolve in water but expand significantly when exposed to water. Because of the excellent absorption capacity, nontoxicity, biocompatible and biodegradable qualities, biopolymer-based superabsorbents are becoming increasingly popular. The superabsorbent hydrogels have the ability to respond to a variety of external stimuli, such as temperature, light, pH, solvent composition, and enzyme. They are synthesized using solution, bulk, suspension or inverse suspension polymerization, and irradiation polymerization methods. Natural and synthetic polymers have been modified to develop smart materials, and these polymeric networks of SABs have subsequently been tailored and developed to satisfy the needs of many applications. SABs can be employed in a variety of applications such as drug delivery, skin care, tissue engineering, agriculture, wastewater treatment, bioremediation, food industries, biosensors, thermal energy storage, and many more. Biodegradable and high-performance superabsorbent hydrogels will become the focus of research as environmental problems become more significant. Wastewater treatment can be integrated with the synthesis of superabsorbents, which is good for the environment. Finally, coupling synthetic monomers with bio-based polymers will enhance the performance of quality of SAPs.

KEYWORDS

- **superabsorbent**
- **environmental degradation**
- **acryamide**
- **carcinogenic**
- **biopolymers**
- **hybrid materials**

REFERENCES

1. Rhim, J. W.; Ng, P. K. W. Natural Biopolymer-based Nanocomposite Films for Packaging Applications. *Criti. Rev. Food Sci. Nutri.* **2007,** *47* (4), 411–433.
2. Tharanathan, R. N. *Biodegradable Films and Composite Coatings: Past, Present and Future. Trends Food Sci. Technol.* **2003,** *14,* 71–78.
3. Sinha, S. Derivatization of Milk Protein Using Poly (Acrylamide): Its Characterization and Application. In *Macromolecular Symposia*, Vol. 381; Wiley-VCH Verlag GmbH & Co., 2018, p 1800120.

4. Michalek, J.; Hobzova, R.; Pradny, M.; Duskova, M.; Hydrogel Contact Lenses. In *Biomedical Applications of Hydrogels Handbook*; Ottenbrite, R. M., Park, K., Okano, T., Eds.; Springer: New York, USA, 2010; pp 303–316.
5. Rosiak, J. M.; Yoshii, F. Hydrogels and Their Medical Applications. *Nucl. Instrum. Methods Phys. Res. B* **1999,** *151,* 56–64.
6. Yue, Y.; Shen, X.; Wang, P. Fabrication and Characterization of Microstructure and pH Sensitive Interpenetrating Networks Hydrogel Films and Application in Drug Delivery Field. *Eur. Polym. J.* **2009,** *45,* 309–315.
7. Andreozzi, L.; Castelvetro, V.; Ciardelli, G.; Corsi, L.; Faetti, M.; Fatarella, E.; Zulli, F. Free Radical Generation Upon Plasma Treatment of Cotton Fibers and their Initiation Efficiency in Surface-Graft-Polymerrization, *J. Colloid Interface Sci.* **2005,** *289,* 455–465.
8. Ibrahim, M. M.; Porreca, F.; Lai, J.; Albrecht, P. J.; Rice, F. L.; Khodorova, A.; Davar, G.; Makriyannis, A.; Vanderah, T. W.; Mata, H. P. Malan, T. P. CB2 Cannabinoid Receptor Activation Produces Antinociception by Stimulating Peripheral Release of Endogenous Opioid. *Proc. Natl. Acad. Sci. USA.* **2005,** *102,* 3093–3098.
9. Bhattacharya, A.; Rawlins, J. W.; Ray, P. Eds. *Polymer Grafting and Crosslinking*. John wiley & Sons: New York, 2008.
10. Nayak, P. L.; Lenka, S. Redox Polymerization Initiated by Metal Ions. *J. Macromol. Sci. C. Polym. Rev.* **1980,** *19* (1), 83–134.
11. Misra, G. S.; Bajpai, U. D. Redox Polymerization (Review). *Progress Poly. Sci.* **1982,** *8,* 61–131.
12. Temel, O.; Ismail, C. Synthesis of Block Copolymers via Redox Polymerization Process: A Critical Review. *Iranian Polym. J.* **2007,** *16* (8), 561–581.
13. Mino, G.; Kaizerman, S. *A New* Method for the Preparation of Graft Copolymers. *J. Polym. Sci.* **1958,** *31,* 242–243.
14. McDowall, D. J.; Gupta, B. S.; Stannett, V. T. Grafting of Vinyl Monomers to Cellulose by Ceric ion Inhibition. *Prog. Polym. Sci.* **1984,** *10,* 1–50.
15. *Zheng, Y. A.; Li, P.; Zhang, J. P.* Study on Superabsorbent Composite XVI: Synthesis, Characterization and Swelling Behaviors of poly(sodiumacrylate)/Vermiculite *Superabsorbent Composites. Eur. Polym. J.* ***2007,*** *43,* 1691–1698.
16. Buchholz, F. L.; Graham, A. T. *Modern Superabsorbent Polymer Technology*. Wiley: New York, 1997.
17. Lenzi, F.; Sannino, A.; Borriello, A.; Porro, F.; Capitani, D.; Mensitieri, G. Probing the Degree of Crosslinking of a Cellulose based Superabsorbing Hydrogel Through Traditional and NMR Techniques. *Polymer.* **2003,** *44,* 1577–1588.
18. Yang, L.; Chu, J. S.; Fix, J. A. Colon-Specific Drug Delivery: New Approaches and In Vitro/In Vivo Evaluation. *Int J Pharm.* **2002,** *235,* 1–15.
19. Hacker, M. C.; Mikos, A. G. Synthetic Polymers. In *Principles of Regenerative Medicine*, 2nd ed.; Academic Press, San Diego, 2011; chapter 33; pp 587–622.
20. U.S. Environmental Protection Agency (U.S. EPA). Integrated Risk Information System (IRIS) Online. Coversheet for Acrylamide. In *Office of Health and Environmental Assessment*. U.S. EPA: Cincinnati, 1994.
21. Staples, T. L.; Henton, D. E.; Buchholz, F. L. Chemistryof Superabsorbent Polyacrylates. In *Modern SuperabsorbentPolymer Technology*; Buchholz, F. L., Graham, A. T., Eds.; Wiley-VCH: New York, 1998; chapter 2.

22. Khan, U.; O'Neill, A.; Lotya, M.; De, S.; Coleman, J. N. High-Concentration Solvent Exfoliation of Graphene. *Small* **2010,** *6,* 864–871.
23. Berger, J.; Reist, M.; Mayer, J. M.; Felt, O.; Peppas, N. A.; Gurny, R. Structure and Interactions in Covalently and Ionically Crosslinked Chitosan Hydrogels for Biomedical Applications. *Eur. J. Pharm. Biopharm.* **2004,** *57* (1), 19–34.
24. Zhong, Q.; Daubert, C. R. Kinetics of Rennet Casein Gelation at Different Cooling Rates. *J. Colloid Interface Sci.* **2004,** *279* (1), 88–94.
25. Hoare, T. R.; Kohane, D. S. Hydrogels in Drug Delivery: Progress and Challenges. Polym. **2008,** *49* (8), 1993–2007.
26. Rivarola, C. R.; Biasutti, M. A.; Barbero, C. A. A Visible Light Photoinitiator System to Produce Acrylamide Based Smart Hydrogels: $Ru(bpy)_3^{+2}$ as Photopolymerization Initiator and Molecular Probe of Hydrogel Microenvironments. *Polymer* **2009,** *50,* 3145–3152.
27. Hernandez, Y.; Nicolosi, V.; Lotya, M.; Blighe, F. M.; Sun, Z. Y.; De, S.; McGovern, I. T.; Holland, B.; Byrne, M.; Gun'ko, Y. K.; Boland, J. J.; Niraj, P.; Duesberg, G.; Krishnamurthy, S.; Goodhue, R.; Hutchison, J.; Scardaci, V.; Ferrari, A. C.; Coleman, J. N. High-Yield Production of Graphene by Liquid-Phase Exfoliation of Graphite. *Nat. Nanotechnol.* **2008,** *3,* 563–568.
28. Lotya, M.; Hernandez, Y.; King, P. J.; Smith, R. J.; Nicolosi, V.; Karlsson, L. S.; Blighe, F. M.; De, S.; Wang, Z.; McGovern, I. T.;Duesberg, G. S.; Coleman, J. N.; Liquid Phase Production of Graphene by Exfoliation of Graphite in Surfactant/Water Solutions. *J. Am. Chem. Soc.* **2009,** *131,* 3611–3620.
29. Kopecek, J.; Yang, J. Peptide-Directed Self-Assembly of Hydrogels. *ActaBiomaterialia* **2009,** *5* (3), 805–816.
30. Ebewele, R. O. *Polymer Science and Technology*. Boca Raton: CRC Press, 2000; pp 8, 182, 363, 431–432.
31. Karadaǒ, E.; Saraydin, D.; Gûven, O. Radiation InducedSuperabsorbent Hydrogels. Acrylamide/Itaconic Acid Copolymers. *Macromol. Mater. Eng.* **2001,** *286,* 34–42.
32. Garnett, J. L.; Ng, L.-T.; Virngkhou, V. Grafting of MethylMethacrylate to Cellulose and Polypropylene with UV and Ionising Radiation in the Presence of Additives Including CT Complexes. *Radiat. Phys. Chem.* **1999,** *56,* 387–403.
33. United States Department of Agriculture. U.S. Patent 3,981,100, 1961.
34. Crini, G. Recent Developments in Polysaccharide-Based Materials Used as Adsorbents in Wastewater Treatment. *Prog. Polym. Sci.* **2005,** *30,* 38–70.
35. Singh, R. P.; Dhania, G.; Sharma, A.; Jaiwal, P. K. Biotechnological Approaches to Improve Phytoremediation Efficiency for Environment Contaminants. In *Environmental Bioremediation Technologies*; Singh, S. N., Tripahti, R. D., Eds.; Springer, 2007; pp 223–258.
36. Kim, S. J.; Lee, C. K.; Kim, S. I. Electrical/pH Responsive Properties of poly(2-Acrylamido-2-Methylpropane Sulfonic Acid)/Hyaluronic Acid Hydrogels. *J Appl Pol Sci.* **2004,** *92,* 1731–1736.
37. Fortunati, E.; Armentano, I.; Zhou, Q.; Puglia, D.; Terenzi, A.; Berglund, L. A.; Kenny, J. M.; Effects of Modified Cellulose Nanocrystals on the Barrier and Migration Properties of PLA Nano-Biocomposites. *Polym. Degrad. Stab.* **2012,** *97,* 2027–2036.
38. Turbak, A. F; Snyder, F. W.; Sandberg, K. R.; Microfibrillated Cellulose, A New Cellulose Product: Properties, Uses, and Commercial Potential. *J. Appl. Polym. Sci. Appl. Polym. Symp.* **1983,** *3,* 815–827.

39. Flory, P. J.; Phase Equilibria in Polymer Systems: Swelling of Network Structures. In: *Principles of Polymer Chemistry*. Cornell University: Ithaca, NY, 1953.
40. Peppas, N. A. Hydrogels in Pharmaceutical Formulations. *Eur. J. Pharm. Biopharm.* **2000,** *50,* 27–46.
41. Kashyap, N.; Kumar, N.; Kumar, M. N. V. R. Hydrogels for Pharmaceutical and Biomedical Applications. *Crit. Rev. Ther. Drug Carrier Syst.* **2005,** *22* (2), 107–150.
42. Singh, A.; Sharma, P. K.; Garg, V. K.; Garg, G. Hydrogels : A Review. *Int. J. Pharma. Sci. Rev. Res.* **2010,** *4* (2), 97–105.
43. Parhi, R. Cross-linked Hydrogel for Pharmaceutical Applications: A Review. *Adv. Pharm. Bull.* **2017,** *7* (4), 515–530.
44. Mishra, B.; Upadhyay, M.; Adena, S. K. R. Hydrogels: An Introduction to a Controlled Drug Delivery Device, Synthesis and Application in Drug Delivery and Tissue Engineering. *Austin J. Biomed. Eng.* **2017,** *4* (1).
45. Ravindra, S.; Mulaba-Bafubiandi, A. F.; Rajineekanth, V.; Varaprasad, K.; Raju, K. M. Synthesis of Surfactant-Modified Poly (acrylamide-copotassium Acrylate) Hydrogels and Its in Vitro Release Studies. *Polym. Plast. Technol. Eng.* **2012,** *51* (13), 1355–1360.
46. Mahkam, M. Synthesis, Characterization and Evaluation of Poly [Glucose Acrylate-methacrylic Acid] Hydrogels for Colon-specific Drug Delivery. *E-Polymers* **2008** (159), 1–8.
47. Gupta, P.; Vermani, K.; Garg, S. Hydrogels: From Controlled Release to pH-Responsive Drug Delivery. *DDT* **2002,** *7* (10), 569–579.
48. Cabral, J.; Moratti, S. C. Hydrogels for Biomedical Applications. *Future Med. Chem.* **2011,** *3* (15), 1877–1888.
49. Coury, A.; et al. Hydrogels in Tissue Engineering. **2006,** *79,* 1.
50. El-sherbiny, I. M.; Yacoub, M. H. Review Article Hydrogel Scaffolds for Tissue Engineering: Progress and Challenges. *GLobal Cardiol. Sci. Pract.* **2013,** *38,* 316–342.
51. Lee, J.; Kim, H. Emerging Properties of Hydrogels in Tissue Engineering. *J. Tissue Eng.* **2018,** *9,* 1–4.
52. Koehler, J.; Brandl, F. P.; Goepferich, A. M. Hydrogel Wound Dressings for Bioactive Treatment of Acute and Chronic. *Eur. Polym. J.* **2017,** 1–38.
53. Madaghiele, M.; Demitri, C.; Sannino, A.; Ambrosio, L. Polymeric Hydrogels for Burn Polymeric Hydrogels for Wound Care: Advanced Skin Wound Dressings and Regenerative Templates. *Burns Trauma* **2014,** *2* (4), 153–161.
54. Gupta, B.; Agarwal, R.; Alam, M. S. Hydrogels for Wound Healing Applications. In *Biomedical Hydrogels: Biochemistry, Manufacture and Medical Applications*, 2011; pp 184–227.
55. Bashari, A.; Shakeri, M. Cellulose-based Hydrogels for Personal Care Products. *Polym. Adv. Technol.* **2018,** *29,* 2853–2867.
56. Sannino, A.; Demitri, C.; Madaghiele, M. Biodegradable Cellulose-based Hydrogels: Design and Applications. *Materials* **2009,** *2,* 353–373.
57. Quattrone, A. Thermosensitive Hydrogel Mask Significantly Improves Skin Moisture and Skin Tone; Bialteral Clinical Trial. *Cosmetics* **2017,** *4* (17), 18.
58. Parente, M. E.; Ochoa Andrade, A.; Ares, G.; Russo, F.; Jiménez-Kairuz, A. Bioadhesive Hydrogels for Cosmetic Applications. *Int. J. Cosmet. Sci.* 2015**,** *37* (5), 511–518.
59. Iris, S.; Christian, H.; Christina, K. Hydrogel for Natural Cosmetic Purposes. US 2013/00, 19, 2013; pp 1–4.

60. Dabhi, R.; Bhatt, N.; Pandit, B. Effect on the Absorption Rate of Agricultural Super Absorbent Polymers under the Mixer of Soil and Different Quality of Irrigation Water. *Int. J. Emerging Technol. Adv. Eng.– IJETAE.* **2014,** *3* (1), 1402–1406.
61. Sayyari, M.; Ghanbari, F. Effects of Super Absorbent Polymer A 200 on the Growth, Yield and Some Physiological Responses in Sweet Pepper (capsicum Annuum L.) Under Various Irrigation Regimes. *Int. J. Agric. Food Res.* **1929,** *1* (1), 1–11.
62. Wu, L.; Liu, M. Preparation and Properties of Chitosan-coated NPK Compound Fertilizer with Controlled-Release and Water-Retention. *Carbohydr. Polym.* **2008,** *72* (2), 240–247.
63. Levy, R.; et al. Encapsulated Systems for Controlled Release and Pest Management. *Symp. A Q. J. Mod. Foreign Liter.* 1993, pp 202–212.
64. Singh, B.; Sharma, D. K.; Negi, S.; Dhiman, A. Synthesis and Characterization of Agar-starch Based Hydrogels for Slow Herbicide Delivery Applications. *Int. J. Plast. Technol.* **2015,** *19* (2), 263–274.
65. Shewan, H. M.; Stokes, J. R. Review of Techniques to Manufacture Micro-hydrogel Particles for the Food Industry and Their Applications. *J. Food Eng.* **2013,** *119* (4), 781–792.
66. Ali, A.; Ahmed, S. Recent Advances in Edible Polymer Based Hydrogels as a Sustainable Alternative to Conventional Polymers. *J. Agric. Food Chem.* **2018,** *66,* 6940–6967.
67. Van Tran, V.; Park, D.; Lee, Y. Hydrogel Applications for Adsorption of Contaminants in Water and Wastewater Treatment. Environ. *Sci. Pollut. Res.* **2018,** *25,* 24569–24599.

CHAPTER 2

Applications of Green Nanocomposites in Polycyclic Aromatic Hydrocarbons (PAHs) Bioremediation: Current Status and Future Direction

PARAMITA CHAKRAVARTY[1], HEMEN DEKA[1], and DEVASISH CHOWDHURY[2]

[1]Ecology and Environmental Remediation Laboratory, Department of Botany, Gauhati University, Guwahati, Assam, India

[2]Institute of Advanced Study in Science and Technology, Paschim Boragaon, Guwahati, Assam, India

ABSTRACT

Polycyclic aromatic hydrocarbons, that is, PAHs are considered major contaminants of environmental pollution. The elimination of these hazardous pollutants in an ecofriendly, renewable, and economically feasible manner is one of the great concern of present times. Nanomaterials are more reactive having large surface area compared to their bulk counterpart, so they have multiple modes of uses including remediation of environmental pollutants. The mechanism behind the mitigation of organic/inorganic contaminants using nanomaterials/nanocomposites can be categorized into three heads: adsorption, transformation, and catalysis. Nanomaterials/nanoparticles increase the degradation percentage of pollutants in the presence of microbes by means of increasing the microbial growth, stabilizing remediating mediators, and inducing the production of some specific enzymes which may help in the bioremediations of pollutants. Moreover, the nanomaterials also reduce the hydrophobicity and increase the conductivity of contaminated environment

Bio-Based Polymers and Composites: Properties, Durability, and Applications. Amadou Belal Gueye, Sabu Thomas, Nandakumar Kalarikkal, & Modou Fall (Eds.)

which enhances microbial degradation. It can also improve soil/water conditions, can act as a catalyst in development and metabolic processes of microbes, and can build adsorption capacity against pollutants which ultimately make a successful bioremediation approach. In the present chapter, the applicability of different green nanomaterials/nanocomposites in remediation of environmental pollutants especially PAHs has been highlighted.

2.1 INTRODUCTION

Polycyclic aromatic hydrocarbons (PAHs) contamination is still a major concern throughout the globe. The PAHs pollution is now considered a key concern of the 21st century as well as being an important international challenge. The profound impacts of these pollutants have attracted worldwide attention due to their far-reaching implications on the environment. In most cases, the oil fields and oil refineries are located on the periphery of human settlements including crop fields and low-lying areas; so the problem of PAHs pollution becomes a major concern.

2.1.1 POLYCYCLIC AROMATIC HYDROCARBONS: AN OVERVIEW

PAHs are well-known organic compounds consisting of at least two single/combined aromatic rings of hydrogen and carbon atoms arranged either linearly or in a cluster of angular arrangement.[72] PAHs are the major group of environmental pollutants that can be formed through both natural and/or anthropogenic processes. The products of incomplete combustion such as forest and brush fires, volcanic eruptions, etc. contribute to natural sources of PAHs formation while automobile emissions, petroleum catalytic cracking, cigarette smoke, industrial combustion of fossil fuels, etc. are considered manmade or anthropogenic sources of PAHs formation.[2,10] PAHs are found everywhere in the atmosphere. The distribution and contents of the PAHs pollutants in the environment fluctuate widely and depend on various factors including source and meteorological conditions.[24] PAHs are hydrophobic in nature which do not dissolve generally in water or volatilize to the atmosphere (except naphthalene).

2.1.1.1 PROBLEMS OF PAHS

The PAHs are comparatively more stable compounds and natural biodegradation of these compounds is very slow and even difficult. PAHs are chemically stable at ambient temperature due to their large negative resonance energy which enables them to persist longer in the environment.[70] The problem is further compounded by their low or total insolubility in water,[14] which enables PAHs to easily adsorb onto the organic matter of solid particles leading to a decline in their bioavailability.[91] This impedes the biological degradation of PAHs and enables them to escape the aqueous microhabitats of degrader organisms and accumulate in the environment. This agglomeration generally makes the atmosphere noxious and inhospitable, characterized by the lack of the nutrients, water, or appropriate electron acceptors which are vital for any successful bioremediation approach.[84] As reported by earlier workers, the presence of PAHs on agricultural land renders the soil's biologically active layer toxic by enhancing soil hydrophobicity and preventing normal oxygen exchange between soil and atmosphere, thus reducing plant growth and making agricultural land unproductive.[68,30,69] The uptake of these carcinogenic, as well as mutagenic contaminants by plants, results in food chain contaminations, while the downward movement of these pollutants causes groundwater pollution and ultimately it harms human health and the environment (www.sciencedaily.com/releases/2017/07/170710103446.htm).[130]

2.1.1.2 TYPES OF PAHS AND THEIR TOXICITY

The PAH pollutants show recalcitrant nature in soil due to their complex/branched structure. As reported by Abdel-Shafy and Mansour[2] PAHs molecules with one to six fused aromatic rings can be considered "small" PAHs, while "large" PAHs possess more than six aromatic rings. The availability of small PAHs is higher in comparison to small PAHs. According to another classification on the basis of molecular weight, PAHs are sorted into two types, low-molecular-weight (LMW) PAHs and high-molecular-weight (HMW) PAHs. The LMW-PAHs such as naphthalene, acenaphthene, fluorene, and phenanthrene consist of one to three aromatic rings, while HMW PAHs such as pyrene, chrysene, benzo[a]pyrene, and dibenz[a,h]anthracene are composed of four or more than four aromatic rings.[72] In general, the HMW PAHs have low solubility in water than LMW PAHs. Several PAHs were identified, among which 16 compounds have been marked as hazardous contaminants by United State Environmental Protection Agency (USEPA), European Union (EU), and European Union Scientific Committee on Food (SCF) (Table 2.1).

It is well known that the PAH pollutants exert serious deleterious effects on the contaminated environment. These contaminants have negatively affected the entire ecosystem as they can accumulate in soil and water and

TABLE 2.1 List of 16 PAH Pollutants According to Recommendations by the European Union (EU), the EU Scientific Committee for Food (SCF), and the United State Environmental protection Agency (USEPA) with Their Structures, Molecular Weight, Water Solubility, Carcinogenicity, and Toxic Factor.

List	Common name	Structure	Molecular weight (g/mol)	Water solubility (μM)	Carcino-genicity	Toxic equivalent factor
USEPA, SCF, EU	Benzo[*a*]pyrene	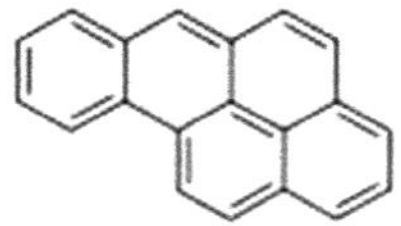	252.32	0.016	High	1.0
USEPA	Acenaphthene	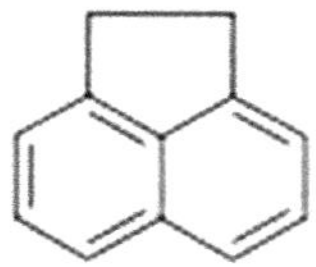	154.21	23	Low	0.001
USEPA	Acenaphthylene	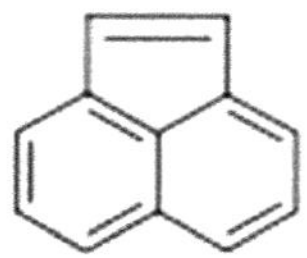	152.20	26	Low	0.001
USEPA	Anthracene	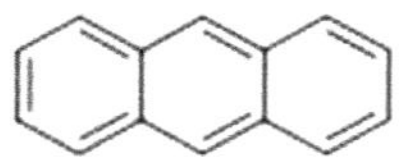	178.23	0.409	Low	0.01
USEPA,SCF, EU	Benzo[*a*] Anthracene	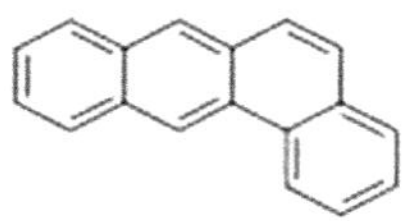	228.29	0.048	High	0.1

TABLE 2.1 *(Continued)*

List	Common name	Structure	Molecular weight (g/mol)	Water solubility (µM)	Carcino-genicity	Toxic equivalent factor
USEPA,SCF, EU	Benzo[*b*] Fluoranthene	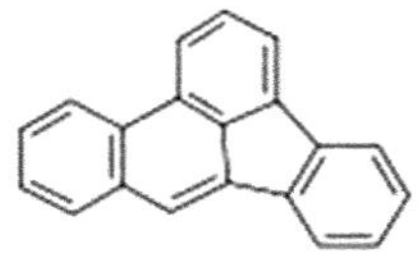	252.32	0.006	High	0.1
USEPA	Fluorene	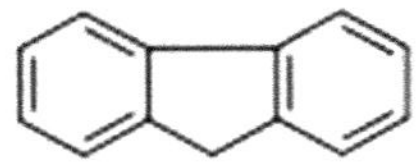	166.22	12	Low	0.001
USEPA	Pyrene	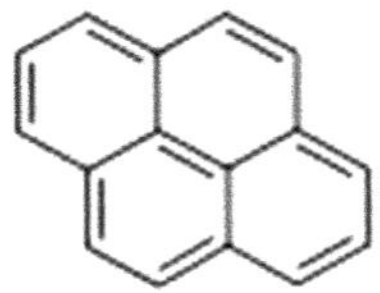	202.26	0.667	Low	0.001
USEPA,SCF, EU	Benzo[*k*] Fluoranthene	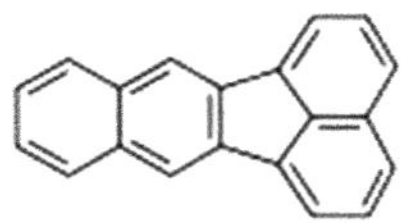	252.32	0.003	High	0.1
USEPA,SCF, EU	Benzo[*g,h,i*] Perylene	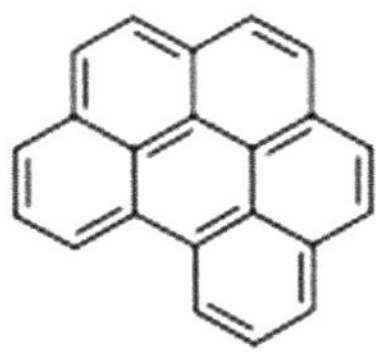	276.34	0.020	High	0.01

TABLE 2.1 *(Continued)*

List	Common name	Structure	Molecular weight (g/mol)	Water solubility (μM)	Carcino-genicity	Toxic equivalent factor
USEPA,SCF, EU	Chrysene	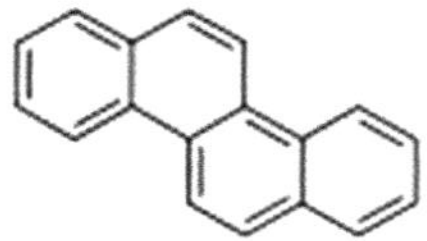	228.29	0.013	Low	0.01
USEPA	Fluoranthene	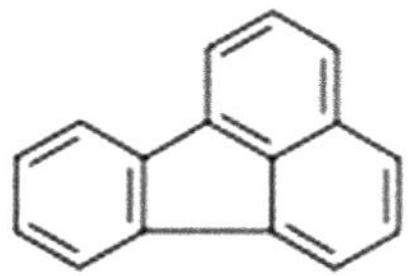	202.26	1.28	Low	0.001
USEPA,SCF, EU	Dibenzo[*a,h*] Anthracene	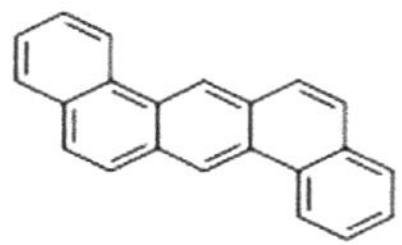	278.35	0.001800	High	1.0
USEPA	Phenanthrene	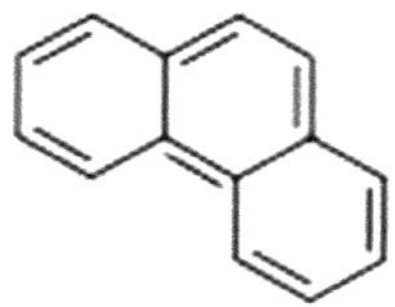	178.23	7.20	Low	0.001
USEPA	Naphthalene	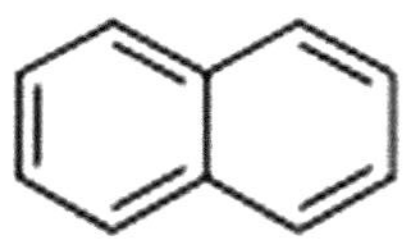	128.17	240	Low	0.001

TABLE 2.1 *(Continued)*

List	Common name	Structure	Molecular weight (g/mol)	Water solubility (µM)	Carcino-genicity	Toxic equivalent factor
USEPA,SCF, EU	Indeno[1,2,3-*cd*]pyrene	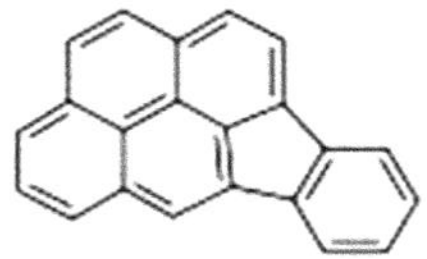	276.34	0.000 700	High	0.1

Source: Refs. [23, 50, 90, 118]

are transferred to living organisms including human beings via the food chain causing noticeable damage to the soil fertility, crop production, as well as to human health and aquatic lives, etc.[17] As reported by Borah et al.[12] and Ying et al.[124] the PAHs and other hydrocarbons alter the physical, chemical, and biological conditions of the soil systems and make soil unfit for vegetation. Moreover, due to the carcinogenicity, mutagenicity, and teratogenicity of the PAH pollutants, they exerted their toxic effects upon mammalian health also.[7] More than a hundred years ago,[32] phenanthrene had been addressed to affect human skin as it acts as a photosensitizer and mild allergen.[10] Abhaand Singh[3] stated that long-term contact to naphthene-aromatics and poly-aromatic compounds often leads to tumors, cancer, and failure of the nervous system. Benzo[*a*]pyrene is documented as one of the major 16 priority PAHs pollutants by the USEPA due to its acute carcinogenicity among all the identified PAHs.[44] It has been suggested that with increasing number of benzene rings the toxicity of a PAH also increases.[7] Considering these facts, researchers all over the world have already paid large attention to the remediation of PAHs' polluted habitats. It is seen that hydrocarbon contamination is most crucial in developing countries like India as the industries of these countries frequently discharge crude waste and residues into an open environment.[8] These have encouraged scientists for searching a variety of sustainable methods to successfully remediate hydrocarbon-polluted environments.

2.1.2 NANOMATERIALS AND NANOCOMPOSITES

2.1.2.1 THE CENTRAL CONCEPT

The nanomaterials are those tiny particles which have the sizes from 1 and 100 nm.[92] The physical and chemical properties of nanoparticles are distinct from the corresponding sourced materials as they possess very small size and greater surface-volume ratio.[92] Nowadays, the synthesis of green or biogenic nanomaterials from plants, biogenic wastes, microbes, etc. has proven ecologically sound and cost effective due to no involvement of harsh chemicals and solvents. The synthesis of nanoparticles by plants is getting more attention from researchers as it offers a single-step biosynthesis route.[4] Green synthesis of nanoparticles by bacteria and fungi is used widely as it is cheaper, easy to synthesize with simpler biomass handling and downstream processing and more specifically for its greater bioaccumulation capacity.[126]

Some of the bio-based nanoparticles along with their green sources are present in Table 2.2.

TABLE 2.2 List of Some Nanomaterials Synthesized from Green Sources

Green sources	Nanomaterials	References
Plants		
Ficus bangalansis, Allium sepa, Aloe vera, Cycas sp., Pinus sp., Ginkobiloba, Alfa alfa, Magnolia sp., Ocimum sanctum, Euphorbia hirta, Azadirachta indica, Trifolium sp., Muska melon, Derisindica	Ag, Au, Cu, Cd, Pd, Pt, ZnO, CNT, Ti, Fe	Saxena et al.,[89] Tripathi et al. 2013, Song et al.,[103] Elia et al.,[28] Mahajan et al.[59]
Bacteria		
Cupriavidus necator, Pseudomonas meridiana, Bacillus indicus, Escherichia coli, Enterobacter cloa, Bacillus megaterium, Bacillus licheniformis, Bacillus subtilis	Ag, Au, Cd, Ti, ZnO	Shivaji et al.,[100] Du et al.,[27] Tripathi et al.,[112] Kirthi et al.[48]
Fungi		
Aspergilus fumigetus, Aspergillus flavus, Thrichoderma longibrachiatum, Fuserium sp., Penicillium sp., Candida albicans, Rhizopus sp., Yeast, *Neurospora crassa, Aspergilus Fuserium oxysporum*	Ag, Au, Fe, Ti, Cd	Velhal et al.,[114] Philip et al.,[77] Thakker et al.,[106] Tripathi et al.,[113] Kumar et al.[49]

Composites are naturally occurring or artificially produced with combination of two or more different constituent materials each having distinct physical or chemical properties. The composites are more superior in characteristics in comparison to their constituent materials.[92] Composites with nanoscale (10 to 100 nm) morphology such as nanoparticles, nanotubes, or lamellar nanostructure to at least one phase out of its multiple phases are considered to be nanocomposites..[47,88] Nanocomposites were designed and proposed to create a new building block with nanoscale properties to generate more advanced materials having enhanced physical properties and flexibility. As reported by Sen,[92] the nanocomposites can provide more surface appearance, improve electrical conductivity, and increase chemical resistance.

2.1.2.2 TYPES OF NANOCOMPOSITES

Nanocomposites material can be classified into two broad categories on the basis of the presence or absence of polymeric material in the composite, that is, one is polymer-based nanocomposites and another is non-polymer-based nanocomposites.

2.1.2.2.1 Polymer-Based Nanocomposite

The polymer-based nanocomposites (PBNC) are those polymers with nanoparticles or nanofillers scattered in the polymer matrix.[25] At least one of the dimension (1D) should be in the scale of 1–50 nm and possess various shapes similar to spheroids, platelets, fibers, etc.[92] PBNCs are class of materials because of their distinct properties; however, the complex interfacial areas in between the polymer matrices can be considered one of the interesting properties of PBNCs due to the formation of small-scale large specific area.[92]

The polymer-based nanocomposites further can be classified into the following categories[92]:

i. *Polymer/ceramic-based nanocomposites*
ii. *Inorganic/organic polymer-based nanocomposites*
iii. *Inorganic/organic hybrid-based nanocomposites*
iv. *Polymer/Layered silicate-based nanocomposites*

2.1.2.2.2 Non-Polymer-Based Nanocomposites

Non-polymer-based nanocomposites (NPBCs), also known as inorganic nanocomposites, are those composites which have no polymer or polymer-derived materials in their configuration. The NPBCs are further categorized into the following three classes[46]:

i. ***Metal-based nanocomposites (MBN):*** The bimetallic nanoparticles in the form of alloy/core-shell structures are further explored because of their enriched catalytic characteristics.
ii. ***Ceramic-based nanocomposites (CBN):*** The CBN are those composites which have two or more solid phases and at least one of those phases should contain the nanoscale dimensions of less than 50–100 nm.
iii. ***Ceramic-ceramic-based nanocomposites (CCBN):*** These nanocomposites are considered effective for prosthesis that can lessen the rate of surgical treatment and increase the flexibility of the patients.

2.1.2.3 APPLICATIONS OF NANOCOMPOSITES

The nanocomposites have been *receiving* cumulative *attention* due to their several applications. They are widely used in various domains like electro-conductive scaffolds, photonic gadgets, *photovoltaic cell*, solid coats, biosensors, remediation of air, water, soil, biomedical purposes, agricultural domain, and others. Renu et al.[86] had stated the plant-based mediated silver or poly-d,l-PLGA nanocomposites can be used in tomography, biosensors, antimicrobial-based wound healing etc. Again, chitin (derived from fungi) and chitosan (derived from crustaceans) are widely used for natural wound healing.[43] Graphene oxide and rGO are also applied in various fields such as biomedical fields, food industry, environmentology, etc.[102] Similarly, the nanoclay is considered a familiar natural fiber nanofiller among nanocomposites.[102] Wypych[122] documented the applications of jute or polyethylene or nanoclay nanocomposites which are synthesized from jute in the presence of propionic anhydride. Another researcher Nasrollahzadeh et al.[63] reported the biosynthesis of Cu-eggshell, Fe_3O_4-eggshell, and Cu-Fe_3O_4-eggshell nanocomposites from the leaves of *Orchismascula* that revealed high catalytic action against the reduction of several dyes in aqueous medium. As a whole, the bio-based nanocomposites are environmental friendly and renewable which have the properties of sustainability, degradability, minimizing, and inhibiting the contamination.[102] It has been reported that green-synthesized hierarchical ZnO–rGO nanocomposites can indicate the photocatalyst activity for the photodegradation of azure B dye.[80] As outlined by Sabherwal et al.,[87] the bio functionalized carbon-nanocomposites, mainly the bioreceptor-functionalized nanocomposites, are used for the generation of recognition systems in medical and/or ecological study. Nanocomposites are nowadays getting attraction in different industries including automobile industries (engine covers, door covers, and timing belt covers) due to their enhanced mechanical properties.[92] The nanocomposites can also be applied in vacuum cleaners (as blades), mower hoods, and mobile phones' covering, etc.[92] In today's technology, applications of the nanocomposite are quite widely used in purification, enrichment of active ingredients, recycling of industrial byproducts, and reducing of the toxicity of wastes.

2.2 OVERVIEW OF PAHS REMEDIATION TECHNOLOGIES

Several chemical and physical methods have been developed for decontamination of PAHs-polluted environment.[41,76] However, the biological methods are receiving much more attentiveness as an alternative for PAH remediation as it is a cost-effective and ecologically sound technique for remediation of PAHs pollution.[72] Among the biological methods, the bioremediation practices are considered very much effective which can further be divided two categories: a) ex-situ and b) in-situ. The ex-situ bioremediation approaches comprise the physical elimination of the contaminants to other sites from the original place through excavation in case of soil pollution and pumping in case of water pollution.[72] On the other hand, the in-situ practices involve treatment of the pollutants in site, that is in original place of pollution.[72] The in-situ bioremediation approaches have been getting more attention as it is a cost-effective method for nominal site disruption, permanent waste elimination etc.[95]

Bioremediation technology can be widely sorted into two categories viz. microbial-remediation and phyto-remediation. In phyto-remediation plants are used to accumulate, sequester, or metabolize environmental pollutants like PAHs. The PAHs toxicity can also be treated by microbial degradation through anaerobic and aerobic environments.[8] Microbes such as some bacteria, e.g., *Pseudomonas* sp., *Mycobacterium* sp., *Xanthomonas* sp., *Rhodococcus* sp., use PAHs as their sole carbon source for the growth/development and metabolic proccss and thus subsequently induce mineralization of such pollutants.[123] Bacterial-aided remediation exhibits oxidation reaction that is performed by the enzyme oxygenase (mono-oxygenase and di-oxygenases) incorporating two atoms of oxygen in aromatic rings of PAHs and thereby producing *cis* dihydrodiols that are again transformed into simpler structures.[42] Some common intermediates produced during the bioremediation of PAHs are catechol, gentisic acid, and protocatechuic acid that are later transformed into aldehydes and acids.[15] Like bacteria, some fungal sp. such as *Caenorhabditis elegans*, *Candida* sp., *Penicillium janthinellum*, etc. are also used in bioremediation of PAHs. Both lignolytic and non-lignolytic fungi oxidize PAHs through cytochrome P-450 monooxygenase and lignin degrading enzymes.[78] Some algal sp. have also the ability to degrade PAHs. As reported by Suresh Kumar et al.[105] the microalgae *Haematococcus pluvialis* and *Parachlorella Kessler* have the potential to degrade both HMW and LMW PAHs with the help of the enzyme cytochrome P-450 monooxygenase (CYP). Nevertheless, in microbial remediation practices, sometimes adequate microbial biomass is not produced, which ultimately hampers the effective degradation of tightly bound hydrophobic PAHs in soil and this is very true in the case of HMW PAHs pollutants.[75] Furthermore, the specificity of the bio-remediation practices (developed in recent past) to a single PAH compound limits the overall success at the sites which are polluted by multiple PAHs. Therefore, value addition to the existing technique is necessary.

2.3 WHY ARE NANOCOMPOSITES AGAINST PAHS?

Nanotechnology is an emergent ground for making the nanoscale products which have proficient reactive capacity as well as large surface area. Due to these distinctive features, nanomaterials depict greater capacity to remediate contaminated areas such as PAHs polluted and/or other contaminant-polluted areas. In remediation of polluted areas through microbes, the nanoparticles serve as a stimulator for enhanced microbial growth, prompting the production of bio-surfactants and other remediating enzymes to improve the solubility/bioavailability of PAHs.[51] The mechanism behind the removal of organic and inorganic pollutants through nano-bioremediation can be categorized into three main heads: a) Adsorption, b) Transformation, and c) Catalysis.

a. ***Adsorption:*** Adsorption is a surface-related phenomenon. It involves the binding of pollutants to the surface of the adsorbing substances. The interaction involved in adsorption is ionic interaction which is exothermic in nature. The concept of enthalpy can be understood by the adsorption phenomenon by calculating heat generated by 1-mole adsorbate in the adsorbent. When temperature and pressure are kept constant, adsorption occurs spontaneously, further decreasing Gibb's free energy.[11]

b. ***Transformation:*** Transformation is the result of mainly oxidation and reduction reactions. These reactions can lead to a reduction in PAHs as well as metal toxicity, enhancement in metal precipitation and PAHs degradation.[111] ZVI nanoparticles can transform heavy metals like Cd, Zn, Cr, Pb, Ni, etc.[101]

c. ***Catalysis:*** Catalysis is a crucial part of various processes such as environmental and industrial processes. Catalytic technologies can be used in the production and emission of CO, NO, and hydrocarbons.

2.4 APPLICATION POTENTIAL OF NANOCOMPOSITES IN ENVIRONMENTAL POLLUTION REMEDIATION

Effective removal of contaminants from the environment is a substantial and challenging task. Recently, green nanomaterials (NMs)-assisted remediation has been proved as ecologically sound and cost-effective techniques for remediation of PAH pollutants from a contaminated environment. The commonly used nanomaterials for environmental remediation are biopolymers, single enzyme nanoparticles (SENs), zero-valent metals, metallic oxides, etc. The use of the nanocomposites in remediation of organic and inorganic contaminants is discussed below:

2.4.1 IN PAHS REMEDIATION

Recent researches established that nanocomposites have a great role in the removal of inorganic and organic compounds, such as PAHs, from domestic and industrial effluents.[63] Inbaraj et al.[41] carried out the green synthesis of a magnetic-activated carbon by using green tea leaves (MNPs-GTAC), aiming at its application for the remediation of BaA, BbF, CHR, and BaP, from wastewater and reaching 80% removal in 5 min. Shanker et al.[97] presented in their studies a green synthesis method of an iron hexacyanoferrate nanostructure (FeHCF) using *Sapindus-mukorossi* as a natural surfactant. Another worker[79] reported that nanomaterials were applied to the photocatalytic degradation and adsorption of PAHs from wastewater (ANT, PHEN, CHR, FLU, and BaP). FeHCFs generated a significant reduction in PAHs (70–90% of reduction).[79] Hassan et al.[39] studied the green synthesis of iron oxide nanoparticles (IONPs) by using an extract from pomegranate peel. The nanomaterial obtained was applied in the adsorption of BaP and PYR from synthetic wastewater. IONPs showed a 98.5% reduction for PRY and 99% for BaP. The studies developed by Inbaraj et al.,[41], Shanker et al.[97], and Hassan et al.[39] successfully performed the green synthesis of magnetic nanoparticles using natural extracts. The green magnetic nanoparticles synthesized in the studies were applied in the adsorption of PAHs from wastewater, and the results indicated high removal efficiency of PAHs in all studies. On account of the negative impacts of PAHs in the surrounding environment, some studies have been developed aiming at the application of green graphene-based nanomaterials for the remediation of these contaminants from wastewater as well as soil.[41] The results obtained by Cheng et al.[19] also showed that graphene oxide (GO) synthesized by the green route is an environmental-friendly and auspicious adsorbent in the field of PAHs remediation from wastewater. A functional composite BBGO (developed by mixing GO with brilliant blue) was applied as an effective adsorbent to remediate two types of PAH pollutants that include anthracenemethanol (AC) and FLT from an artificial discharge, and the adsorption results indicated high adsorption capacity for AC and FLT, a 72% reduction for AC was obtained after 11 days and a 93.2% reduction for FLT was obtained after 1.4 h.[79] Some of the nanomaterials used in PAHs remediation have been listed in Table 2.3.

2.4.2 IN OTHER POLLUTANT REMEDIATION

Nanocomposites play a great role in the remediation of other pollutants also like heavy metals, polychorynatedbiphenyls (PCBs), etc. The nanocomposites possess some distinctive properties to stabilize the soil rapidly as well as proficiently. Attapulgite-based nanocomposites can be used as repairing agents restore mining areas, crop-lands polluted by means of HMs and other pollutants (both organic and inorganic) in an ecological manner (Medina-Pérez, 2018). Another mixture containing 5 to 20 parts of mass-modified carbon nanotubes, 10 to 60 parts of mass-modified clay mineral, and 10 to 75 parts of mass-lime was employed by Chen et al.[18] as a remediating agent for heavy metal lead-cadmium and lead-cadmium sulfide composite contaminated soil which resulted in 75-90% reduction of cadmium. Fang and Yang[31] patented a composite material that is used for in-situ remediation of lead-contaminated soil. As reported by Medina-Pérez (2018), an ecological bag filled with multifunctional nanocomposite material was also patented for the elimination of HMs, total phosphorus, and total nitrogen from the water and soil matrix. Another research carried out by earlier worker Shao et al.[96] using grafted methyl methacrylate on multi-walled carbon nanotubes revealed enhanced removal of PCB from water. Shao et al.[98] again reported that the composite of β-cyclodextrin and multi-walled carbon nanotubes revealed a considerable PCB remediation efficacy (>95%). A study performed by Ren et al.,[85] highlighted the expediency of graphene-based nanomaterials as an excellent adsorbent for PCB28 and as a promising nanomaterial for application in environmental remediation. Some of the nanomaterials applied for decontaminations of pollutants such as heavy metals (HMs), PCBs, and others are listed in Table 2.4.

2.5 LIMITATIONS

Nanocomposites/nanomaterials have made a significant influence in the world's economy due to their outstanding potential applications. However the high surface area-to-mass ratio, crystalline structure, and reactivity

TABLE 2.3 List of Some Nanomaterials Used in PAHs Remediation.

Nanomaterial	Target pollutant	Percentage of removal	Means of removal	References
Multiwalled-carbon nanotube (MWCNTs)	PAH	99	Extraction	Paszkiewicz et al.[71]
TiO_2/SiO_2–carbon	PAH	>90	Adsorption	Rasheed et al.[81]
Nanotubes sulfonated graphene oxide	PAH	95	Adsorption	Zhao et al.[128]
Graphene oxide	PAH	~99	Adsorption	Wang et al.[116]
Nano-reduced graphene oxide-hybridized polymeric high-internal phase emulsions (RGO/polyHIPEs)	PAH	50–90	Adsorption	Huang et al.[40]
Magnetic graphene oxide	PAH	100	Adsorption	Huang et al.[38]
Fe_3O_4–carbon nanotubes	PAH	99	Magnetic solid-phase Extraction	Abdar et al.[1]
TiO_2@ZnHCF	Phenanthrene	95	Photocatalytic degradation	Rachna et al.[82]
Fe_3O_4–carbon nanotubes	PAH	–	Adsorption	Zhang and Fang[127]
Graphene oxide–Fe_3O_4	Naphthalene	–	Extraction	Nazari et al.[64]
Fe_3O_4–carbon nanosheets	PAH	100	Hydrothermal	Zhang and Fang[127]
Graphene/Fe_3O_4 grafted onto polythiophene(G/Fe_3O_4@PT)	PAH	95	MSPE	Mehdinia et al.[62]
Fe_3O_4–FeOOH	Anthracene	–	Photodegradation	Gupta et al.[36]
Graphitic carbon nitride (g-C_3N_4/Fe_3O_4)	PAH	100	Adsorption	Nian et al.[66]
Fe_3O_4–chitosan	PAH	96	Sorption	Nisticò et al.[67]
Graphitic carbon nitride (g-C_3N_4/Fe_3O_4)	PAH	80–99.8	MSPE	Wang et al.[117]
Fe_3O_4–wood biochar	PAH	90	Adsorption	Dong et al.[26]
TiO_2nanodots fixed on carbon nanotubes (TiO_2/Co@NCT)	PAH	98.48	Photocatalytic remediation	Li et al.[53]
Nanometer TiO_2	PAH	95	Photocatalytic remediation	Soni et al.[104]
Sr–TiO_2	Phenanthrene	100	Photocatalytic remediation	Liu et al.[57]

TABLE 2.3 *(Continued)*

Nanomaterial	Target pollutant	Percentage of removal	Means of removal	References
Graphite oxide–TiO_2–$Sr(OH)_2$/$SrCO_3$	Phenanthrene	100	Photocatalytic degradation	Fu et al.[33]
TiO_2–graphene oxide	PAH	80	Photocatalytic degradation	Bai et al.[5]
Iron-based chitosan nanocomposites	Phenanthrene	92	Photocatalytic Remediation	Rani et al.[83]
Fullerenes C60	Naphthalene	100	Adsorption	Cheng et al.[20]
Graphitic carbon nitride (g-C_3N_4)	Nitro-PAH	100	Adsorption	Lin et al.[56]
Nanoscale zero iron coupled with silica oxides (Fe@SiO_2@ PDA)	Phenanthrene Anthracene	100	Adsorption	Li et al.[52,53]
Bismide perylene dopamine (PBI-DA)	PAH	83–100	Adsorption	Gutierrez et al.[37]
Magnetic nanoparticles	PAH	100	Adsorption	Torabian et al.[110]

Source: Adapted from Ref. [13, 16].

TABLE 2.4 List of Nanomaterials Used in Decontaminations of Environmental Pollutants

Nanomaterial	Target pollutant	Percentage of removal	Means of removal	References
Methyl methacrylate on multiwalled carbon nanotube (MWCNT-gpMMA)	PCB	95	Adsorption	Shao et al.[96]
Single-walled carbon nanotubes (SWCNTs)	PCB	100	Adsorption	Mahdavian and Mousavi[60]
Multiwalled carbon nanotubes grafted Cyclodextrin (MWCNT-g-CD)	PCB	>95	Adsorption	Shao et al.[98]
Graphene oxide	PCB-28	99	Adsorption	Ren et al.[85]
Amino-functionalized polypropylene	PCB	85	Adsorption	Tian et al.[109]
Nonwoven graphene oxide (PP-g-DMAEMA/GO)				
Metal organic nanotube (Fe_3O_4@Co-MONT)	PCB	100	MSPE	Li et al.[54]
Metal grafted graphene oxide (Fe_3O_4@GO)	PCB 28	100	MSPE	Zeng et al.[125]
Fe_3O_4 with ammonium chloride dispersed	PCB	99.1	MSPE	Gan et al.[34]
On graphene oxide sheets (Fe_3O_4@PDDA/GOx@DNA)				
Nano-Fe/Pd bimetallic	PCB	100	Adsorption	Choi et al.[21]
Fe_3O_4 beta-cyclodextrin	PCB	100	Adsorption	Wang et al.[116]
Chitosan-coated iron-oxide nanocomposites	As(III)	13	Adsorption	Neeraj et al.[65]
Chitosan and Alginate nanocomposites	Cr(VI)	-	Adsorption	Gokila et al.[35]
Dead yeast biomass and titania nanoparticles	Cr(VI)	99.92	adsorption	Choudhury et al.[22]
Zero-valent copper-chitosan nanocomposites	Cr(VI)	-	adsorption	Wu et al.[121]
Chitosan-g-poly (acrylamide)/ZnS (ChPA/ZS) nanocomposite	MO and CR	CR-75 MO -69	Photocatalytic degradation	Pathania et al.[73]

TABLE 2.4 *(Continued)*

Nanomaterial	Target pollutant	Percentage of removal	Means of removal	References
Chitosan-TiO_2 nanocomposite	RhB and CR	RhB-65 CR-57	Photocatalytic degradation	Karthikeyan et al.[45]
Gelatin-Zr(IV) Phosphate nanocomposite (GT/ZPNC)	Methylene blue (MB) &	MB-87.81 & FG-89.91	Photocatalytic degradation	Thakur et al.[107]
Guar gum/Al_2O_3 nanocomposite	MG	~80–90	Photocatalytic degradation	Pathania et al.[74]
Hybrid TiO_2/ microcrystalline	MB	40–90	Adsorption	Virkutyte et al.[115]
Alginate/carboxymethyl cellulose TiO_2 nanoparticles (TiO_2-NPs) and graphene oxide(GO)	CR	98	Photocatalytic degradation	Thomas et al.[108]
Organo-bentonite/Fe_3O_4(OB-Fe_3O_4 PSA)	Th(IV)	-	Adsorption	Wu et al.[120]
Bamboo charcoal iron oxide (BC@ Fe_3O_4)	PCB	98.4	SPME	Zhao et al.[129]
Bamboo charcoal iron oxide (BC@Fe)	PCB	90	SPME	Liu et al.[58]
Carbon-modified titanium oxide nanoparticles (CMn- TiO_2)	PCB	93	Photocatalytic degradation	Shaban et al.[94]
Nanoscale zero-valent iron (NZVI)	PCB	100	Adsorption	EL-Temsah et al.[29]
Velvet-like magnetic carbon nitride nanocomposites (V-g-C_3N_4)	PCB	100	Adsorption	Li et al.[52]
Gold nanoparticles scattered on graphene oxide sheets (RGOAuNp)	PCB 77	100	Adsorption	Wu et al.[119]
Nanoclays	PCB	77	Adsorption	Shariat and Raihan[99]
Nanoscale zero-valent iron (NZVI)	PCB	97.5	Degradation	Sevcu et al.[93]

Source: Adapted from Ref. [6, 13].

of nanocomposites may enable them to interact with the cell components and produce lethal effect. This possesses a serious concern for nanocomposites as an emerging technology due to their major health and safety hazards when released to the environment. Moreover, they do show certain antagonistic effect when combined to microbes and possess potential risk of forming colloidal reactions with contaminants which will get sunk into the soil beyond the extent of remediation. Therefore, in-depth study and proper research on toxicological effect and irregular cluster properties of nanocomposites are utmost important before its wide-scale synthesis and applications.

2.6 CONCLUSION AND FUTURE DIRECTION

The rapid development of nanoscience over the past decades is responsible for rising in nanomaterials-assisted remediation research. The rapid development has made it possible to interpret ideas from research laboratories to consumer use with a wide range of accessibility and low cost. The main advantage of using biogenic nanomaterials is the cost-effective and eco-friendly way of removal of toxic pollutants from the environment. The constrained situation within the microbial cells, viz. cytoplasmic vascular compartments, and periplasmic space modulate the size and shape of nanomaterials and need further research for efficient utilization. It is always important to develop our remediation strategies for sustainable development in the future; in this regard nanotechnology is a great boon. Apart from health hazards exerted by nanomaterials, the role played by nanomaterials in bioremediation is vital. It is one of the important responsibilities of researchers to minimize the potential risk of health hazards of nanomaterials and use nanomaterials for environmental remediation and sustainable development.

KEYWORDS

- **nanomaterials**
- **nanocomposites**
- **PAHs remediation**
- **green synthesis**

REFERENCES

1. Abdar, A.; Sarafraz-Yazdi, A.; Amiri, A. andBagheri, N. Magnetic Solid-Phase Extraction of Polycyclic Aromatic Hydrocarbons in Water Samples by Fe_3O_4@polypyrrole/Carbon Nanotubes. . *J. Sep. Sci.* **2016,** 39, 2746–2753.
2. Abdel-Shafy, H. I.; Mansour, M. S. A Review on Polycyclic Aromatic Hydrocarbons: Source, Environmental Impact, Effect on Human Health and Remediation. *Egypt. J. Pet.* **2016,** *25,* 107–123.
3. Abha, S.; Singh, C. S. Hydrocarbon Pollution: Effects on Living Organisms, Remediation of Contaminated Environments, and Effects of Heavy Metals Co-Contamination on Bioremediation. In *Introduction to Enhanced Oil Recovery (EOR) Processes and Bioremediation of OilContaminated Site*; Romero-Zerón, L., Ed.; INTECH press: Croatia, 2012; pp 186–206.
4. Ahmed, S.; Ahmad, M.; Swami, B. L.; Ikram, S. A Review on Plants Extract Mediated Synthesis of Silver Nanoparticles for Antimicrobial Applications: A Green Expertise. *J. Adv. Res.* **2016,** *7* (1), 17–28.
5. Bai, H.; Zhou, J.; Zhang, H.; Tang, G. Enhanced Adsorbability and Photocatalytic Activity of TiO_2-Graphene Composite for Polycyclic Aromatic Hydrocarbons Removal in Aqueous Phase. *Colloids Surf. B.* **2017,** 150, 68–77.
6. Bajpai, A.; Sharma, M.; Gond, L. Nanocomposites for Environmental Pollution Remediation. In *Sustainable Polymer Composites and Nanocomposites*. Springer: Cham, 2019; pp 1407–1440.
7. Bamforth, S. M.; Singleton, I. Bioremediation of Polycyclic Aromatic Hydrocarbons: Current Knowledge and Future Directions. *J. Chem. Technol. Biotechno.* **2005,** *80* (7), 723–736.
8. Baek, K. H.; Kim, H. S.; Oh, H. M.; Yoon, B. D.; Kim, J.; Lee, I. S. Effects of Crude Oil, Oil Components, and Bioremediation on Plant Growth. *J. Environ. Sci. Health A* **2004,** *39* (9), 2465–2472.
9. Beless, B.; Rifai, H. S.; Rodrigues, D. F. Efficacy of Carbonaceous Materials for Sorbing Polychlorinated Biphenyls from Aqueous Solution. *Environ. Sci. Technol.* **2014,** *48,* 10372–10379.
10. Bisht, S.; Pandey, P.; Bhargava, B.; Sharma, S.; Kumar, V.; Sharma, K. D. Bioremediation of Polyaromatic Hydrocarbons (PAHs) Using Rhizosphere Technology. *Braz. J. Microbiol.* **2015,** *46,* 7–21.
11. Bhargava, S.; Sawant, K. Drought Stress Adaptation: Metabolic Adjustment and Regulation of Gene Expression. *Plant Breed.* **2013,** *132* (1), 21–32.
12. Borah, G.; Nath, N.; Deka, H. Effects on Anatomy of Some Abundantly Growing Herbs in the Effluents Contaminated Soil of Oil Refinery. *Environ. Sci. Pollut. Res.* **2021,** *28* (9), 11549–11557.
13. Borji, H.; Ayoub, G. M.; Al-Hindi, M.; Malaeb, L.; Hamdan, H. Z. Nanotechnology to Remove Polychlorinated Biphenyls and Polycyclic Aromatic Hydrocarbons from Water: Review. *Environ. Chem. Lett.* **2020,** *18* (3), 729–746.
14. Cameotra, S. S.; Bollag, J. M. Biosurfactant-Enhanced Bioremediation of Polycyclic Aromatic Hydrocarbons. *Crit. Rev. Environ. Sci. Technol.* **2003,** *33* (2), 111–126.
15. Cébron, A.; Norini, M. P.; Beguiristain, T.; Leyval, C. Real-Time PCR Quantification of PAH-Ring Hydroxylatingdioxygenase (PAH-RHDα) Genes from Gram Positive and

Gram Negative Bacteria in Soil and Sediment Samples. *J. Microbiol. Meth.* **2008,** *73* (2), 148–159.
16. Chauhan, H. A.; Rafatullah, M.; Ahmed Ali, K.; Siddiqui, M. R.; Khan, M. A.; Alshareef, S. A. Metal-Based Nanocomposite Materials for Efficient Photocatalytic Degradation of Phenanthrene from Aqueous Solutions. *Polymers* **2021,** 13 (14), 2374.
17. Chakravarty, P.; Deka, H. Enzymatic defense of Cyperus brevifolius in hydrocarbons stress environment and changes in soil properties. *Sci. Rep.* **2021,** *11* (1), 1–12.
18. Chen, M.; Wang, X.; Wang, R.; Zhang, J.; Li, X. Remediating Agent Used for Heavy Metal Lead-Cadmium and Lead-Cadmium Sulfide Composite Contaminated Soil, Preferably Mine Soil or Agricultural Land, Comprises Modified Carbon Nanotubes, Modified Clay Mineral and Lime. Patent CN104893732-A, 2015.
19. Cheng, H.; Bian, Y.; Wang, F.; Jiang, X.; Ji, R.; Gu, C.; Yang, X.; Song, Y. Green Conversion of Crop Residues Into Porous Carbons and Their Application to Efficiently Remove Polycyclic Aromatic Hydrocarbons from Water: Sorption Kinetics, Isotherms and Mechanism. *Bioresour. Technol.* **2019,** *284*, 1–8.
20. Cheng, X.; Kan, A. T.; Tomson, M. B. Naphthalene Adsorption and Desorption from Aqueous C60 Fullerene. *J. Chem. Eng. Data* **2004,** *49* (3), 675–683.
21. Choi, H.; Al-Abed, S. R.; Agarwal. S.; Dionysiou, D. D. Synthesis of Reactive Nano-Fe/Pd Bimetallic System-Impregnated Activated Carbon for the Simultaneous Adsorption and Dechlorination of PCBs. *Chem. Mater.* **2008,** *20,* 3649–3655.
22. Choudhury, P. R.; Bhattacharya, P.; Ghosh, S.; Majumdar, S.; Saha, S.; Sahoo, G. C. Removal of Cr (VI) by Synthesized Titania Embedded Dead Yeast Nanocomposite: Optimization and Modeling by Response Surface Methodology. *J. Environ. Chem. Eng.* **2017,** *5* (1), 214–221.
23. Dandajeh, H. A.; Ladommatos, N.; Hellier, P.; Eveleigh, A. Influence of Carbon Number of C1–C7 Hydrocarbons on PAH Formation. *Fuel* **2018,** 228, 140–151.
24. Dat, N. D.; Chang, M. B. Review on Characteristics of PAHs in Atmosphere, Anthropogenic Sources and Control Technologies. *Sci. Total Environ.* **2017,** *609,* 682–693.
25. Delides, C. G. Everyday Life Applications of Polymer Nanocomposites. no. October, 2016.
26. Dong, C. D.; Chen, C. W.; Kao, C. M.; Chien, C. C.; Hung, C. M. Woodbiochar-Supported Magnetite Nanoparticles for Remediation of PAH-Contaminated Estuary Sediment. *Catalysts* **2018,** *8,* 73.
27. Du, L.; Jiang, H.; Liu, X.; Wang, E. Biosynthesis of Gold Nanoparticles Assisted by *Escherichia coli* DH5α and Its Application on Direct Electrochemistry of Hemoglobin. *Electrochem. Commun.* **2007,** *9* (5), 1165–1170.
28. Elia, P.; Zach, R.; Hazan, S.; Kolusheva, S.; Porat, Z. E.; Zeiri, Y. Green Synthesis of Gold Nanoparticles Using Plant Extracts as Reducing Agents. *Int. J. Nanomedicine* **2014,** *9,* 4007.
29. El-Temsah, Y. S.; Sevcu, A.; Bobcikova, K.; Cernik, M.; Joner, E. J. DDT Degradation Efficiency and Ecotoxicological Effects of Two Types of Nano-Sized Zero-Valent Iron (nZVI) in Water and Soil. *Chemosphere* **2016,** *144,* 2221–2228.
30. Essien, O. E.; John, I. A. Impact of Crude-Oil Spillage Pollution and Chemical Remediation on Agricultural Soil Properties and Crop Growth. *J. Appl. Sci. Environ. Manag.* **2010,** *14,* 147–154.
31. Fang, Z.; Yang, Z. Preparation of Composite Material Used for Remediation of Lead-Contaminated Soil, Involves Dispersing Nano Hydroxyapatite in Water, Mixing

Potassium Chloride and Agricultural Waste and Performing Pyrolysis Process. Patent CN105131960-A, 2015.
32. Fawell, J. K.; Hunt, S. The Polyaromatic hydrocarbons. In *Environmental Toxicology: Organic Pollutants*; Fawell, J. K.; Hunt, S., Eds.; Ellis Harwood: West Sussex, 1988; pp 241–269.
33. Fu, J.; Kyzas, G. Z.; Cai, Z.; Deliyanni, E. A.; Liu, W.; Zhao, D. Photocatalytic Degradation of Phenanthrene by Graphite Oxide-TiO_2–$Sr(OH)_2$/$SrCO_3$ Nanocomposite Under Solar Irradiation: Effects of Water Quality Parameters and Predictive Modeling. *Chem. Eng. J.* **2018,** *335,* 290–300.
34. Gan, N.; Zhang, J.; Lin, S.; Long, N.; Li, T.; Cao, Y. A Novel Magnetic Graphene Oxide Composite Absorbent for Removing Trace Residues of Polybrominateddiphenyl Ethers in Water. *Materials (Basel)* **2014,** *7,* 6028–6044.
35. Gokila, S.; Gomathi, T.; Sudha, P. N.; Anil, S. Removal of the Heavy Metal Ion Chromium (VI) Using Chitosan and Alginate Nanocomposites. *Int. J. Biol. Macromol.* **2017,** 104, 1459–1468.
36. Gupta, H.; Kumar, R.; Park, H. S.; Jeon, B. H. Photocatalytic Efficiency of Iron Oxide Nanoparticles for the Degradation of Priority Pollutant Anthracene. *Geosyst. Eng.* **2016,** *20,* 21–27.
37. Gutierrez, M. S.; Duel, P.; Hierro, F.; Morey, J.; Pina, M. N. A Very Highly Efficient Magnetic Nanomaterial for the Removal of PAHs from Aqueous Media. *Small* **2018,** *14,* 1702573.
38. Huang, D.; Xu, B.; Wu, J.; Brookes, P.C. andXu, J. Adsorption and Desorption of Phenanthrene by Magnetic Graphene Nanomaterials from Water: Roles of pH, Heavy Metal Ions and Natural Organic Matter. *Chem. Eng. J.* **2019,** *368,* 390–399.
39. Hassan, S. S. M.; Abdel-Shafy, H. I.; Mansour M. S. M. Removal of Pyrene and Benzo(a) Pyrene Micropollutant from Water via Adsorption by Green Synthesized Iron Oxide Nanoparticles. *Adv. Nat. Sci. Nanosci. Nanotechnol.* **2018,** *9,* 015006.
40. Huang, Y.; Zhang, W.; Ruan, G.; Li, X.; Cong, Y.; Du, F.; Li, J. Reduced Graphene Oxide-Hybridized Polymeric High-Internal Phase Emulsions for Highly Efficient Removal of Polycyclic Aromatic Hydrocarbons from Water Matrix. *Langmuir* **2018,** *34,* 3661–3668.
41. Inbaraj, B. S.; Sridhar, K.; Chen, B. H. Removal of Polycyclic Aromatic Hydrocarbons from Water by Magnetic Activated Carbon Nanocomposite from Green Tea Waste. *J. Hazard. Mater.* **2021,** *415,* 125701.
42. Jouanneau, Y.; Martin, F.; Krivobok, S.; Willison, J. C. Ring-Hydroxylatingdioxygenases Involved in PAH Biodegradation: Structure, Function and Biodiversity. In *Microbial Bioremediation of Non Metals: Current Research.* Caister Academic Press: Norflok, UK, 2011; pp 149–175.
43. Jones, M.; Kujundzic, M.; John, S.; Bismarck, A. Crab vs. Mushroom: A Review of Crustaceanand Fungal Chitin in Wound Treatment. *Mar. Drugs.* **2020,** *18* (1), 64.
44. Juhasz, A. L.; Naidu, R. Bioremediation of High Molecular Weight Polycyclic Aromatic Hydrocarbons: A Review of Themicrobial Degradation of Benzo[*a*]Pyrene. *Int. Biodeterior. Biodegrad.* **2000,** *45,* 57–88.
45. Karthikeyan, K. T.; Nithya, A.; Jothivenkatachalam, K. Photocatalytic and Antimicrobial Activities ofChitosan-TiO_2 Nanocomposite. *Int. J. Biol. Macromol.* **2017,** *104,* 1762–1773.
46. Khandoker, N.; Hawkins, S. C.; Ibrahim, R.; Huynh, C. P.; Deng, F. Tensile Strength of Spinnable Multiwall Carbon Nanotubes. *Procedia Eng.* **2011,** *10,* 2572–2578.

47. Khan, I.; Saeed, K.; Khan, I. Nanoparticles: Properties, Applications and Toxicities. *Arab. J. Chem.* **2019,** *12* (7), 908–931.
48. Kirthi, A. V.; Rahuman, A. A.; Rajakumar, G.; Marimuthu, S.; Santhoshkumar, T.; Jayaseelan, C.; Elango, G.; Zahir, A. A.; Kamaraj, C.; Bagavan, A. Biosynthesis of Titanium Dioxide Nanoparticles Using Bacterium Bacillus Subtilis. *Mater. Lett.* **2011,** *65* (17–18), 2745–2747.
49. Kumar, S. A.; Ansary, A. A.; Ahmad, A.; Khan, M. I. Extracellular Biosynthesis of CdSequantum Dots by the Fungus, Fusariumoxysporum. *J. Biome. Nanotech.* **2007,** *3* (2), 190–194.
50. Kumar, V.; Kothiyal, N. C.; Saruchi; Masih, A. Environmental Fate and Behavior of Some PAHs at Roadside Ambient Air in a Fast Developing City Environment of Northern India. *J. Chinese Adv. Mat. Soc.* **2014,** *2* (2), 82–98.
51. Kumari, B.; Singh, D. P. A Review on Multifaceted Application of Nanoparticles in the Field of Bioremediation of Petroleum Hydrocarbons. *Ecol. Eng.* **2016,** *97,* 98–105.
52. Li, D.; Zhu, J.; Wang, M.; Bi, W.; Huang, X.; Chen, D. D. Extraction of Trace Polychlorinated Biphenyls in Environmental Waters by Welldispersed Velvet-Like Magnetic Carbon Nitride Nanocomposites. *J. Chromatogr. A* **2017a,** *1491*, 27–35.
53. Li, J.; Zhou, Q.; Liu, Y.; Lei, M. Recyclable Nanoscale Zero-Valent Iron-Based Magnetic Polydopamine Coated Nanomaterials for the Adsorption and Removal of Phenanthrene and Anthracene. *Sci. Technol. Adv. Mater.* **2017b,** *18*, 3–16.
54. Li, Q. L.; Wang, L. L.; Wang, X.; Wang, M. L.; Zhao, R. S. Magnetic Metal-Organic Nanotubes: an Adsorbent for Magnetic Solid-Phase Extraction of Polychlorinated Biphenyls from Environmental and Biological Samples. *J. Chromatogr. A* **2016,** *1449,* 39–47.
55. Li, X.; Yu, J.; Li, G.; Liu, H.; Wang, A.; Yang, L.; Zhou, W.; Chu, B.; Liu, S. TiO_2 Nanodots Anchored on Nitrogen-Doped Carbon Nanotubes Encapsulated Cobalt Nanoparticles As Photocatalysts with Photo-Enhanced Catalytic Activity Towards the Pollutant Removal. *J. Colloid. Interface. Sci.* **2018,** *526,* 158–166.
56. Lin, Z.; Zheng, J.; Lin, G.; Tang, Z.; Yang, X.; Cai, Z. Negative Ion Laser Desorption/Ionization Time-of-Flight Mass Spectrometric Analysis of Small Molecules Using Graphitic Carbon Nitride Nanosheetmatrix. *Anal. Chem.* **2015,** 87, 8005–8012.
57. Liu, S.; Yu, J.; Jaroniec, M. Anatase TiO_2 with Dominant Highenergy 001 Facets: Synthesis, Properties, and Applications. *Chem. Mater.* **2011,** *23,* 4085–4093.
58. Liu, Y. L.; Chen, X. F.; Wang, X.; Zhou, J. B.; Zhao, R. S. Sensitive Determination of Polychlorinated Biphenyls in Environmental Water Samples by Headspace Solid-Phase Microextraction with Bamboo Charcoal@iron Oxide Black Fibers Prior to Gas Chromatography with Tandem Mass Spectrometry. *J. Sep. Sci.* **2014,** *37,* 1496–1502.
59. Mahajan, R.; Bhadwal, A. S.; Kumar, N.; Madhusudanan, M.; Pudake, R. N.; Tripathi, R. M. Green Synthesis of Highly Stable Carbon Nanodots and Their Photocatalytic Performance. *IET Nanobiotechnol.* **2016,** *11* (4), 360–364.
60. Mahdavian, L.; Mousavi, S. E. Diameter of Single-Walled Carbon Nanotubes (SWCNTs) as an Effective Factor in the Detection and Degradation of PCBs. Polycycl. *Aromat. Compd.* **2016,** *39,* 14–22.
61. Medina-Pérez, G.; Fernández-Luqueño, F.; Vazquez-Nuñez, E.; López-Valdez, F.; Prieto-Mendez, J.; Madariaga-Navarrete, A.; Miranda-Arámbula, M. Remediating Polluted Soils Using Nanotechnologies: Environmental Benefits and Risks. *Pol. J. Environ. Stud.* **2019,** *28* (3).

62. Mehdinia, A.; Khodaee, N.; Jabbari, A. Fabrication of Graphene/Fe_3O_4@polythiophene Nanocomposite and Its Application in the Magnetic Solid-Phase Extraction of Polycyclic Aromatic Hydrocarbons from Environmental Water Samples. *Anal. Chim. Acta* **2015,** *868,* 1–9.
63. Nasrollahzadeh, M.; Sajadi, S. M.; andArezo, H. Waste Chicken Eggshell as A Natural Valuable Resource and Environmentally Benign Support for Biosynthesis of Catalyticallyactive Cu/Eggshell, Fe_3O_4/Eggshell and Cu/Fe_3O_4/Eggshell Nanocomposites. *Appl. Catal. Environ.* **2016,** *191,* 209–227.
64. Nazari, N.; Masrournia, M.; EsHaghi, Z.; Bozorgmehr, M. Simultaneous Extraction and Preconcentration of Aniline, Phenol, and Naphthalene Using Magnetite-Graphene Oxide Composites Before Gas Chromatography Determination. *J Sep. Sci.* **2016,** *39,* 3046–3053.
65. Neeraj, G.; Krishna, S.; Senthil Kumar, P.; Cabana, H.; kumar, V. V. Adsorptive Potential of Dispersible Chitosan Coated Iron-Oxide Nanocomposites Toward the Elimination of Arsenic from Aqueous Solution. *Process Saf. Environ. Prot.* **2016,** *104,* 185–195.
66. Nian, Q.; Wang, X.; Wang, M.; Zuo, G. A Hybrid Material Composedof Graphitic Carbon Nitride and Magnetite (Fe_3O_4) for Magnetic Solid-Phase Extraction of Trace Levels of Hydroxylated Polycyclic Aromatic Hydrocarbons. *Microchim. Acta.* **2019,** *186,* 497..
67. Nisticò, R.; Franzoso, F.; Cesano, F.; Scarano, D.; Magnacca, G.; Parolo, M. E.; Carlos, L. Chitosan-Derived Iron Oxide Systems for Magnetically Guided and Efficient Water Purification Processes from Polycyclic Aromatic Hydrocarbons. *ACS Sustain. Chem. Eng.* **2016,** *5,* 793–801.
68. Odukoya, J.; Lambert, R.; Sakrabani, R. Understanding the Impacts of Crude Oil and Its Induced Abiotic Stresses on Agrifood Production: A Review. *Horticulturae* **2019,** *5,* 47.
69. Onuoha, S. C.; Olugbue, V. U.; Uraku, J. A.; Uchendu, D. O. Biodegradation Potentials of Hydrocarbon Degraders from Waste Lubricating Oil-Spilled Soils in Ebonyi State, Nigeria. *Int. J. Agric. Biol.* **2011,** *13,* 586–590.
70. Partila, A. M. Biodegradation of Polycyclic Aromatic Hydrocarbons in Petroleum Oil Contaminating the Environment. Ph.D. Thesis, faculty of science, Cairo University, 2013.
71. Paszkiewicz, M.; Sikorska, C.; Leszczynska, D.; Stepnowski, P. Helical Multi-Walled Carbon Nanotubes As an Efficient Material for the Dispersive Solid-Phase Extraction of Low and High Molecularweight Polycyclic Aromatic Hydrocarbons from Water Samples: Theoretical Study. *Water Air Soil Pollut.* **2018,** *229,* 253.
72. Patel, A. B.; Shaikh, S.; Jain, K. R.; Desai, C.; Madamwar, D. Polycyclic Aromatic Hydrocarbons: Sources, Toxicity, and Remediation Approaches. *Front. Microbiol.* **2020**, 2675.
73. Pathania, D.; Gupta, D.; Ala'a, H.; Sharma, G.; Kumar, A.; Naushad, M.; Ahamad, T.; Alshehri, S. M. Photocatalytic Degradation of Highly Toxic Dyes Using Chitosan-g-Poly (Acrylamide)/ZnS in Presence of Solar Irradiation. *J. Photochem. Photobiol. A* **2016a,** *329*, 61–68.
74. Pathania, D.; Katwal, R.; Sharma, G.; Naushad, M.; Khan, M.R.; Ala'a, H. 2016b. Novel Guar Gum/Al_2O_3 Nanocomposite as an Effective Photocatalyst for the Degradation of Malachite Green Dye. *Int. J. Biol. Macromol. 87*, 366–374.
75. Patowary, R.; Patowary, K.; Kalita, M. C.; Deka, S. Application of Biosurfactant for Enhancement of Bioremediation Process of Crude Oil Contaminated Soil. *Int. Biodeterior. Biodegrad.* **2018,** *129,* 50–60.

76. Patra, D. K.; Pradhan, C.; Patra, H. K. Toxic Metal Decontamination by Phytoremediation Approach: Concept, Challenges, Opportunities and Future Perspectives. *Environ. Technol. Innov.* **2020,** *18,* 100672.
77. Philip, D. Biosynthesis of Au, Ag and Au–Ag Nanoparticles Using Edible Mushroom Extract. *Spectrochim. Acta - A: Mol. Biomol. Spectrosc.* **2009,** *73* (2), 37 4–381.
78. Pozdnyakova, N. N. Involvement of the Ligninolytic System of White-Rot and Litter-Decomposing Fungi in the Degradation of Polycyclic Aromatic Hydrocarbons. *Biotechnol. Res. Int.* **2012,** 243217.
79. Queiroz, R. N.; Prediger, P.; Vieira, M. G. A. Adsorption of Polycyclic Aromatic Hydrocarbons from Wastewater Using Graphene-based Nanomaterials Synthesized by Conventional Chemistry and Green Synthesis: A Critical Review. *J. Hazard. Mater.* **2022,** *422,* 126904.
80. Rabieh, S.; Nassimi, K.; Bagheri, M. Synthesis of Hierarchical ZnO-Reduced Grapheneoxide Nanocomposites with Enhanced Adsorption–Photocatalytic Performance. *Mater. Lett.* **2016,** *162,* 28–31.
81. Rasheed, T.; Adeel, M.; Nabeel, F.; Bilal, M.; Iqbal, H. M. N. TiO_2/SiO_2 Decorated Carbon Nanostructured Materials as a Multifunctional Platform for Emerging Pollutants Removal. *Sci. Total Environ.* **2019,** *688,* 299–311.
82. Rachna; Rani, M.; Shanker, U.; Degradation of Tricyclic Polyaromatic Hydrocarbons in Water, Soil and River Sediment with a Novel TiO_2 based Heterogeneous Nanocomposite. *J. Environ. Manag.* **2019,** *248,* 109340.
83. Rani, M.; Rachna,Shanker, U. Metal Oxide-Chitosan Based Nanocomposites for Efficient Degradation of Carcinogenic PAHs. *J. Environ. Chem. Eng.* **2020,** *8,* 103810.
84. Rao, M. A.; Scelza, R.; Scotti, R.; Gianfreda, L. Role of Enzymes in the Remediation of Polluted Environments. *J. Soil Sci. Plant Nutr.* **2010,** *10* (3), 333–353.
85. Ren, W.; Chang, H.; Wang, Y.; Teng, Y.; Ma, W.; Luo, Y. Effect of Sulfonated Graphene on Uptake, Translocation, and Metabolism of 2, 4,4′-Trichlorobiphenyl in Maize Seedlings. *Environ. Sci.Pollut. Res.* **2018**. DOI: 10.1007/s11356-018-2203-z
86. Renu, S.; Shivashangari, K. S.; Ravikumar, V. Incorporated Plant Extract Fabricated Silver/Poly-D,l-Lactide-Co-Glycolide Nanocomposites for Antimicrobial Based Wound Healing. *Spectrochim. Acta - A: Mol. Biomol. Spectrosc.* **2020,** *228,* 117673.
87. Sabherwal, P.; Mutreja, R.; Suri, C. R. Biofunctionalized Carbon Nanocomposites: Newgeneration Diagnostic Tools. *TrAC Trends Anal. Chem.* **2016,** *82,* 12–21.
88. Salavati-Niasari, M.; Ghanbari, D. Polymeric Nanocomposite Materials. In *Advances in Diverse Industrial Applications of Nanocomposites*. IntechOpen, 2011.
89. Saxena, A.; Tripathi, R. M.; Zafar, F.; Singh, P. Green Synthesis of Silver Nanoparticles Using Aqueous Solution of Ficusbenghalensis Leaf Extract and Characterization of their Antibacterial Activity. *Mater. Lett.* **2012,** *67* (1), 91–94.
90. Schirmer, K.; Dixon, D. G.; Greenberg, B. M.; Bols, N. C. Ability of 16 Priority PAHs to be Directly Cytotoxic to a Cell Line from the Rainbow Trout Gill. *Toxicology* **1998,** *127* (1–3), 129–141.
91. Semple, K. T.; Doick, K. J.; Jones, K. C.; Burauel, P.; Craven, A.; Harms, H. Defining Bioavailability and Bioaccessibility of Contaminated Soil and Sediment is Complicated. *Environ. Sci. Technol.* **2004,** *38,* 228A–31A.
92. Sen, M.; Nanocomposite Materials. In *Nanotechnology and the Environment*. IntechOpen, 2020. DOI: 10.5772/intechopen.93047

93. Sevcu, A.; El-Temsah, Y. S.; Filip, J.; Joner, E. J.; Bobcikova, K.; Cernik, M.; Zero-Valent Iron Particles for PCB Degradation and An Evaluation of their Effects on Bacteria, Plants, and Soil Organisms. *Environ. Sci. Pollut. Res.* **2017,** *24,* 21191–21202.
94. Shaban, Y. A.; El Sayed, M. A.; El Maradny, A. A.; Al Farawati, R. K.; Al Zobidi, M. I.; Khan, S. U. M. Photocatalytic Removal of Polychlorinated Biphenyls (PCBs) Using Carbon-Modified Titanium Oxide Nanoparticles. *Appl. Surf. Sci.* **2016,** *365,* 108–113.
95. Shah, M. P. Environmental Bioremediation: A Low Cost Nature's Natural Biotechnology for Environmental Clean-Up. *J. Pet. Environ. Biotechnol.* **2014,** *5,* 4.
96. Shao, D.; Hu, J.; Jiang, Z.; Wang, X. Removal of 4,4-Dichlorinated Biphenyl from Aqueous Solution Using Methyl Methacrylate Grafted Multiwalled Carbon Nanotubes. *Chemosphere* **2011,** *82,* 751–758.
97. Shanker, U.; Jassal, V.; Rani, M. Green Synthesis of Iron Hexacyanoferrate Nanoparticles: Potential Candidate for the Degradation of Toxic PAHs. *J. Environ. Chem. Eng.* **2017,** *5,* 4108–4120.
98. Shao, D.; Sheng, G.; Chen, C.; Wang, X.; Nagatsu, M. Removal of Polychlorinated Biphenyls from Aqueous Solutions Using Beta-Cyclodextrin Grafted Multiwalled Carbon Nanotubes. *Chemosphere* **2010,** *79,* 679–685.
99. Shariat, M.; Raihan, T. M. *Nano Clay as an Adsorbent for Removes PCB from Soil.* In Paper Presented at the Conference: International Conference on Engineering and Built Environment, 2015.
100. Shivaji, S.; Madhu, S.; Singh, S. Extracellular Synthesis of Antibacterial Silver Nanoparticles Using Psychrophilic Bacteria. *Process Biochem.* **2011,** 46 (9), 1800–1807.
101. Singh, R.; Misra, V.; Singh, R. P.; Removal of Hexavalent Chromium from Contaminated Ground Water Using Zero-Valent Iron Nanoparticles. *Environ. Monit. Assess.* **2012,** *184* (6), 3643–3651.
102. Singh, R. P.; Singh, P.; Singh, K. R. Introduction to Composite Materials: Nanocomposites and Their Potential Applications. In *Composite Materials*. CRC Press, 2021, pp 1–28.
103. Song, J.; Kim, B. Rapid Biological Synthesis of Silver Nanoparticles Using Plant Leaf Extracts. *Bioproc. Biosyst. Eng.* **2009,** *32* (1), 79–84.
104. Soni, H.; Kumar, N.; Patel, K.; Kumar, R. N. Investigation on the Heterogeneous Photocatalytic Remediation of Pyrene and Phenanthrene in Solutions Using Nanometer TiO_2 Under UV Irradiation. *Polycycl. Aromat. Compd.* **2017**.
105. Suresh Kumar, P.; Thomas, J.; Poornima, V. Structural Insights on Bioremediation of Polycyclic Aromatic Hydrocarbons Using Microalgae: A Modelling-Based Computational Study. *Environ. Monit. Assess.* **2018,** *190* (2), 92.
106. Thakker, J. N.; Dalwadi, P.; Dhandhukia, P. C. Biosynthesis of Gold Nanoparticles Using *Fusarium oxysporum*f.sp. cubense JT1, a plant pathogenic fungus. *ISRN Biotechnol.* **2013,** *5.*
107. Thakur, M.; Sharma, G.; Ahamad, T.; Ghfar, A. A.; Pathania, D.; Naushad, M. Efficient Photocatalytic Degradation of Toxic Dyes from Aqueous Environment Using Gelatin-Zr(IV) Phosphate Nanocomposite and Its Antimicrobial Activity. *Colloids Surf. B Biointerfaces.* **2017,** *157,* 456–463.
108. Thomas, M.; Natarajan, T .S.; Sheikh, M. U. D.; Bano, M.; Khan, F. Self-Organized Graphene Oxide and TiO_2 Nanoparticles Incorporatedalginate/Carboxymethyl Cellulose Nanocomposites with Efficient Photocatalytic Activity Under Direct Sunlight. *J. Photochem. Photobiol. A* **2017,** *346,* 113–125.

109. Tian, J.; Wei, J.; Zhang, H.; Kong, Z.; Zhu, Y.; Qin, Z. Grapheneoxide-Functionalized Dual-Scale Channels Architecture for High Throughput Removal of Organic Pollutants from Water. *Chem. Eng. J.* **2019,** 359, 852–862.
110. Torabian, A.; Panahi, H. A.; Hendi, G. R. N. B.; Mehrdadi, N. Synthesis, Modifcation and Graft Polymerization of Magnetic Nano Particles for PAH Removal in Contaminated Water. *J. Environ. Health. Eng.* **2014,** *12,* 105.
111. Torimiro, N.; Daramola, O. B.; Oshibanjo, O. D.; Otuyelu, F. O.; Akinsanola, B. A.; Yusuf, O. O. Ore, O. T.; Omole, R. K. Ecorestoration of Heavy Metals and Toxic Chemicals in Polluted Environment Using Microbe-Mediated Nanomaterials. *Int. J. Environ. Bioremed. Biodegrad.* **2021,** *9,* 8–21.
112. Tripathi, R. M.; Bhadwal, A. S.; Gupta, R. K.; Singh, P.; Shrivastav, A.; Shrivastav, B. R.; ZnO Nanoflowers: Novel Biogenic Synthesis and Enhanced Photocatalytic Activity. *J. Photochem. Photobiol. B Biol.* **2014,** *141,* 288–295.
113. Tripathi, R. M.; Kumar, N.; Bhadwal, A. S.; Gupta, R. K.; Shrivastav, B. R.; Shrivastav, A. Facile and Rapid Biomimetic Approach for Synthesis of HApnanofibers and Evaluation of their Photocatalyticactivity. *Mater. Lett.* **2015,** 140, 64–67.
114. Velhal, S. G.; Kulkarni, S. D.; Latpate, R. V. Fungal Mediated Silver Nanoparticles Synthesis Using Robust Experimental Design and Its Application in Cotton Fabric. *Int. Nano Lett.* **2016,** *6* (4), 257–264.
115. Virkutyte, J.; Jegatheesan, V.; Varma, R. S. Visible Light Activated TiO2/Microcrystalline Cellulose Nanocatalyst to Destroy Organic Contaminants in Water. *Bioresour. Technol.* **2012,** *113,* 288–293.
116. Wang, F.; Haftka, J. J.; Sinnige, T. L.; Hermens, J. L.; Chen, W. Adsorption of Polar, Nonpolar, and Substituted Aromatics to Colloidal Graphene Oxide Nanoparticles. *Environ. Pollut.* **2014,** *186,* 226–233.
117. Wang, M.; Liu, P.; Wang, Y.; Zhou, D.; Ma, C.; Zhang, D.; Zhan, J. Core-Shell Superparamagnetic Fe_3O_4@beta-CD Composites for Host-Guest Adsorption of Polychlorinated Biphenyls (PCBs). *J. Colloid. Interface. Sci.* **2015,** *447,* 1–7.
118. Wenzl, T.; Simon, R.; Anklam, E.; Kleiner, J. Analytical Methods for Polycyclic Aromatic Hydrocarbons (PAHs) in Food and the Environment Needed for New Food Legislation in the European Union. *TrAC Trends Anal. Chem.* **2006,** *25* (7), 716–725.
119. Wu, L.; Lu, X.; Fu, X.; Wu, L.; Liu, H. Gold Nanoparticles Dotted Reduction Graphene Oxide Nanocomposite Based Electrochemical Aptasensor for Selective, Rapid, Sensitive and Congener-Specific PCB77 Detection. *Sci. Rep.* **2017,** *7,* 5191.
120. Wu, L.; Ye, Y.; Liu, F.; Tan, C.; Liu, H.; Wang, S.; Wang, J.; Yi, W.; Wu, W. Organo-Bentonite-Fe3O4 Poly(Sodium Acrylate) Magnetic Superabsorbent Nanocomposite: Synthesis, Characterization, and Thorium(IV) Adsorption. *Appl. Clay. Sci.* **2013,** *83,* 405–414.
121. Wu, S. J.; Liou, T. H.; Mi, F. L. Synthesis of Zero-Valent Copper-Chitosan Nanocomposites and their Application for Treatment of Hexavalent Chromium. *Bioresour. Technol.* **2009,** *100,* 4348–4353.
122. Wypych, F. Chemical Modification of Clay Surfaces. In *Clay Surfaces: Fundamentals, Applications*; Wypych, F., Satyanarayana, K. G., Eds.; Amsterdam: Academic Press, 2004.
123. Xu, X.; Liu, W.; Tian, S.; Wang, W.; Qi, Q.; Jiang, P.; Gao, X.; Li, F.; Li, H.; Yu, H. Petroleum Hydrocarbon-Degrading Bacteria for the Remediation of Oil Pollution Under Aerobic Conditions: A Perspective Analysis. *Front. Microbiol.* **2018,** *9,* 2885.

124. Ying, L.; Shaogang, L.; Xiaoyang, C. Assessment of Heavy Metal Pollution and Human Health Risk in Urban Soils of a Coal Mining City in East China.*Hum. Ecol. Risk Assess.* **2016,** *22* (6), 1359–1374.
125. Zeng, S.; Gan, N.; Weideman-Mera, R.; Cao, Y.; Li, T.; Sang, W. Enrichment of Polychlorinated Biphenyl 28 from Aqueous Solutions Using Fe3O4 Grafted Graphene Oxide. .*Chem. Eng. J.* **2013,** *218*, 108–115.
126. Zhang, D.; Ma, X. L.; Gu, Y.; Huang, H.; Zhang, G. W. Green Synthesis of Metallic Nanoparticles and their Potential Applications to Treat Cancer. *Front. Chem.* **2020,** *799.*
127. Zhang, L.; Fang, M. Nanomaterials in Pollution Trace Detection and Environmental Improvement. *Nano Today* **2010,** *5,* 128–142.
128. Zhao, G.; Jiang, L.; He, Y.; Li, J.; Dong, H.; Wang, X.; Hu, W. Sulfonated Graphene for Persistent Aromatic Pollutant Management. *Adv. Mater.* **2011,** *23*, 3959–3963.
129. Zhao, R. S.; Liu, Y. L.; Chen, X. F.; Yuan, J. P.; Bai, A. Y.; Zhou, J. B. Preconcentration and Determination of Polybrominateddiphenyl Ethers in Environmental Water Samples by Solid-Phase Microextraction with Fe_3O_4-Coated Bamboo Charcoal Fibers prior to Gas Chromatography-Mass Spectrometry. *Anal. Chim. Acta.* **2013,** *769*, 65–71.

Web References: Springer

130. "Oil spill impacts may perturb entire food webs: Study identifies a major new ecological damage pathway following oil spills." ScienceDaily. 10 July 2017.www.sciencedaily.com/releases/2017/07/170710103446.htm

CHAPTER 3

Lignin- and Chitosan-Based Biosorbents for Waste Water Treatment

JYOTHY G. VIJAYAN

Department of Chemistry, M. S. Ramaiah University of Applied Sciences, Bengaluru, Karnataka, India

ABSTRACT

Water pollution through heavy metal ions and organic dyes is the major cause of environmental concern. Conventional methods used to remove contaminants are not much economical and also have disposal issues. Adsorption by low cost adsorbents is most efficient for the treatment of waste water. Biochars and agricultural wastes are considered potential low-cost adsorbents for wastewater treatment. Biopolymers such as cellulose, chitosan, and lignin are considered very efficient adsorbents for water treatment. Lignin is collected from agricultural waste and has high adsorption capacity than other biosorbents. Main advantages of lignin and chitosan-based biosorbents are their easy availability, low cost, high efficiency, and high adsorption rate, respectively. This review discusses about the synthesis and application of lignin and chitosan-based materials in waste water treatment. This review concerns the investigation of the potential of lignin and chitosan for the effective removal of heavy metal ions and organic dyes. Lignin and chitosan are considered the abundant biopolymer available in the nature and can be used in water treatment. The importance of using biopolymers and their future aspects are also discussed.

Bio-Based Polymers and Composites: Properties, Durability, and Applications. Amadou Belal Gueye, Sabu Thomas, Nandakumar Kalarikkal, & Modou Fall (Eds.)

3.1 INTRODUCTION

Increasing population affects the availability of water and food. Water from different industrial units, agricultural and commercial units disposes large quantity of effluent into the water which is hazardous and carcinogenic in nature.[1] This scenario makes the present situation more serious now. Different industries such as metal mining, leather, cosmetic, pharmaceutical, electroplating, paint, and paper discharge different organic and inorganic pollutants into the water.[2-6] Poisonous heavy metal ions and organic dye cause serious threat to the human health system. This leads finally to finding a solution to search a cost-effective method for waste water treatment and sustainable availability of pure water. Conventional methods such as coagulation, precipitation, adsorption, flocculation, and ionic exchange are comparatively more expensive and cause more sludge to dispose.[7] Heavy metal ions and dyes cannot degrade itself and create severe environmental pollution and also cause health issues to the human system. These organic pollutants cause serious threat to the ecosystem. Textile effluents cause poisonous wastes which affect the quality of the drinking water. Textile effluents contain hazardous chemical compounds that influence the purity of the water and need to be processed cautiously before disposal.[8] Organic dyes are the major reasons for water pollution. Textile dyes are synthetic organic compounds with aromatic structures. These dyes are resistant to light, heat, and oxidizing agents. The colorful textile dyes in water bodies reduce light penetration into it causing a serious impact on the aquatic life. Organic dyes are carcinogenic, mutagenic, and toxic in nature.[9] Different processes are used to eliminate color from the industrial effluents.

Decolorization of dyes in aerobic and anaerobic condition with the help of fungi or bacteria is applied in biological processes. Physical methods include reverse osmosis, nanofiltration, ultrafiltration, flotation, chemical oxidation and reduction, chemical precipitation, chemical complexation etc are used conventionally for WWT.[10] Mixed methods such as mix of physical, chemical, and biological are also applied traditionally. Recently need for search of material as biosorbents aroused. The criteria were that the material should possess high sorption efficiency and is applicable in waste water treatment. Different materials were listed based on the surface charge of the adsorbent on contact with water, for example, coal, flyash, activated carbon, etc. Special attention was concentrated on lignin and chitosan-based materials due to their low cost, high porosity, mechanical resistance, high surface area, and hydrophilicity.[11] Chitosan is present mainly in crustacean shells such as crab, shrimp, and lobster. Modification of chitosan is very important for the enhancement of adsorption and widens their application in different fields[12] for the elimination of different contaminants at low cost. Functionalization varies the physicochemical properties of chitosan and also increases the adsorption efficiency.[13] Chitosan without modification, that is flakes or in powder form, is also used for water treatment. Chitosan is generated from chitin in solid form which exhibits high crystallinity. These are called chitosan flakes. It is mainly used to remove dyes from industrial waste water. Lignin can be surface functionalized easily and used in different continuous and noncontinuous processes.[14] According to the literature removal of organic dyes and heavy metal ions by using lignin or modified lignin-based biosorbents has been applied in waste water treatment and its effect is studied. Lignin and chitosan are among some of the abundant biopolymers on earth. Both are natural, nontoxic, cheap, biodegradable, and biocompatible in nature.[15]

This chapter details about the synthesis, regeneration, reuse, and application of lignin and chitosan-based adsorbents in wastewater treatment.

3.2 LIGNIN: EMERGENCE OF A NEW BIOSORBENT

Lignin is a major waste byproduct from agriculture. Lignin is a natural polymer like cellulose, chitin, and chitosan.[16] It is the second most abundantly present biopolymer after cellulose. It covers almost 30% of the terrestrial plant cell wall. Properties of lignin include biodegradability, cost effectiveness, antioxidant, antimicrobial etc. Lignin is classified based on its crosslinking, rigidity, and variance of substitution.[17] Lignin guaiacyl is dominant in softwoods. It is also known as kraft lignin. It constitutes 85% of the total world lignin manufacturing. But in hardwood lignin syringy and lignin guaiacyl are dominant. Soda lignin is the major constituent in black liquor. It is produced from surface-free processes. According to its structure lignin possesses hydroxyl group which helps to interact with water. Lignin is the most abundant natural polymer present in lignocellulosic biomass along with cellulose and hemicellulose.[18] It is widely identified as amorphous phenolic polymer containing phenyl propane constituents. Lignin from hardwood is made up of equal mix of coniferyl and sinopyl units. Lignin is used for the adsorption of different kinds of organic or inorganic pollutants from waste water. It is mainly due to the low cost, biocompatibility, abundance, and physico-chemical properties. Surface modification of lignin causes the enhancement of the properties of the lignin toward water treatment. Lignin is heterogeneous due to the influence of extraction and isolation process. Neat

TABLE 3.1 Lignin-Based Adsorbents and Properties.

Alkaline lignin	Ligno sulfonate	Organosolv lignin	Hydrolysis lignin
• Alkaline pulping is the initial method for separating lignin from lignocellulose. • This process generates free phenolic hydroxyl groups. It increases the hydrophilicity of the alkaline lignin. Finally it turned alkaline lignin into black liquor • Absorption ability of alkaline lignin is associated with surface phenolic hydroxyl groups through ion exchange mechanism.	• Lignosulfonate is gained through sulfite pulping process. Contains 7 elemental sulfate • Lignosulfonate is water soluble and is not applicable for the adsorption of heavy metal ions.	• Organosolv lignin is prepared from biomass with solvents methanol, ethanol, acetic or formic acid. It is an ecofriendly synthesis. It does not need toxic sulfites or high temperature or pressure. It is the most purest lignin form which contains very small amount of sulfate. • Organosolv lignin in the hardwood has higher adsorption capacity for metal ions and dyes.	• In hydrolysis, the removal of lignin from plants, removal of C-O bonds in lignin takes place by replacing C-C bond • It is cost effective • Usage of acid makes the reaction more toxic and corrosive and needs anticorrosive vessels

lignin cannot be used in the adsorption process or the use of elimination of heavy metal ions due to its 3D and coiled structure.[19] It exhibits less adsorption capacity toward heavy metal ions. Generally, methylation, demethylation, hydrothermal depolymerization etc are used for the functionalization of lignin.[20]

3.3 FUNCTIONALIZATION OF LIGNIN

Lignin is a crosslinked aromatic polymer. It contains phenyl propane units in their structure. It is insoluble in water. Lignin in its pure form is not suitable for effective application as adsorbents. It is due to its low adsorption capacity and nonspecific selectivity toward pollutants. Modification of lignin enhances physico-chemical properties of the lignin which includes hydrophobicity, chemical stability, and hydrophilicity.[21] Lignin-based materials are modified and used as new adsorbents due to their hydrogen bond network formation and copolymerization. Modification of lignin is now classified based on 3 main functional group categories.

3.3.1 MODIFIED LIGNIN WITH OXYGEN CONTAINING FUNCTIONAL GROUPS

Lignin contains some fraction of oxygen-containing functional groups including carboxyl and hydroxyl groups (Table 3.2).

TABLE 3.2 Adsorption by Oxygen-Modified Lignin.

Compounds	Adsorbed Metal ions	Quantity of adsorption (mg/g)	References
DALAC	Cu(II)	399	[22]
Lignosulfonate	Pb(II)	195	[23]
Carboxymethyl lignin	Cd(II)	68	[24]
Lignin gel	Cd(II)	44	[25]

3.3.2 MODIFIED LIGNIN WITH NITROGEN CONTAINING FUNCTIONAL GROUPS

Addition of $-NH_2$ groups on lignin to tailor the structure and create functional groups is achieved by amination (Table 3.3).

TABLE 3.3 Adsorption by Nitrogen-Modified Lignin.

Compounds	Adsorbed Metal ions	Quantity of adsorption (mg/g)	References
Lignin	Pb(II)	60	[26]
Bagasse lignin	Cu(II)	56	[27]
Lignin PNMA	Hg(II)	180	[28]
Lignin	Pb(II)	72	[29]
Lignin	Pb(II)	70	[30]

3.3.3 MODIFIED LIGNIN WITH SULFUR CONTAINING FUNCTIONAL GROUPS

Sulfur is a soft base and has strong affinity toward heavy metal ions. Surface containing functional groups are widely used in the design and development of advanced adsorbents[31] (Tables 3.4 and 3.5).

TABLE 3.4 Adsorption by Sulfur-modified Lignin.

Compounds	Adsorbed Metal ions	Quantity of adsorption (mg/g)	References
Sulfomethylated Lignin	Cu(II)	54	[32]
Mesoporous lignin with sulfur trioxide	Pb(II)	952	[33]
Lignin Xanthate resin	Pb(II)	65	[34]
Lignin dithiocarbamate	Cu(II)	176	[35]
Lignin adsorbent	Cd(II)	72	[36]

TABLE 3.5 Adsorption by Modified Lignin-Based Composites.

Compounds	Adsorbed metal ions	Quantity of adsorption (mg/g)	References
Si/Lignin	Ni(II)	77	[37]
TiO_2/lignin	Pb(II)	60	[38]
MgO-SiO_2-Lignin	Cu(II)	84	[39]
Bentonite/Lignin	Pb(II)	315	[40]
TiO_2/SiO_2/lignin	Pb(II)	60	[41]

Chemical modification is again classified into ***fragmentation*** (hydrolysis. Pyrolysis, hydrogenation, oxidation, gasification, enzymatic oxidation, and microbial conversion), ***hydroxyl modification*** (sulfonation, amination, nitration, alkylation, halogenation, and hydroxyalkylation) ***New chemically active structure*** (etherification, esterification, silylation, oxidation, reduction, phenolation, and reaction with isocyanate).[42-45]

3.3.4 REGENERATION AND REUSE OF LIGNIN-BASED ADSORBENTS

Low-cost lignin and its compound composites were subjected to desorption-adsorption process analysis.[46] These composites can remove organic dyes and heavy metal ions with a pH range of 5–7. Reuse application of lignin and lignin-based composites for withdrawing the toxicity of pollutions is reported in the literature. In steam treatment by using lignin-based adsorbents high thermal stream reuse is conducted.[47] Recovery of steam can take place at high or low temperature. Steam temperature of organic compounds is in the range of 30–50°C, that is, above the adsorbents boiling temperature. In this method vapor temperature depends upon the organic and inorganic adsorbent solvency.

3.3.5 FUNCTIONALIZED LIGNIN IN THE SYNTHESIS OF POLYMERIC MATERIALS

For enhancing the reactivity of lignin it has to be processed under multistep chemical modification. The main cost-effective way is to blend it with polymers through melting and mixing. Identifying the interactions of polymerization on miscibility is very important to study the system with versatile properties. The main advantage is that when adding lignin to the polymers, the property of lignin such as poor miscibility, dispersion etc causes a decrease in the mechanical property of the product. It has to be well monitored while adding higher amount of lignin to the polymers.[48] Alkylation and esterification helps to produce thermoplastic blends and reduces lignin–lignin interactions. Acetylation helps to improve the ductility and thermal properties.[49]

3.3.6 COVALENT INCORPORATION OF FUNCTIONALIZED LIGNIN

Lignin is the most abundant amorphous, 3D crosslinked phenolic polymer. Lignin acts as a binding agent between the wood cell wall. Lignin is produced by processes which involve kraft, sulfite, and organosolv reactions. Functionalized lignin into the polymer forms different polymer matrices.[50]

3.3.6.1 POLYURETHANE

PU is widely applied in the making of flexible/rigid forms and nonporous materials such as elastomers, adhesives, and coatings.[51] Lignin is covalently incorporated into polymer matrix to enhance the physical strength of PU materials. Functionalization processes such as carboxylation, esterification, amination, and demethoxylation are the most commonly used methods to increase the reactivity of lignin.[52] Modified lignin was used as fillers in the vegetable oil-based PU composites. It was proved that lignin urethane was more compatible than lignin butyrate in the association with the PU matrix.[53]

3.3.6.2 LIGNIN–PHENOL FORMALDEHYDE

Phenol formaldehyde is synthesized via step growth polymerization between phenol and formaldehyde. These are good wood adhesives. Methods like methylation, demethylation, phenolation, sulphonation, and depolymerization are used to enhance the reactivity of lignin for generating PF resins. Autocatalyzed phenolation was done for sulfite lignin and acid-catalyzed phenolation was proved to be not efficient for kraft lignin.[54]

3.3.6.3 EPOXY

Epoxy resins are also called prepolymers containing reactive epoxide groups. Two main methods are used for preparing covalently incorporated technical lignin epoxy resins—direct epoxidation of lignin and premodification of lignin before oxidation. Mechanical property of epoxy resins is not efficient compared to the pure neat lignin. Amination process includes aliphatic and aromatic amine as curing agents or hardeners to form crosslinked thermosets with epoxy resins. Amination involves the introduction of amine groups into the aromatic rings giving modified lignin into epoxy and cured ones. Mannich amination increases the dispersion of lignin in the epoxy matrices, which leads to the comparable thermomechanical properties into the neat epoxy ones.[55]

3.4 NANOLIGNIN

Due to the renewability and biodegradability, lignin becomes the main precursor for developing green nanoscale materials. They are called nanolignin or colloidal lignin. Nano lignin is synthesized by methods like precipitation, atomization, ultrasonication, and homogenization. Methods used to prepare nanolignin are divided into wet and dry. Wet NPs were produced by dissolving lignin in organic solvent and precipitating in an ant solvent. Dry lignin nanoparticles are prepared via well-controlled drying of dilute lignin solutions. Application of lignin NPs in antibacterial, UV protection, cosmetics, anticorrosion, drug carriers, emulsions etc is reported in the literature. Nanolignin is added as fillers in the preparation of biocomposites to enhance thermal stability and mechanical performance of the material.[56-58] It also increases the antimicrobial activity of the materials.

3.5 LIGNIN COMPOUNDS AS EFFICIENT ADSORBENTS FOR THE ELIMINATION OF HEAVY METAL IONS

Adsorption is an effective way to remove contaminants such as heavy metal ions from waste water. Need of low-cost and renewable green bioadsorbent has become a topic of great interest in the past. Now great efforts are used to synthesize cost-effective adsorbents derived from biomass due to its versatile functionality. Modified lignin is incorporated into polymer matrix for the use of removal of pollutants from contaminated water (Table 3.6).[59-61]

TABLE 3.6 Adsorption of Heavy Metal Ions by Different Lignin-Based Compounds.

Lignin	Type of lignin	Metal ion adsorbed	Adsorption capacity	Reference
Wheat straw	Hydrolysis	Cu(II)	4	[62]
Wood	Alkaline	Cr(II)	18	[63]
Eucalyptus	Kraft	Cd(II)	137	[64]
Hardwood	Organosolv	Cu(II)	3	[65]
Softwood	Organosolv	Cd(II)	1	[66]
Wheat straw	Alkaline	Cu(II)	26	[67]

3.5.1 MODIFIED LIGNIN-BASED ADSORBENTS

Lignin in its pure form is not suitable for metal ion adsorption due to its nonspecific selectivity for certain metal ions. Modification of lignin causes the tuning of specific physicochemical properties such as adsorption capacity, hydrophilicity, etc. New lignin-based adsorbents are prepared via hybridization, blending, crosslinking, hydrogen bond network formation etc.[68]

Generally, lignin is modified with oxygen, sulfur and nitrogen containing functional groups are more efficient (Tables 3.7, 3.8, and 3.9).

TABLE 3.7 Adsorption by Functionalized Lignin.

Lignin	Type of modification	Method of modification	Metal ions	Adsorption capacity	References
Wood	Oxygen containing	Grafting	Pb(II)	371	[69]
Wheat straw	Oxygen containing	Carboxymethylation	Cu(II)	399	[70]
Bagasse	Oxygen containing	Carboxymethylation	Cd(II)	68	[71]
Alkaline	Nitrogen containing	Grafting	Pb(II)	61	[72]
Bagasse	Nitrogen containing	Grafting	Cu(II)	72	[73]
Enzymatic	Nitrogen containing	Copolymerization	Ag(I)	1557	[74]
Alkaline	Sulfur containing	Grafting	Pb(II)	54	[75]
Rice straw	Sulfur containing	Microthermal expansion	Pb(II)	950	[76]
Bamboo	Sulfur containing	Click reaction	Cd(II)	72	[77]
Alkaline	Sulfur containing	Dithio carbamate	Pb(II)	188	[78]

TABLE 3.8 Adsorption by Lignin-Based Composites.

Lignin composites	Adsorbed metal ions	Adsorption capacity	References
TiO_2/lgnin	Pb(II)	36	[79]
Si/lignin	Ni(II)	77	[80]
SiO_2/lignin	Pb(II)	60	[81]
CNT/lignin	Pb(II)	235	[82]
MgO/SiO_2/lignin	Cu(II)	90	[83]

TABLE 3.9 Adsorption of Dyes by Lignin-Based Materials.

Lignin	Adsorbate	Adsorption capacity	References
Industrial	Brilliant red	10	[84]
Natural	Methyl orange	300	[85]
Organosolv	Mehylene blue	40	[86]
hydrolyzed	Porocion blue	74	[87]
Acetic acid based	Methylene blue	63	[88]

3.6 CHITOSAN-BASED ADSORBENTS

Nanosized chitosan exhibits more efficiency than microsized chitosan material because of the latter faster adsorption rate and adsorption capacity.[89] Due to their small size, absence of diffusion and large surface area chitosan and nanochitosan is used for water treatment. Nanochitosan is modified by different procedures such as amination, carboxylation, and phosphorylation. Chitosan NPs are aminated to enhance the adsorption capacity and selectivity of adsorption. $-NH_2$ groups present in chitosan are the reason for electrostatic interaction with the contaminated species. Modification agents used in amination include ethylene diamine, hexane diamine, diethylene triaamine etc.[90] Chitosan is also modified by using the parameters such as carboxylation, CNT, clay, etc.

3.7 CHITOSAN: STEPS AND SYNTHESIS

To synthesize chitosan, various methods have been used such as deprotenization, demineralization, decolorization, and deacetylation.[91]

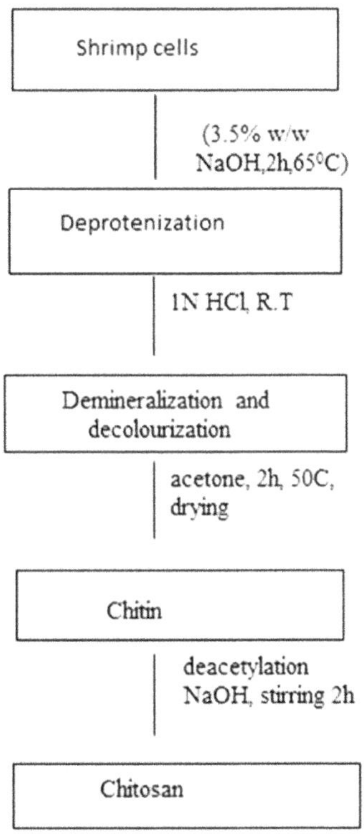

3.8 MODIFIED CHITOSAN: DIFFERENT FORMS

Chitosan is a highly basic polysaccharide. They are efficient to form chelate metal ions, films, etc. Chitosan is readily soluble in dilute acetic acid solutions. Presence of NH_2 and −OH groups in chitosan molecule provides a suitable base for many chemical species namely monomers, polymers, and nanoparticles.

3.8.1 NANOCHITOSAN

It is an ecofriendly biopolymer with wide application in the elimination of contaminants from waste water. The nanometric scale supports their high performance due to their large surface area and better adsorption performance. Nanochitosan is prepared by the methods like coagulation, precipitation, crosslinking, etc.[92] Nanochitosan is also used as an effective antibacterial agent. It is also used for the elimination of dyes from wastewater.

3.8.2 CHITOSAN BIOPOLYMER BLEND

Many biopolymers are used to make blended adsorbents with chitosan for the efficient removal of pollutants. Biopolymer includes alginate, cellulose that are added with chitosan for the contamination process.[93] Chitosan has been blended with nanoparticles for water treatment, for example, chitosan/ZnO bionanocomposite. It is used for the removal of acid black from waste water. Chitosan hydrogels are very efficient for the use of the removal of pollutants. It is because of their properties such as softness and efficiency to store water. Among hydrogels, pH-sensitive hydrogels are good conductors because of the presence of functional groups on it which are ionic in nature and are suitable for the separation. Chitosan beads exhibit excellent properties due to the presence of hydroxyl groups in it. It also exhibits properties such as acid stability, higher surface area, pore size, chemical stability etc. Modification of chitosan is done through physical and chemical methods.

Physical modification is done by blending or physical mixing of two polymers to create new novel material with enhanced properties. Blending is very cheaper to carry out and is also time consuming. Chitosan membranes are efficient due to its high mechanical and physical properties. Chitosan adsorptive membranes are used to remove heavy metal ions from contaminated water.[94]

Chemical modification is carried out by introducing different functional groups into the chitosan structure. Photosensitizer modification is the most promising chemical modification which is less explored till date. Photosensitizer is a chemical compound used to absorb light energy. Use of chitosan as photosensitizer is the promising one and relevant for synergetic operation with light-derived catalyst to enhance contaminant degradation.[95] Chitosan and its composites are used for the removal of dyes and heavy metal ions. They are potential biosorbents due to their amino and hydroxyl functional groups.

3.9 CHITOSAN-BASED NANOCOMPOSITES FOR WATER TREATMENT

3.9.1 MAGNETIC CHITOSAN NANOCOMPOSITES

These were used for the adsorption of heavy metal ions such as Cd^{2+}, Cu^{2+}, Pb^{2+} etc. The process includes addition of nanocomposites with the contaminant solution by sonication and allows to adsorb heavy metal ions from it. Ethylene diamine modified chitosan NPs were found to be very efficient for the removal of palladium and platinum metal ions. Magnetic chitosan composites are used efficiently in waste water treatment due to their high surface area, surface selectivity, and regeneration by the simple application of external magnetic field, good selectivity, and chelation capacity.[96] This external magnetic field helps in easy recovery and reuse of target contaminants from waste water.

3.9.2 CHITOSAN/HYDROXYAPATITE HA COMPOSITES

HA materials are mainly applied for deflouridation application. Biocomposites made up of nanoHA with chitosan are used for the removal of pollutants. Addition of HA into the polymer matrix helps to increase the mechanical strength and stability of the polymer.[97]

3.9.3 CHITOSAN/NANOCLAY COMPOSITES

Chitosan/nanoclay composites are used for the removal of organic dye from waste water. Clay increases the thermal stability of the complexes. Montmorillonite is considered as most effective for adding with chitosan.[98]

It is able to adsorb metal ions and dyes from polluted water. Carbonaceous materials have been added with chitosan to increase the strength, stability, and specific area of chitosan-based adsorbents. GO and CNT possess high SA and chemical stability.[99] But the disadvantage with the material is that it is solved by adding chitosan into it, as shown in Table 3.10. Biochar is also added with chitosan for better applications. The main advantage of biochar is its low cost and high surface area.[100] Metal oxides like titania NPs and other inorganic materials (Zeolite) are added with chitosan for WWT.

TABLE 3.10 Adsorption of Heavy Metal Ions by Chitosan Composites.

Composites	Adsorbate	Adsorption efficiency	Reference
Chitosan/ethylene diamine	Pt(II)	171	[101]
Chitosan/magnetic NPs	Ni (II)	833	[102]
Chitosan/HA	Cr (II)	284	[103]
Chitosan/ HA	Cd (II)	124	[104]
Chitosan/ CNT	Methyl orange	417	[105]
Chitosan/ZnO	Acid black 26	53	[106]
Chitosan/Clay NPs	Direct blue 78	35	[107]

3.10 CHEMICAL MODIFICATIONS OF CHITOSAN

Chemical modifications of chitosan include the processes such as crosslinking, grafting, combining with other adsorbents, and ion templating. Structural modifications of chitosan include the synthesis of hydrogels/aerogels, nano/micro particles, nanofibers, and magnetic adsorbents.

3.10.1 CROSSLINKING

Main limitation of chitosan is its low mechanical strength and poor chemical stability. It hinders the reuse of chitosan in water treatment. Crosslinkers help to build a bridge between polymer matrix and functional groups of chitosan. Another main factor is to control the degree of crosslinking to make the balance between adsorption capacity and strength of the adsorbent. Cross-linked[108] chitosan exhibits better mechanical and chemical properties than pure chitosan adsorbents. Mostly crosslinking would lower the adsorption capacity of the chitosan. But functionalization of chitosan helps to create more adsorption site and increases the adsorption capacity. In chitosan–polymer composite synthesis some crosslinkers are not preferred to use. Those crosslinkers are highly toxic and hinder the large-scale production.

3.10.2 GRAFTING

Grafting is a general and cheap method to increase the adsorption capacity of the adsorbent. Grafting is the process where covalent binding of functional groups onto the chitosan helps to increase the adsorption capacity of the material.[109] Grafting carboxyl groups such as acrylic acid, succinic acid, and methacrylic acid onto the surface of chitosan helps to use the product in different applications. EDTA, cyclodextrin, etc is also used in grafting processes. It helps to adsorb metal ions from waste water.

3.10.3 COMBINING CHITOSAN WITH OTHER ADSORBENTS

Combining chitosan with other polymers helps in adsorption processes. Chitosan is able to adsorb more contaminants due to the presence of hydroxyl and amine groups on it. Due to low cost and high mechanical and chemical stability, clay minerals are a good option to add with chitosan. Carbonaceous materials have been added with chitosan to increase the strength, surface area, and stability of chitosan-based adsorbents. GO and CNT possess high surface area and stability. But, the disadvantage is with the agglomeration which is solved by adding chitosan into it. Biochar is also added with chitosan to prepare the composites. Main advantage of biochar is its low cost and high S.A. Metal oxide nanoparticles and other inorganic materials e.g. zeolite are added with chitosan and studied the applications. Nanoporous structure, chemical and thermal stability of chitosan-based materials helps them to be used as an effective adsorbent (Tables 3.11 and 3.12).

3.10.4 ION TEMPLATING

Ion templating method helps to synthesize adsorbents which have higher selectivity for targeted metal ions. It involves a) Addition of template ion to polymerizable ligand, b) polymerization of the complex, and c) removal of template ion from the complex.[110]

TABLE 3.11 Adsorption of Dyes by Chitosan-Based Composites.

Composites	Dye	Adsorption capacity	Reference
Chitosan/CNT	Direct blue 71	61	[111]
Chitosan/silica	Acid red 88	26	[112]
Chitosan/GO	Methyl orange	398	[113]
Chitosan/montmorillonite	Ramzol blue	310	[114]
Chitosan NPs	Acid orange 10	1.8	[115]
Chitosan /Ethylene diamine	Acid orange 7	1215	[116]
Chitosan /Fe(II)	Acid red 73	295	[117]

TABLE 3.12 Adsorption of Heavy Metal Ions by Chitosan-Based Composites.

Composites	Dye	Adsorption capacity	Reference
Chitosan/GO/PEG	Cu(II)	78	[118]
Chitosan/thiourea/titania	Cd(II)	256	[119]
Chitosan/HA	Cd(II)	123	[120]
Chitosan/EDTA/GO	Cu(II)	207	[121]
Chitosan / PEO	Pb(II)	263	[122]
Chitosan /HA	Cd(II)	124	[123]
Chitosan /Fe_3O_4	Co(II)	28	[124]
Chitosan/HA	Cd(II)	124	[125]

3.11 FUTURE ASPECTS AND CONCLUSIONS

From the past few decades lignin-based materials are studied for waste water treatment. It is due to the use of lignin as a cost-effective adsorbent which is generated from renewable biomass. Lignin-based adsorbents are efficient due to abundance, biocompatibility, biodegradability, low cost, chemical stability etc. Lignin-based materials are ecofriendly and efficient to capture metal ions through chelation, electrostatic interactions, and ion exchange. Lignin can be functionalized by using versatile physical and chemical methods to enhance adsorption capacity and selectivity for the targeted contaminants. Chitosan is used for the removal of pollutants from waste water due to their low cost, biocompatibility, and ecofriendliness. Commercialization of chitosan-based material at large scale leads to gaining a sustainable renewable material for future water treatment. Waste water treatment by chitosan-based adsorbents can be effectively used in the place of expensive activated carbon-based adsorbents.

It is very essential to commercialize novel lignin and chitosan-based adsorbents. Special highlights should be given on the modification strategies of lignin and chitosan macromolecules such as grafting, crosslinkage, copolymerization, and hybridization to design novel advanced lignin and chitosan-based adsorbents.[126-130] This review mainly concentrates on the synthesis, functionality, and application of lignin, modified lignin, chitosan, and modified chitosan-based composites. Lignin is considered a vital material to replace petroleum and petroleum-based products. Aromatic structure of lignin helps it to apply different areas such as the adsorption of gases, organic dyes, metal ions etc. Three processes such as carbonization, modification, and activation of lignin are very important to enhance the adsorption capacity and selectivity. Future efforts are essential to identify lignin and chitosan composites with adsorption properties.

KEYWORDS

- **biosorbents**
- **organic dyes**
- **heavy metal ions**
- **biosorption**
- **adsorption**
- **biochar**
- **adsorption capacity**

REFERENCES

1. Carneiro, P. A.; Umbuzeiro, G. A.; Oliveira, D. P.; Zanoni, M. V. B. Assessment of Water Contamination Caused by a Mutagenic Textile Effluent/Dyehouse Effluent Bearing Disperse Dyes. *J. Hazard. Mater.* **2010,** *174*(1–3), 694–699.
2. Chukwujike, I. C.; Ewulonu, C. M.; Nwanonenyi, S. C.; Uche, I. C. Adsorption Treatment of Industrial Paint Effluent for the Removal Of Pollutants by Local Clays. *Int. J. Res. Sci. Eng.* **2015,** *6* (10), 690–703.
3. Lladó, J.; Gil, R. R.; Lao-Luque, C.; Solé-Sardans, M.; Fuente, E.; Ruiz, B. Highly Microporous Activated Carbons Derived from Biocollagenic Wastes of the Leather Industry as Adsorbents of Aromatic Organic Pollutants in Water. *J. Environ. Chem. Eng.* **2017,** *5* (3), 2090–2100.

4. Rahman, M. L.; Fui, C. J.; Sarjadi, M. S.; Arshad, S. E.; Musta, B.; Abdullah, M. H.; ... O'Reilly, E. J. Poly (Amidoxime) Ligand Derived from Waste Palm Fiber for the Removal of Heavy Metals from Electroplating Wastewater. *Environ. Sci. Pollut. Res.* **2020,** *27* (27), 34541–34556.
5. Islam, M. T.; Dominguez, A.; Turley, R. S.; Kim, H.; Sultana, K. A.; Shuvo, M. A. I.; ... Noveron, J. C. Development of Photocatalytic Paint based on TiO2 and Photopolymer Resin for the Degradation of Organic Pollutants in Water. *Sci. Total Environ.* **2020,** *704,* 135406.
6. Ma, G.; Chen, J. In *Nitrogen and Phosphorus Pollutants in Cosmetics Wastewater and Its Treatment Process of a Certain Brand.* IOP Conference Series: Earth and Environmental Science, IOP Publishing, 2018, Vol. 113(1); p 012051.
7. Adams, C.; Wang, Y.; Loftin, K.; Meyer, M. Removal of Antibiotics from Surface and Distilled Water in Conventional Water Treatment Processes. *J. Environ. Eng.* **2002,** *128* (3), 253–260.
8. Olukanni, O. D.; Osuntoki, A. A.; Gbenle, G. O. Textile Effluent Biodegradation Potentials of Textile Effluent-Adapted and Non-Adapted Bacteria. *Afr. J. Biotechnol.* **2006,** *5* (20).
9. Chatterjee, S.; Chatterjee, S.; Chatterjee, B. P.; Guha, A. K. Adsorptive Removal of Congo Red, a Carcinogenic Textile Dye by Chitosan Hydrobeads: Binding Mechanism, Equilibrium and Kinetics. *Colloids Surf. A Physicochem. Eng. Aspects,* **2007,** *299* (1–3), 146–152.
10. Stackelberg, P. E.; Gibs, J.; Furlong, E. T.; Meyer, M. T.; Zaugg, S. D.; Lippincott, R. L. Efficiency of Conventional Drinking-Water-Treatment Processes in Removal of Pharmaceuticals and Other Organic Compounds. *Sci. Total Environ.* **2007,** *377* (2–3), 255–272.
11. Wang, K.; Loo, L. S.; Goh, K. L. A Facile Method for Processing Lignin Reinforced Chitosan Biopolymer Microfibres: Optimising the Fibre Mechanical Properties Through Lignin Type and Concentration. *Mater. Res. Express* **2016,** *3* (3), 035301.
12. Mittal, H.; Ray, S. S.; Kaith, B. S.; Bhatia, J. K.; Sharma, J.; Alhassan, S. M. Recent Progress in the Structural Modification of Chitosan for Applications in Diversified Biomedical Fields. *Eur. Polym. J.* **2018,** *109,* 402–434.
13. Sobahi, T. R.; Abdelaal, M. Y.; Makki, M. S. Chemical Modification of Chitosan for Metal Ion Removal. *Arab. J. Chem.* **2014,** *7* (5), 741–746.
14. Pokhrel, S.; Yadav, P. N. Functionalization Of Chitosan Polymer and their Applications. *J. Macromol. Sci. Phys. A* **2019,** *56* (5), 450–475.
15. Laschimke, R. Investigation of the Wetting Behaviour of Natural Lignin-a Contribution to the Cohesion Theory of Water Transport in Plants. *Thermochim. Acta* **1989,** *151,* 35–56.
16. Padalkar, S.; Capadona, J. R.; Rowan, S. J.; Weder, C.; Won, Y. H.; Stanciu, L. A.; Moon, R. J. Natural Biopolymers: Novel Templates for the Synthesis of Nanostructures. *Langmuir* **2010,** *26* (11), 8497–8502.
17. Saito, T.; Brown, R. H.; Hunt, M. A.; Pickel, D. L.; Pickel, J. M.; Messman, J. M.; Naskar, A. K. Turning Renewable Resources into Value-Added Polymer: Development of Lignin-based Thermoplastic. *Green Chem.* **2012,** *14* (12), 3295–3303.
18. Ververis, C.; Georghiou, K.; Christodoulakis, N.; Santas, P.; Santas, R. Fiber Dimensions, Lignin and Cellulose Content of Various Plant Materials and their Suitability for Paper Production. *Ind Crops Prod.* **2004,** *19* (3), 245–254.

19. Wagner, M.; Rębiś, T.; Inganäs, O. Enhancing Charge Storage of Conjugated Polymer Electrodes with Phenolic Acids. *J. Power Sources* **2016,** *302,* 324–330.
20. Ekeberg, D.; Gretland, K. S.; Gustafsson, J.; Bråten, S. M.; Fredheim, G. E. Characterisation of Lignosulphonates and Kraft Lignin by Hydrophobic Interaction Chromatography. *Anal. Chim. Acta* **2006,** *565* (1), 121–128.
21. Sammond, D. W.; Yarbrough, J. M.; Mansfield, E.; Bomble, Y. J.; Hobdey, S. E.; Decker, S. R.; Crowley, M. F. Predicting Enzyme Adsorption to Lignin Films by Calculating Enzyme Surface Hydrophobicity. *J. Bio. Chem.* **2014,** *289* (30), 20960–20969.
22. Tian, J.; Ren, S.; Fang, G.; Ma, Y.; Ai, Q. Preparation and Performance of Dimethyl-Acetoxy-(2-Carboxymethyl ether)-Lignin Ammonium Chloride Amphoteric Surfactant. *BioResources* **2014,** *9* (4), 6290–6303.
23. Liang, F. B.; Song, Y. L.; Huang, C. P.; Li, Y. X.; Chen, B. H. Synthesis of Novel Lignin-based Ion-Exchange Resin and Its Utilization in Heavy Metals Removal. *Ind. Eng. Chem. Res.* **2013,** *52* (3), 1267–1274.
24. Peternele, W. S.; Winkler-Hechenleitner, A. A.; Pineda, E. A. G. Adsorption of Cd (II) and Pb (II) onto Functionalized Formic Lignin from Sugar Cane Bagasse. *Bioresour. Techn.* **1999,** *68* (1), 95–100.
25. Parajuli, D.; Inoue, K.; Ohto, K.; Oshima, T.; Murota, A.; Funaoka, M.; Makino, K. Adsorption of Heavy Metals on Crosslinked Lignocatechol: A Modified Lignin Gel. *React. Funct. Polym.* **2005,** *62* (2), 129–139.
26. Liu, X.; Zhu, H.; Qin, C.; Zhou, J.; Zhao, J. R.; Wang, S. Adsorption of Heavy Metal Ion from Aqueous Single Metal Solution by Aminated Epoxy-Lignin. *BioResources*. **2013,** *8* (2), 2257–2269.
27. Lü, Q. F.; Luo, J. J.; Lin, T. T.; Zhang, Y. Z. Novel Lignin–Poly (N-Methylaniline) Composite Sorbent for Silver Ion Removal and Recovery. *ACS Sustain. Chem. Eng.* **2014,** *2* (3), 465–471.
28. Li, Z.; Xiao, D.; Kong, Y.; Ge, Y. Enhancing Lead Adsorption Capacity by Controlling the Chain Length of Alkyl Amine Grafted Lignin. *BioResources* **2015,** *10* (2), 2425–2432.
29. Ge, Y.; Qin, L.; Li, Z. Lignin Microspheres: An Effective and Recyclable Natural Polymer-based Adsorbent for Lead Ion Removal. *Mater. Des.*. **2016,** *95,* 141–147.
30. Ge, Y.; Qin, L.; Li, Z. Lignin Microspheres: An Effective and Recyclable Natural Polymer-based Adsorbent for Lead Ion Removal. *Mater.Des.* **2016,** *95,* 141–147.
31. Zhang, X.; Jeremic, D.; Kim, Y.; Street, J.; Shmulsky, R. Effects of Surface Functionalization of Lignin on Synthesis and Properties of Rigid Bio-Based Polyurethanes Foams. *Polymers* **2018,** *10* (7), 706.
32. Li, Z.; Xiao, D.; Kong, Y.; Ge, Y. Enhancing Lead Adsorption Capacity by Controlling the Chain Length of Alkyl Amine Grafted Lignin. *BioResources* **2015,** *10* (2), 2425–2432.
33. Xu, F.; Zhu, T. T.; Rao, Q. Q.; Shui, S. W.; Li, W. W.; He, H. B.; Yao, R. S. Fabrication of Mesoporous Lignin-Based Biosorbent From Rice Straw and Its Application for Heavy-Metal-ion Removal, *Res. J. Environ. Sci.* **2017,** *53,* 132–140.
34. Li, Z.; Kong, Y.; Ge, Y. Synthesis of Porous Lignin Xanthate Resin for Pb2+ Removal from Aqueous Solution. *Chem. Eng. Sci.* **2015,** *270,* 229–234.
35. Ge, Y.; Qin, L.; Li, Z. Lignin Microspheres: An Effective and Recyclable Natural Polymer-based Adsorbent for Lead Ion Removal. *Mater.Design* **2016,** *95,* 141–147.
36. Jin, C.; Zhang, X.; Xin, J.; Liu, G.; Wu, G.; Kong, Z.; Zhang, J. Clickable Synthesis of 1, 2, 4-Triazole Modified Lignin-based Adsorbent for the Selective Removal of Cd (II). *ACS Sustain. Chem. Eng.* **2017,** *5* (5), 4086–4093.

37. Klapiszewski, Ł.; Bartczak, P.; Wysokowski, M.; Jankowska, M.; Kabat, K.; Jesionowski, T. Silica Conjugated with Kraft Lignin and Its Use as a Novel 'green' sorbent for Hazardous Metal Ions Removal. *Chem. Eng. Sci.* **2015,** *260,* 684–693.
38. Klapiszewski, Ł.; Siwińska-Stefańska, K.; Kołodyńska, D. Preparation and Characterization of Novel TiO2/Lignin and TiO2-SiO2/Lignin Hybrids and their Use as Functional Biosorbents for Pb (II). *Chem. Eng. Sci.* **2017,** *314,* 169–181.
39. Ciesielczyk, F.; Bartczak, P.; Klapiszewski, Ł.; Jesionowski, T. Treatment of Model and Galvanic Waste Solutions of Copper (II) Ions Using a Lignin/Inorganic Oxide Hybrid as an Effective Sorbent. *J. Hazard. Mater.* **2017,** *328,* 150–159.
40. Yao, Q.; Xie, J.; Liu, J.; Kang, H.; Liu, Y. Adsorption of Lead Ions Using a Modified Lignin Hydrogel. *J. Polym. Res.* **2014,** *21* (6), 1–16.
41. Klapiszewski, Ł.; Siwińska-Stefańska, K.; Kołodyńska, D. Preparation and Characterization of Novel TiO2/Lignin and TiO2-SiO2/Lignin Hybrids and their Use as Functional Biosorbents for Pb (II). *Chem. Eng. Sci.* **2017,** *314,* 169–181.
42. Gonçalves, A. R.; Soto-Oviedo, M. A. Production of Chelating Agents Through the Enzymatic Oxidation of Acetosolv Sugarcane Bagasse Lignin. *Appl. Biochem. Biotechn.* **2002,** *98* (1), 365–371.
43. Gonçalves, A. R.; Benar, P. Hydroxymethylation and Oxidation of Organosolv Lignins and Utilization of the Products. *Bioresour. Technol.* **2001,** *79* (2), 103–111.
44. Chen, C.; Zhu, M.; Li, M.; Fan, Y.; Sun, R. C. Epoxidation and Etherification of Alkaline Lignin to Prepare Water-Soluble Derivatives and Its Performance in Improvement of Enzymatic Hydrolysis Efficiency. *Biotechnol. Biofuels* **2016,** *9* (1), 1–15.
45. An, L.; Si, C.; Wang, G.; Choi, C. S.; Yu, Y. H.; Bae, J. H.;Kim, Y. S. Efficient and Green Approach for the Esterification of Lignin with Oleic Acid Using Surfactant-Combined Microreactors in Water. *BioResources* **2020,** *15* (1), 89–104.
46. Buono, P.; Duval, A.; Verge, P.; Averous, L.; Habibi, Y. New Insights on the Chemical Modification of Lignin: Acetylation versus Silylation. *ACS Sustain. Chem. Eng.* **2016,** *4* (10), 5212–5222.
47. Feghali, E.; Cantat, T. Unprecedented Organocatalytic Reduction of Lignin Model Compounds to Phenols and Primary Alcohols Using Hydrosilanes. *Chem. Commun.* **2014,** *50* (7), 862–865.
48. Guo, X.; Zhang, S.; Shan, X. Q. Adsorption of Metal Ions on Lignin. *J. Hazard. Mater.* **2008,** *151* (1), 134–142.
49. Li, J.; Henriksson, G.; Gellerstedt, G. Lignin Depolymerization/Repolymerization and Its Critical Role for Delignification of Aspen Wood by Steam Explosion. *Bioresour. Technol.* **2007,** *98* (16), 3061–3068.
50. Kun, D.; Pukánszky, B. Polymer/Lignin Blends: Interactions, Properties, Applications. *Eur. Polym. J.* **2017,** *93,* 618–641.
51. Yetiş, F.; Liu, X.; Sampson, W. W.; Gong, R. H. Acetylation of Lignin Containing Microfibrillated Cellulose and Its Reinforcing Effect for Polylactic Acid. *Eur. Polym. J.* **2020,** *134,* 109803.
52. Wang, Y. Y.; Meng, X.; Pu, Y.; J Ragauskas, A. Recent Advances in the Application of Functionalized Lignin in Value-added Polymeric Materials. *Polymers* **2020,** *12* (10), 2277.
53. Griffini, G.; Passoni, V.; Suriano, R.; Levi, M.; Turri, S. Polyurethane Coatings based on Chemically Unmodified Fractionated Lignin. *ACS Sustain. Chem. Eng.* **2015,** *3* (6), 1145–1154.

54. Konduri, M. K.; Kong, F.; Fatehi, P. Production of Carboxymethylated Lignin and Its Application as a Dispersant. *Eur. Polym. J.* **2015,** *70,* 371–383.
55. Li, H.; Liang, Y.; Li, P.; He, C. Conversion of Biomass Lignin to High-Value Polyurethane: A Review. *J. Bioresour.* **2020**.
56. Chakar, F. S.; Ragauskas, A. J. Review of Current and Future Softwood Kraft Lignin Process Chemistry. *Ind. Crops Prod.* **2004,** *20* (2), 131–141.
57. Khalil, H. A.; Marliana, M. M.; Alshammari, T. Material Properties of Epoxy-Reinforced Biocomposites with Lignin from Empty Fruit Bunch as Curing Agent. *BioResources.* **2011,** *6* (4), 5206–5223.
58. Cui, C.; Sadeghifar, H.; Sen, S.; Argyropoulos, D. S. Toward Thermoplastic Lignin Polymers; Part II: Thermal Polymer Characteristics of Kraft Lignin Derivatives. *BioResources* **2013,** *8* (1), 864–886.
59. Sarkar, S.; Adhikari, B. Thermal Stability of Lignin–Hydroxy-Terminated Polybutadiene Copolyurethanes. *Polym. Degrad. Stab.* **2001,** *73* (1), 169–175.
60. Naseem, A.; Tabasum, S.; Zia, K. M.; Zuber, M.; Ali, M.; Noreen, A. Lignin-Derivatives based Polymers, Blends and Composites: A Review. *Int. J. Biol. Macromol.* **2016,** *93,* 296–313.
61. Thakur, V. K.; Thakur, M. K.; Raghavan, P.; Kessler, M. R. Progress in Green Polymer Composites from Lignin for Multifunctional Applications: A Review. *ACS Sustain. Chem. Eng.* **2014,** *2* (5), 1072–1092.
62. Da Cunha, C.; Deffieux, A.; Fontanille, M. Synthesis and Polymerization of Lignin-based Macromonomers. III. Radical Copolymerization of Lignin-based Macromonomers with Methyl Methacrylate. *J. Appl. Polym. Sci.* **1993,** *48* (5), 819–831.
63. Merdy, P.; Guillon, E.; Aplincourt, M.; Dumonceau, J.; Vezin, H. Copper Sorption on a Straw Lignin: Experiments and EPR Characterization. *J. Colloid Interface Sci.* **2002,** *245* (1), 24–31.
64. Wu, Y.; Zhang, S.; Guo, X.; Huang, H. Adsorption of Chromium (III) on Lignin. *Bioresour. Technol.* **2008,** *99* (16), 7709–7715.
65. Mohan, D.; Pittman, C. U. Jr; Steele, P. H. Single, Binary and Multi-Component Adsorption of Copper and Cadmium from Aqueous Solutions on Kraft Lignin—A Biosorbent. *J. Colloid Interface Sci.* **2006,** *297* (2), 489–504.
66. Harmita, H.; Karthikeyan, K. G.; Pan, X. Copper and Cadmium Sorption onto Kraft and Organosolv Lignins. *Bioresour. Technol.* **2009,** *100* (24), 6183–6191.
67. Harmita, H.; Karthikeyan, K. G.; Pan, X. (200). Copper and Cadmium Sorption onto Kraft and Organosolv Lignins. *Bioresour. Technol. 100* (24), 6183–6191.
68. Demirbas, A. Adsorption of Lead and Cadmium Ions in Aqueous Solutions onto Modified Lignin from Alkali Glycerol Delignication. *J. Hazard. Mater.* **2004,** *109* (1–3), 221–226.
69. Klapiszewski, Ł.; Siwińska-Stefańska, K.; Kołodyńska, D. Preparation and Characterization of Novel TiO2/Lignin and TiO2-SiO2/Lignin Hybrids and their Use as Functional Biosorbents for Pb (II). *Chem. Eng. Sci.* **2017),** *314,* 169–181.
70. Parajuli, D.; Inoue, K.; Ohto, K.; Oshima, T.; Murota, A.; Funaoka, M.; Makino, K. Adsorption of Heavy Metals on Crosslinked Lignocatechol: A Modified Lignin Gel. *React. Funct. Polym.* **2005,** *62* (2), 129–139.
71. Dizhbite, T.; Jashina, L.; Dobele, G.; Andersone, A.; Evtuguin, D.; Bikovens, O.; Telysheva, G. Polyoxometalate (POM)-aided Modification of Lignin from Wheat Straw Biorefinery. *Holzforschung* **2013,** *67* (5), 539–547.

72. Peternele, W. S.; Winkler-Hechenleitner, A. A.; Pineda, E. A. G. Adsorption of Cd (II) and Pb (II) onto Functionalized Formic Lignin from Sugar Cane Bagasse. *Bioresour. Technol.* **1999,** *68* (1), 95–100.
73. Ge, Y.; Song, Q.; Li, Z. A Mannich base biosorbent Derived from Alkaline Lignin for Lead Removal from Aqueous Solution. *J. Ind. Eng. Chem.* **2015,** *23,* 228–234.
74. Liu, X.; Zhu, H.; Qin, C.; Zhou, J.; Zhao, J. R.; Wang, S. Adsorption of Heavy Metal Ion from Aqueous Single Metal Solution by Aminated Epoxy-Lignin. *BioResources* **2013,** *8* (2), 2257–2269.
75. Lü, Q. F.; Luo, J. J.; Lin, T. T.; Zhang, Y. Z. Novel Lignin–Poly (N-methylaniline) Composite Sorbent for Silver Ion Removal and Recovery. *ACS Sustain. Chem. Eng.* **2014,** *2* (3), 465–471.
76. Ge, Y.; Li, Z.; Kong, Y.; Song, Q.; Wang, K. Heavy Metal Ions Retention by Bi-Functionalized Lignin: Synthesis, Applications, and Adsorption Mechanisms. *J. Ind. Eng. Chem.* **2014,** *20* (6), 4429–4436.
77. Xu, F.; Zhu, T. T.; Rao, Q. Q.; Shui, S. W.; Li, W. W.; He, H. B.; Yao, R. S. Fabrication of Mesoporous Lignin-based Biosorbent from Rice Straw and Its Application for Heavy-Metal-ion Removal. *Res. J. Environ. Sci.* **2017,** *53,* 132–140.
78. Jin, C.; Zhang, X.; Xin, J.; Liu, G.; Wu, G.; Kong, Z.; Zhang, J. Clickable Synthesis of 1, 2, 4-Triazole Modified Lignin-based Adsorbent for the Selective Removal of Cd (II). *ACS Sustain. Chem. Eng.* **2017,** *5* (5), 4086–4093.
79. Ge, Y.; Xiao, D.; Li, Z.; Cui, X. Dithiocarbamate Functionalized Lignin for Efficient Removal of Metallic Ions and the Usage of the Metal-Loaded Bio-Sorbents as Potential Free Radical Scavengers. *J. Mater. Chem. A* **2014,** *2* (7), 2136–2145.
80. Klapiszewski, Ł.; Siwińska-Stefańska, K.; Kołodyńska, D. Preparation and Characterization of Novel TiO2/Lignin and TiO2-SiO2/Lignin Hybrids and their Use as Functional Biosorbents for Pb (II). *Chem. Eng. Sci.* **2017,** *314,* 169–181.
81. Klapiszewski, Ł.; Bartczak, P.; Wysokowski, M.; Jankowska, M.; Kabat, K.; Jesionowski, T. Silica Conjugated with Kraft Lignin and Its Use as a Novel 'Green' Sorbent for Hazardous Metal Ions Removal. *Chem. Eng. Sci.* **2015,** *260,* 684–693.
82. Klapiszewski, Ł.; Siwińska-Stefańska, K.; Kołodyńska, D. Preparation and Characterization of Novel TiO2/Lignin and TiO2-SiO2/Lignin Hybrids and their Use as Functional Biosorbents for Pb (II). *Chem. Eng. Sci.* **2017,** *314,* 169–181.
83. Li, Z.; Chen, J.; Ge, Y. Removal of Lead Ion and Oil Droplet from Aqueous Solution by Lignin-Grafted Carbon Nanotubes. *Chem. Eng. Sci.* **2017,** *308,* 809–817.
84. Ciesielczyk, F.; Bartczak, P.; Klapiszewski, Ł.; Jesionowski, T. Treatment of Model and Galvanic Waste Solutions of Copper (II) Ions Using a Lignin/Inorganic Oxide Hybrid as an Effective Sorbent. *J. Hazard. Mater.* **2017,** *328,* 150–159.
85. Ge, Y.; Qin, L.; Li, Z. Lignin Microspheres: An Effective and Recyclable Natural Polymer-based Adsorbent for Lead Ion Removal. *Mater. Des.* **2016,** *95,* 141–147.
86. Suteu, D.; Malutan, T.; Bilba, D. Removal of Reactive dye Brilliant Red HE-3B from Aqueous Solutions by Industrial Lignin: Equilibrium and Kinetics Modeling. *Desalination* **2010,** *255* (1–3), 84–90.
87. Mahmoudi, K.; Hamdi, N.; Kriaa, A.; Srasra, E. Adsorption of Methyl Orange Using Activated Carbon Prepared from Lignin by ZnCl 2 Treatment. *Russ. J. Phys. Chem. A* **2012,** *86* (8), 1294–1300.
88. Zhang, B.; Liu, Y.; Ma, Z.; Guo, M.; Shen, B. Adsorption of AuCl 4—from Acidic Chloride Solution by Chemically Modified Lignin Based on Rice Straw. *Macromol. Res.* **2018,** *26* (2), 121–127.

89. Adebayo, M. A.; Prola, L. D.; Lima, E. C.; Puchana-Rosero, M. J.; Cataluña, R.; Saucier, C.;Ruggiero, R. Adsorption of Procion Blue MX-R dye from Aqueous Solutions by Lignin Chemically Modified with Aluminium and Manganese. *J. Hazard. Mater.* **2014,** *268,* 43–50.
90. Feng, Q.; Cheng, H.; Chen, F.; Zhou, X.; Wang, P.; Xie, Y. Investigation of Cationic dye Adsorption from Water Onto Acetic Acid Lignin. *J. Wood Chem. Technol.* **2016,** *36* (3), 173–181.
91. Jeon, C.; Park, K. H. Adsorption and Desorption Characteristics of Mercury (II) Ions Using Aminated Chitosan Bead. *Water Res.* **2005,** *39* (16), 3938–3944.
92. Li, Q.; Dunn, E. T.; Grandmaison, E. W.; Goosen, M. F. Applications and Properties of Chitosan. *J. Bioact. Compat. Polym.* **1992,** *7* (4), 370–397.
93. Sivakami, M. S.; Gomathi, T.; Venkatesan, J.; Jeong, H. S.; Kim, S. K.; Sudha, P. N. Preparation and Characterization of Nano Chitosan for Treatment Wastewaters. *Int. J. Biol.* **2013,** *57,* 204–212.
94. Chatterjee, T.; Chatterjee, S.; Woo, S. H. Enhanced Coagulation of Bentonite Particles in Water by a Modified Chitosan Biopolymer. *Chem. Eng. Sci.* **2009,** *148* (2–3), 414–419.
95. Wang, B.; Bai, Z.; Jiang, H.; Prinsen, P.; Luque, R.; Zhao, S.; Xuan, J. Selective Heavy Metal Removal and Water Purification by Microfluidically-Generated Chitosan Microspheres: Characteristics, Modeling and Application. *J. Hazard. Mater.* **2019,** *364,* 192–205.
96. Padmaja, P.; Bhatt, R. Environmental Applications of Chitosan and its Derivatives. *J. Mar. Biotechnol.* **2020,** *2,* 1065–1081.
97. Monier, M.; Ayad, D. M.; Wei, Y.; Sarhan, A. A. Preparation and Characterization of Magnetic Chelating Resin based on Chitosan for Adsorption of Cu (II), Co (II), and Ni (II) ions. *React. Funct. Polym.* **2010,** *70* (4), 257–266.
98. Gandhi, M. R.; Kousalya, G. N.; Meenakshi, S. Removal of Copper (II) Using Chitin/Chitosan Nano-Hydroxyapatite Composite. *React. Funct. Polym.* **2011,** *48* (1), 119–124.
99. Shameli, K.; Ahmad, M. B.; Zargar, M.; Yunus, W. M. Z. W.; Ibrahim, N. A.; Shabanzadeh, P.; Moghaddam, M. G. Synthesis and Characterization of Silver/Montmorillonite/Chitosan Bionanocomposites by Chemical Reduction Method and their Antibacterial Activity. *Int. J. Nanomed..* **2011,** *6,* 271.
100. Shieh, Y. T.; Yang, Y. F. Significant Improvements in Mechanical Property and Water Stability of Chitosan by Carbon Nanotubes. *Eur. Polym. J.* **2006,** *42* (12), 3162–3170.
101. Al-Rashed, M. M.; Niknezhad, S.; Jana, S. C. Mechanism and Factors Influencing Formation and Stability of Chitosan/Lignosulfonate Nanoparticles. *Macromol. Chem. Phys.* **2019,** *220* (1), 1800338.
102. Zhou, L.; Xu, J.; Liang, X.; Liu, Z. Adsorption of Platinum (IV) and Palladium (II) from Aqueous Solution by Magnetic Crosslinking Chitosan Nanoparticles Modified with Ethylenediamine. *J. Hazard. Mater.* **2010,** *182* (1), 518–524.
103. Dyal, A.; Loos, K.; Noto, M.; Chang, S. W.; Spagnoli, C.; Shafi, K. V.; Ulman, A.; Cowman, M.; Gross, R. A. Activity of Candida Rugosa Lipase Immobilized on γ-Fe2O3 Magnetic Nanoparticles. *J. Am. Chem. Soc.* **2003,** *125* (7), 1684–1685.
104. Crini, G.; Badot, P.-M. Application of Chitosan, a Natural Aminopolysaccharide, for Dye Removal from Aqueous Solutions by Adsorption Processes Using Batch Studies: A Review of Recent Literature. *Prog. Polym. Sci.* **2008,** *33* (4), 399–447.
105. Salah, T. A.; Mohammad, A. M.; Hassan, M. A.; El-Anadouli, B. E. Development of Nano-Hydroxyapatite/Chitosan Composite for Cadmium Ions Removal in Wastewater Treatment. *J. Taiwan Inst. Chem. Eng.* **2014,** *45* (4), 1571–1577.

106. Mahmoodian, H.; Moradi, O.; Shariatzadeha, B.; Salehf, T. A.; Tyagi, I.; Maity, A.; Asif, M.; Gupta, V. K. Enhanced Removal of Methyl Orange from Aqueous Solutions by Poly HEMA—Chitosan-MWCNT Nano-Composite. *J. Mol. Liq.* **2015,** *202,* 189–198.
107. Salehi, R.; Arami, M.; Mahmoodi, N. M.; Bahrami, H.; Khorramfar, S. Novel Biocompatible Composite (Chitosan–zinc Oxide Nanoparticle): Preparation, Characterization and dye Adsorption Properties. *Colloids Surf. B: Biointerfaces.* **2010,** *80* (1), 86–93.
108. Seyedi, S. M.; Anvaripour, B.; Motavassel, M.; Jadidi, N. Comparative Cadmium Adsorption from Water by Nanochitosan and Chitosan. *Int. J. Eng.* **2013,** *2* (9), 145–148.
109. Beppu, M. M.; Vieira, R. S.; Aimoli, C. G.; Santana, C. C. Crosslinking of Chitosan Membranes Using Glutaraldehyde: Effect on Ion Permeability and Water Absorption. *J. Membr. Sci.* **2007,** *301* (1–2), 126–130.
110. Jayakumar, R.; Prabaharan, M.; Reis, R. L.; Mano, J. Graft Copolymerized Chitosan—Present Status and Applications. *Carbohydr. Polym.* **2005,** *62* (2), 142–158.
111. Sun, S.; Wang, L.; Wang, A. Adsorption Properties of Crosslinked Carboxymethyl-Chitosan Resin with Pb (II) as Template Ions. *J. Hazard. Mater.* **2006,** *136* (3), 930–937.
112. Abbasi, M. Synthesis and Characterization of Magnetic Nanocomposite of Chitosan/SiO2/Carbon Nanotubes and Its Application for Dyes Removal. *J. Clean. Prod.* **2017,** *145,* 105–113.
113. Soltani, R. D. C.; Khataee, A. R.; Safari, M.; Joo, S. W. Preparation of Bio-Silica/Chitosan Nanocomposite for Adsorption of a Textile dye in Aqueous Solutions. *Int. Biodeterior. Biodegrad.* **2013,** *85,* 383–391.
114. Jiang, Y.; Gong, J. L.; Zeng, G. M.; Ou, X. M.; Chang, Y. N.; Deng, C. H.; ... Huang, S. Y. Magnetic Chitosan–Graphene Oxide Composite for Anti-microbial and dye Removal Applications. *Int. J. Biol. Macromol.* **2016,** *82,* 702–710.
115. Karimi, F.; Ayati, A.; Tanhaei, B.; Sanati, A. L.; Afshar, S.; Kardan, A.; ... Karaman, C. Removal of Metal Ions Using a New Magnetic Chitosan Nano-Bio-Adsorbent; A Powerful Approach in Water Treatment. *Environ. Res.* **2022,** *203,* 111753.
116. Le, H. R.; Qu, S.; Mackay, R. E.; Rothwell, R. Fabrication and Mechanical Properties of Chitosan Composite Membrane Containing Hydroxyapatite Particles. *J. Adv. Ceram.* **2012,** *1* (1), 66–71.
117. Bouyahmed, F.; Cai, M.; Reinert, L.; Duclaux, L.; Dey, R. K.; Youcef, H. B.; Delpeux-Ouldriane, S. A Wide Adsorption Range Hybrid Material based on Chitosan, Activated Carbon and Montmorillonite for Water Treatment. *J. Carb. Res.* **2018,** *4* (2), 35.
118. Cortinez, D.; Palma, P.; Castro, R.; Palza, H. A Multifunctional Bi-Phasic Graphene Oxide/Chitosan Paper for Water Treatment. *Sep. Purif. Technol.* **2020,** *235,* 116181.
119. Perez, J. V. D.; Nadres, E. T.; Nguyen, H. N.; Dalida, M. L. P.; Rodrigues, D. F. Response Surface Methodology as a Powerful Tool to Optimize the Synthesis of Polymer-based Graphene Oxide Nanocomposites for Simultaneous Removal of Cationic and Anionic Heavy Metal Contaminants. *RSC adv.* **2017,** *7* (30), 18480–18490.
120. Chen, A.; Zeng, G.; Chen, G.; Hu, X.; Yan, M.; Guan, S.; ... Xie, G. Novel Thiourea-Modified Magnetic Ion-Imprinted Chitosan/TiO2 Composite for Simultaneous Removal of Cadmium and 2, 4-Dichlorophenol. *Chem. Eng. Sci.* **2012,** *191,* 85–94.
121. Mohammad, A. M.; Hassan, M. A.; El-Anadouli, B. E. Development of Nano-Hydroxyapatite/Chitosan Composite for Cadmium Ions Removal in Wastewater Treatment. *J. Taiwan Inst. Chem. Eng.* **2014,** *45* (4), 1571–1577.
122. Shahzad, A.; Miran, W.; Rasool, K.; Nawaz, M.; Jang, J.; Lim, S. R.; Lee, D. S. Heavy Metals Removal by EDTA-Functionalized Chitosan Graphene Oxide Nanocomposites. *RSC Adv.* **2017,** *7* (16), 9764–9771.

123. Chen, A.; Zeng, G.; Chen, G.; Hu, X.; Yan, M.; Guan, S.;Xie, G. Novel Thiourea-Modified Magnetic Ion-Imprinted Chitosan/TiO2 Composite for Simultaneous Removal of Cadmium and 2, 4-Dichlorophenol. *Chem. Eng. Sci.* **2012,** *191,* 85–94.
124. Liu, Z.; Ge, X.; Lu, Y.; Dong, S.; Zhao, Y.; Zeng, M. Effects of Chitosan Molecular Weight and Degree of Deacetylation on the Properties of Gelatine-based Films. *Food Hydrocoll.* **2012,** *26* (1), 311–317.
125. Wang, B.; Zhu, Y.; Bai, Z.; Luque, R.; Xuan, J. Functionalized Chitosan Biosorbents with Ultra-High Performance, Mechanical Strength and Tunable Selectivity for Heavy Metals in Wastewater Treatment. *Chem. Eng. Sci.* **2017,** *325,* 350–359.
126. Hamza, M. F.; Aly, M. M.; Abdel-Rahman, A. A. H.; Ramadan, S.; Raslan, H.; Wang, S.;Guibal, E. Functionalization of Magnetic Chitosan Particles for the Sorption of U (VI), Cu (II) and Zn (II)—Hydrazide Derivative of Glycine-Grafted Chitosan. *Materials* **2017,** *10* (5), 539.
127. Salah, T. A.; Mohammad, A. M.; Hassan, M. A.; El-Anadouli, B. E. Development of Nano-Hydroxyapatite/Chitosan Composite for Cadmium Ions Removal in Wastewater Treatment. *J. Taiwan Inst. Chem. Eng.* **2014,** *45* (4), 1571–1577.
128. Joseph, S. M.; Krishnamoorthy, S.; Paranthaman, R.; Moses, J. A.; Anandharamakrishnan, C. A Review on Source-Specific Chemistry, Functionality, and Applications of Chitin and Chitosan. *Carbohydr. Polym.* **2021,** *2,* 100036.
129. He, H.; Meng, X.; Yue, Q.; Yin, W.; Gao, Y.; Fang, P.; Shen, L. Thiol-ene Click Chemistry Synthesis of a Novel Magnetic Mesoporous Silica/Chitosan Composite for Selective Hg (II) Capture and High Catalytic Activity of Spent Hg (II) Adsorbent. *Chem. Eng. Sci.* **2021,** *405,* 126743.
130. Cavallaro, G.; Micciulla, S.; Chiappisi, L.; Lazzara, G. Chitosan-based Smart Hybrid Materials: A Physico-Chemical Perspective. *J. Mater. Chem. B* **2021,** *9* (3), 594–611.

CHAPTER 4

Advancement of Natural Polymers from Waste and Their Biotechnological Applications

MOHAMMAD AMIR, NAUSHIN BANO, AAMINA SHAHAB, SHAREEN FATIMA RIZVI, and ROOHI

Protein Research Laboratory, Department of Bioengineering, Integral University, Lucknow, Uttar Pradesh, India

ABSTRACT

Polymers are utilized to create a wide range of important industrial items. Apart from social issues, the necessity to replace traditional materials is driving their rapid expansion in production. Every year, about 140 million tons of synthetic polymers are manufactured worldwide. These polymers have substantial stability and don't degrade quickly in ecological degradation cycles. Natural polymers have been isolated from biological systems such as plants, microbes, algae, and mammals. Natural polymers tend to replace synthetic and petroleum-based polymers as they are biodegradable. These natural polymers are not restricted to packaging only; they also have many other biotechnological applications such as nutraceuticals, food, food packaging, pharmaceuticals, agriculture, and bioremediation. For example, collagen is widely used in the pharmaceutical industry to produce microneedles, scaffolds, wound dressings, etc. The only drawback of natural-based polymers is that they are not cost-effective. The increase in production will tend to decrease the cost of the natural polymers.

Bio-Based Polymers and Composites: Properties, Durability, and Applications. Amadou Belal Gueye, Sabu Thomas, Nandakumar Kalarikkal, & Modou Fall (Eds.)

4.1 INTRODUCTION

A wide range of important industrial items is made from polymers. Apart from societal issues, the desire to switch to more environmentally friendly materials fuels the increase in production. J. J. Berzelius, a Swedish chemist, coined the term "polymer." According to him, benzene (C_6H_6) is an ethyne polymer (C_2H_2). Later, this definition underwent a minor change.[1] Polymer science studies natural and synthetic fibers, plastics, coatings, rubbers, sealants, lubricants, and other coatings. Polymers are among the greatest inventions of the 20th century. It was initially proposed in the 1920s, and Nobel Prize winner H. Staudinger is mainly responsible for its widespread acceptance. Several synthetic polymers are listed; some are well-known, for example, nylons or polyesters, while others are less known and used in therapeutic applications for degradable sutures, organs, etc. Plastic materials come in a variety of shapes and sizes and are classified as a commodity (defined by the large volume and cheap cost) or engineering (marked by high volume and high price or higher cost and low volume). Poly(vinyl chloride) (PVC), polypropylene (PP), and polyethylene (PE) belong to the first category, whereas polyimide (PI), polyether ether ketone (PEEK), polycarbonate (PC), and others belong to the second.

The thrust for a greener environment and more sustainable methods has been picked up over the last 2 decades, and much attention has been paid to the mismanagement of natural feedstocks that can be retrieved without competing with natural food supplies as a way to reduce the use of fossil resources. The utilization of sustainable raw materials is one of green chemistry's most important aims, as outlined in the original set of 12 laws known as "The Twelve Principles of Green Chemistry." Environmental regulations, like the Pollution Prevention Act of 1990, imposed by the US Environmental Protection Agency (EPA), the Rio Declaration on Environment and Development, adopted by the United Nations (UN) Conference on Environment and Development 2 years later, and the adoption of the UN objectives of the Vision 2030 have all contributed to this mindset.[2] Natural, manufactured (improved natural), and synthetic fibers have high aspect ratios, high toughness and modulus, and other qualities that vary depending on their use. Every year, about 140 million tons of synthetic polymers are manufactured worldwide. These polymers have high stability and don't degrade quickly in ecological degradation cycles. In wastewater, plastic waste and synthetic polymers have long been recognized as substantial sources of synthetic polymer pollution. This problem must be addressed to promote continued sustainable development around the globe.

Consequently, plastic degradation has been widely investigated over the last 3 decades. Some forms of plastic are degradable, and the mechanisms of degradation have grown clearer with time. This study examines recent developments in this field. Polymers' fundamental principle, which involves numerous assemblies of simple structural elements to produce a three-dimensional superstructure, is found in all biological systems. This covers a wide range of topics.[1]

In addition to being environmentally beneficial and generating low levels of greenhouse gases, natural materials can be derived from various renewable sources like microbes, animals, and plants and their wastes, making them widely available. This is also a benefit when contrasted to typical plastic polymers, which are petroleum-based and nonrenewable. Natural polymeric materials, such as plant- or marine-based polysaccharides and proteins, are inherently biodegradable, meaning they may be broken down by local microbes and returned to the environment. Natural polymers offer significant technical and biological qualities that enable them to be used in food packaging. In general, these materials have strong film-forming and gelling capabilities, making them distinct and long-lasting options for making edible films and coatings.

Biopolymeric films may pose various processing issues when used for this purpose. They have brittleness, poor water and gas barrier qualities, poor mechanical behavior, and all natural characteristics. Alternative techniques to overcome these technological challenges include using hydrophobic chemicals or mixing two biopolymers to generate composites.[3] Polymer-based dendrimers, micelles, nanoparticles, and nanogels are common nanoparticulate structures used in drug delivery systems, ranging from 1 to 1000 nm. Nanocarriers can integrate a drug into the polymer via covalent bonding, resulting in a polymer-drug conjugate. However, the drug can be incorporated into an amphiphilic polymer platform via hydrophobic interactions. The most commonly utilized polymers in nanocarriers include poly(amino acids), polysaccharides, poly(alkyl cyanoacrylate), proteins, polypeptides, and polyesters because of their biocompatibility and biodegradability. Poly(glycolic acid) and poly(lactic acid) (PLA), as well as their copolymers poly(lactic-co-glycolic acid) (PLGA), have been studied broadly in conjunction with poly(ethylene glycol) (PEG). Polyester made of polycaprolactone or polyhydroxyalkanoates (PHA) creates nanosized structures similar to PEG. As a result, PEG has been employed as a biocompatible hydrophilic portion to develop nanoparticles with precise sizes. Moreover, PEG has been demonstrated to offer "stealth," decreasing immune system recognition (i.e., opsonization) and reducing particle aggregation. The nonbiodegradability of this polyether, on the other hand, may limit its use in drug delivery systems. Furthermore, PEG's immunological inertia has not been shown, and the buildup of PEGylated substances in the liver has undetermined long-term toxicological implications.[4]

Synthetic polymers created by humans are virtually as diverse as natural polymers, despite most developments taking place after the Second World War. Newly discovered polymers, such as polyesters and polyamides, were quickly adopted in medical applications, such as synthetic suture materials. Synthetic polymers have grown in prominence in technological and therapeutic applications for various purposes.[5] Most of the synthesis and construction technologies for complex-shaped devices have been developed. The primary purpose of these polymers is to fulfill structural and mechanical needs. Mechanical self-reinforcement is achieved when oriented fibers of the same material are integrated into the matrix.[6] Polymers with reversible shape memory can be spontaneously deformed and then return to their novel shape in the presence of an external stimulus (such as a magnetic field or light) that alters the properties of the polymer. Only a few biological uses include vascular stents, drug delivery devices, clot removal devices, sutures, orthodontic treatment, and aneurysm or ductus arteriosus occlusion.[7,8]

4.2 SYNTHETIC POLYMER

Man-made polymers are known as synthetic polymers. After being refined and cracked, the monomer used in synthetic polymers is often derived from petroleum. Metals and ceramics are among the materials employed in these methods, as are natural and man-made polymers and their micro- and nano-composites.[9] Synthetic materials are used in various regenerative medicine applications, for example, implants, tissue engineering scaffolds, and orthopedic fixation devices.[10] More examples are sutures, medication delivery methods, nonviral gene delivery vectors, and synthetic polymer sensors. Synthetic polymers that are hydrolytically labile and erode (biodegradable polymers) and bioinert polymers that remain unaltered after implantation are examples of nondegradable polymers. Every year, roughly 140 million tons of synthetic polymers are produced worldwide. These polymers are highly stable and do not degrade quickly in the biosphere's decomposition cycles. Synthetic polymer pollution, which includes waste plastics and water-soluble synthetic polymers in wastewater, has long been acknowledged as a significant issue.[11] The properties of synthetic polymers are shown in Figure 4.1.

- Polymers created from chemical substances are known as synthetic polymers.[5]
- Scientists have been able to imitate the structures of natural polymers to create synthetic polymers through scientific research.
- Many of the raw components for synthetic polymers come from petroleum after it has been refined and cracked.[9]

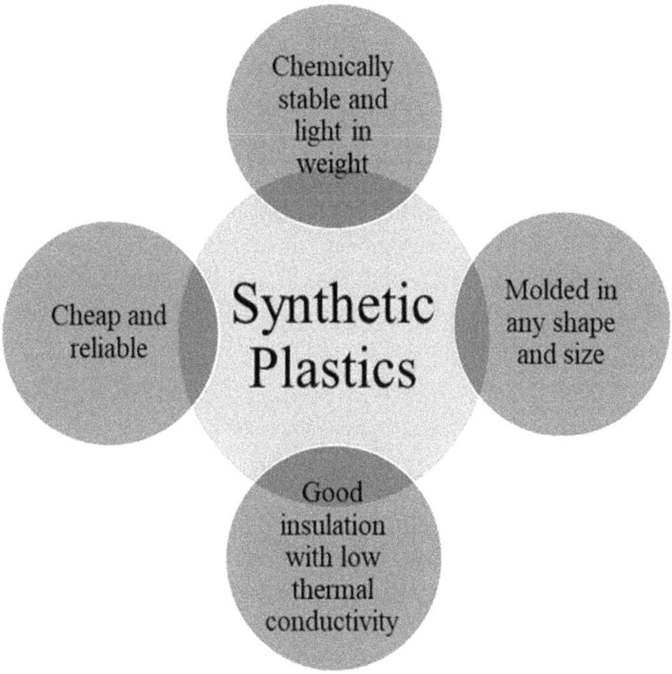

FIGURE 4.1 Properties of synthetic polymers.

4.3 TYPES OF SYNTHETIC POLYMERS

4.3.1 PLASTICS

Plastics are polymers made up of polymer chains that can be partially organic or completely synthetic. To put it another way, while all plastics are polymers, not all polymers are plastics. If a material can be molded or molded into different shapes, it is plastic. Dough, plasticine, and plastics are examples of clay and heated glass. Plastic, on the other hand, is now a generic term for a variety of man-made materials that may be molded into desired forms or drawn into threads. These materials all have one thing in common: they're formed of polymers. Scientists have developed a broad spectrum of human-made plastics with various qualities. Crude oil is the primary source of these polymers.[12]

Plastics are divided into two classes based on how they react to heat.

- Thermoplastics
- Thermosetting plastics.

The process of softening by heating and hardening by chilling in thermoplastics can be repeated to soften the plastic and obtain new shapes. On the other hand, thermosetting polymers can only be softened by heating them once. They can keep their forms and sizes even at very high temperatures once they've been placed in a mold and allowed to harden on cooling. Plastics play a crucial role in our lives nowadays. Plastics' adaptability is what accounts for their wide range of applications. Plastics are used to make carry bags, soft drink bottles, car parts, computer parts, refrigerator parts, telephones, and many other items.[13] Some common man-made plastics and their uses are explained in Table 4.1.

4.3.2 POLYVINYL ALCOHOL

Polyvinyl alcohol (PVA) is a vinyl polymer in which the leading chains are connected only through carbon–carbon bonds. It's the same with water-soluble polymers like polyacrylic acid and polyacrylamide and plastics like polyethylene, polypropylene, and polystyrene. Among the industrially made vinyl polymers, PVA is the only one that has been found to be mineralized by microbes. PVA is a water-soluble thermoplastic polymer. In addition to being a water-soluble polymer, it may be remolded into different shapes, like containers and films (e.g., as a starch substitute in industrial operations). Thus, PVA may be used to develop water-soluble and biodegradable carriers, which could be beneficial in developing fertilizer, pesticide, and herbicide delivery systems.[17]

4.3.3 BAKELITE

Bakelite was the world's first synthetic plastic, and as such, it played a crucial role in ushering humanity into the "plastics age." Bakelite was groundbreaking because it was the first synthetic plastic and because it allowed

TABLE 4.1 Some Common Man-Made Plastics and Their Uses.

Man-made plastics	Types of plastic	Properties	Uses	References
Polythene	Thermoplastic	Strong but flexible, it can be molded into any shape or rolled into a sheet.	Packaging film, waste bags, supermarket bags, and wire and cable insulation.	[14]
Polyvinyl chloride (PVC)	Thermoplastic	PVC is a long-lasting and robust substance.	Construction, transportation, packaging, electrical/electronics, and healthcare are just a few of the industrial, technical, and everyday applications for PVC.	[14]
Polystyrene	Thermoplastic	Easily molded	Packaging, rigid food service containers, medical and optical instruments, electronics, toys, kitchen appliances, and household items.	[15]
Bakelite	Thermosetting	When heated, it becomes liquefiable and flexible, but when cooled, it becomes permanently hard and unyielding. As a result, it's a thermosetting material and a good insulator.	Nonconducting components of radios and other electrical devices including sockets, switches, automotive distribution caps, wire insulation, and brake pads.	[16]

for the production and shaping of products made of it to be combined. From the beginning, Bakelite was employed as a matrix for composites. Finally, Baekeland serves as an example for researchers, particularly those in applied science, of someone who was able to take his innovation "from the laboratory to the factory and from the factory to the public."[16]

4.3.4 FIBERS

Synthetic polymers are man-made fibers derived from petroleum. Rayon, nylon, polyester, acrylic, and other materials are examples.

4.3.5 RAYON

Silk is a beautiful natural material with a lovely texture. Attempts to create a synthetic fiber with a similar texture were successful in the 19th century, when scientists developed rayon. Rayon is sometimes known as synthetic silk. It is less expensive than silk and can be woven similarly to silk fibers. Rayon is a fabric composed of wood pulp. First, the wood pulp is dissolved in an alkaline solution. Fibers are created by passing the thick liquid through microscopic holes. Passing the fibers through a sulfuric acid bath hardens them. After that, the fibers are spun into yarn and sewed into clothing. Rayon is known as regenerated fiber because it is made from cellulose, broken down, and then reformed. This makes the fabric more durable and dyeable in a wide range of hues. Another advantage of rayon is that it absorbs moisture. Rayon's clothing is so cool to wear in the summer. Rayon is used to manufacture suits, ties, blouses, jackets, stockings, pants, and athletic apparel, among other things. Bedspreads, bedsheets, blankets, curtains, tablecloths, upholstery, carpets, and other home furnishings are created from rayon fiber, which has a velvety shine.[18]

4.3.6 NYLON

The first synthetic fiber manufactured entirely of chemicals was nylon. It was first sold in the cities of New York and London, and hence got its name. Nylon thread is more substantial than steel wire because it is very strong and elastic. It's lightweight and wrinkle-free. Nylon clothing lasts a long time because it is also wear-resistant. Because nylon is easy to wash and absorbs little water, clothing made of it dries rapidly. It has a beautiful sheen to it. Due to the strength of nylon fiber, it is used to manufacture ropes (including rock climbing ropes), tents, fishing nets, and parachutes. It also possesses flexibility, making it ideal for socks and stockings. Nylons are also used to manufacture combs, brushes, zips, fasteners, and machine parts.[19]

4.3.7 POLYESTERS

Polyesters are polymers made up of monomers that are linked together by ester bonds. Many different esters can be found naturally, and the enzymes that destroy them, known as esterases, are found in all living creatures. The hydrolysis of ester linkages is usually simple. In fact, PHA and bacterial polyesters have been employed to manufacture biodegradable polymers, and a variety of synthetic polyesters have been discovered to be biodegradable. A number of firms that develop biodegradable polymers are now producing several polyesters on a semicommercial basis.[20] Polyester is also used to make a variety of home furnishings, including bedsheets, curtains, blankets, and pillowcases. Upholstered furniture also uses polyester.[21]

4.3.8 ELASTOMER

Elastomer is a fancy name that just means rubber. Polyisoprene, natural rubber, polybutadiene, polyisobutylene, and polyurethanes are elastomers. During the curing process, the lengthy polymer chains cross-link. The molecular structure of elastomers is similar to that of "spaghetti and meatballs," with the meatballs signifying cross-links.[22]

4.3.9 NATURAL RUBBER

The latex of the rubber tree is used to make natural rubber. Natural rubber is found in around 2000 plant species, but only a few are used commercially. The properties of natural rubber are high tensile strength, wear, abrasion, fatigue resistance, and exceptionally high molecular weight; it has demonstrated resistance to alkalis and mild acids.[23]

4.3.10 POLYBUTADIENE RUBBER

Butadiene (CH2=CH=CH2) is created when butane is dehydrogenated or from cracked petroleum. The properties of polybutadiene rubber (BR) are good abrasion resistance. Elasticity and cold qualities are two of the most important characteristics of a material; because of its lower price and high abrasion resistance, it is often utilized in tires. [24]

4.3.11 SILICONE RUBBER

Silicone rubber is an elastomer manufactured from silicone, a polymer composed of silicon, carbon, hydrogen, and oxygen. Silicone rubber is a form of elastomer (rubber-like material) made of silicone, a polymer of silicon, carbon, hydrogen, and oxygen. Fillers are added to silicone rubbers, which are one- or two-part polymers, to improve their qualities or reduce their price. The properties of silicone rubber are relatively low transition temperatures for glass, high-temperature resistance, low-temperature adaptability, and good electrical resistivity.[25]

4.3.12 BUTYL RUBBER (IIR)

Butyl rubber is a copolymer made up of isobutylene and a little isoprene (0.5–3%), serving as a curing agent. It polymerizes at shallow temperatures. Butyl rubber's properties are: gas permeability is extremely low; exceptional damping qualities (energy absorption); superior ageing stability; adhesion abilities are poor; and a high level of weather resistance.[26]

4.3.13 THERMOPLASTIC ELASTOMER (TPE)

It is a copolymer that consists of materials with both thermoplastic and elastomeric characteristics (typically plastic and rubber). Thermoplastic rubbers are another name for it. The properties of TPE polymers are dipole or hydrogen-linked polymers with a weaker dipole and high elastomer characteristics. Due to vulcanization, it has high strength. Elastomers are extensively employed as sealing agents in buildings, in the preparation of nylon, in tires, in industries, and in gaskets. Styrene block copolymers are used in shoe soles because they are easy to manufacture. They're frequently utilized as glues. These materials have a wide range of applications in the automobile and household appliance industries. Elastomers can be combined or coupled with other materials to improve certain qualities. They can also be easily installed next to various other materials, such as metal, hard plastic, or different types of rubber, with excellent adhesion. The benefits of synthetic polymers have a lot of desirable properties, which is a big plus. They have good strength, flexibility, resistivity, chemical inertness, and other properties. The disadvantages of synthetic polymers are Chemical inertness or resistance to many types of chemical breakdown is a desirable property in many synthetic polymers.[27]

4.4 NATURAL POLYMER

Natural polymers have been isolated from biological systems such as plants, microbes, algae, and mammals.[28] Natural polymers are synthesized either enzymatically or by microbial fermentation due to their biocompatibility, cost-effectiveness, abundance, and lack of toxicity.[29] In comparison to chemical synthesis methodologies based on alkali and acid treatment or copolymer blending, the creation of microbial synthesis and biobased polymers from renewable sources is both scalable, simple, and has a low environmental impact.[30] Agricultural waste is a significant source of raw materials for manufacturing biobased polymers and plasticizers.[31] Bioplastics are made from natural polymers obtained from agricultural waste and are an excellent alternative to hazardous and nonbiodegradable plastics.[32] Since these natural polymer-based bioplastics are highly recyclable and biodegradable, they help reduce the carbon footprint. Bioplastics derived from biopolymers such as PHA, thermoplastic starch (TPS), and PLA are already on the market, albeit in modest quantities.[33] Natural polymers are used extensively in various industries, like drug delivery, cosmetics, tissue engineering, wound healing, and dentistry.[34] Because these polymers are nontoxic, they exhibit biomimetic qualities, allowing for their incorporation into biological systems to examine and enhance cellular processes like adhesion, migration, differentiation, and proliferation. Natural polymers can be classified based on their source: plant, fungal, microbial, and animal-based polymers[35]; their nature: proteinaceous, polysaccharide, polyester, lipids/surfactants, polyisoprenes, and polyphenol-based polymers; or their formation methods: condensation or addition.[36]

4.4.1 PECTIN

Pectin is the most complex polymer in plant cell walls, both physically and functionally. A variety of foods and specialty items use pectin as a gelling and stabilizing polymer, and it also plays a function in plant growth, morphology, and defense. It also benefits human health and has a variety of biological applications. Pectin has the most intricate molecular structure of all the sugar polysaccharide families in nature, accounting for approximately 35% of primary walls in dicots and nongraminaceous monocots, 2–10% of primary walls in grass and other commelinid plants, and up to 5% of walls in woody tissue (Ridley et al., 2001).

4.4.2 GELATIN

Gelatin polymer is a biodegradable and biocompatible substance of 85–92% proteins, mineral salts, and water. It's a type I collagen molecular derivative with a wide range of culinary, cosmetic, and medicinal applications. It's made by irreversibly hydrolyzing collagen's triple helical structure, resulting in random coiled domains, using heat and enzymatic denaturation techniques. As a result, gelatin is less organized than collagen but has a relatively similar molecular composition.[37]

4.4.3 KERATIN

Keratin, which is found in abundance in nails, hair, wool, hooves, horns, and feathers, is one of the most abundant and underutilized sources of protein. It is a polypeptide made up of amino acids with disulfide cysteine intermolecular bonding and polar and nonpolar acids inter- and intramolecular bonding. Keratin has been extensively researched as a biobased material. It's a robust, high-modulus, axially oriented microfibril-in-amorphous-matrix composite material with b-sheet crystallites that are highly crosslinked to cysteine. Primary chains are linear molecules with many hydrophobic residues that are 10.2–10.4 kDa in size.[38]

4.4.4 STARCH

Starch is one of nature's most auspicious natural polymers because of its intrinsic biodegradability, abundant supply, and annual renewability. Since the well-known challenges of oil shortages and the growing interest in relieving the environmental load caused by excessive usage of petrochemically-derived polymers, the development and applications of biodegradable starch-based materials have gotten a lot of attention. The natural polysaccharide raw material starch is among the most plentiful, cheap, and renewable resources. This polymer accounts for more than 60% of grain kernels and is relatively straightforward to separate from the other chemical components.[39]

4.4.5 ALGINATE

When divalent cations like Ca^{2+} are added to alginate, it moderately gels, making it ideal for a wide range of biomedical applications. Alginate is a naturally occurring anionic polymer generated from brown seaweed that has been extensively investigated and used. Brown algae (*Phaeophyceae*) such as *Laminaria hyperborea*, *Laminaria digitata*, *Laminaria japonica*, *Ascophyllum nodosum*, and *Macrocystis pyrifera* are commonly extracted by treating them with aqueous alkali solutions, typically NaOH.[40]

4.4.6 CHITOSAN

Chitosan (5-1,4-linked glucosamine) is a deacetylated derivative of chitin that dissolves in solutions of weak acid but not at neutral pH. It is one of the most prevalent natural polymers. Its unusual polycationic nature makes it useful in various industries, including wastewater purification, transition chelation, coacérvate production for cell entrapment, and seed coating for increased yield. The US Food and Drug Administration has approved chitosan as a flocculating aid for protein recovery from food processing wastes. The resulting protein concentrates are used in livestock feed (chitosan is hypolipidemic in rabbits, hens, and broilers at suitable doses).[41]

4.4.7 LIGNINS

Lignins are natural aromatic polymers found in plant cell wall complexes that are important for plant development and significantly impact the utilization of lignocellulosic biomass. P-hydroxyphenyl (H) units are generated principally from three typical 4-hydroxyphenylpropanoids known as monolignols (i.e., p-coumaryl, coniferyl, and sinapyl alcohols) that differ in the degree of methoxylation of the aromatic ring and result in p-hydroxyphenyl (H) units in lignin polymers. Lignin structural research has traditionally concentrated on how polymerization occurs, the structural consequences of that process, and how structural changes affect lignin processing.[42]

4.4.8 PHB

PHAs were first extracted and described as poly(3-hydroxybutyrate) (P3HB). P3HB is very crystalline, with both amorphous and crystalline phases, thanks to its linear chain structure. It is created industrially through bacterial fermentation and is generated as a carbon reserve in various bacterial strains. In the production of some packaging applications, P3HB also has several benefits over synthetic polymers, including having a higher barrier permeability than both PE and PP, as well as being more rigid and less flexible than PP.[43] Another important feature of P3HB materials is their biodegradability, which occurs in a reasonable amount of time when they come into contact with degrading microorganisms in biologically active environments like soils, freshwater, and aerobic and anaerobic composting, making them a sought-after eco-friendly alternative to synthetic polymers.[44]

4.5 BIOTECHNOLOGICAL APPLICATIONS OF NATURAL POLYMERS

In a time of depleting resources and exponentially increasing pollution, natural polymers have emerged as a ray of hope. Efforts by researchers all over the globe have managed to develop a wide range of biotechnological applications of natural polymers in several fields. Natural polymers have reawakened interest due to worries about environmental contamination and degradation associated with the usage of nonbiodegradable petroleum-based products, the growing expense of petroleum, and the knowledge that petroleum supplies are limited and exhaustible. The government and other scientific, engineering, biomedical, and technical research and development funders have yielded a broad range of natural polymer applications.[45] Polymers sourced from animal and plant waste are extensively used in pharmacy, food and nutraceuticals, agriculture, and cosmetics. Natural polymers include qualities such as biocompatibility and biodegradability that make them a preferable alternative to synthetic polymers in the biomedical, cosmetic, and packaging industries.[46]

4.5.1 NATURAL POLYMERS IN BIOMEDICINE AND PHARMACEUTICAL INDUSTRY

Polymers are enormous macromolecules made up of recurring building units. Covalent chemical bonds link the subunits in most cases. It's possible to employ synthetic or natural polymers in pharmaceutical applications, but natural polymers are preferable because they're inexpensive, readily available, and nontoxic. In addition to undergoing chemical alterations, they are also biocompatible and may one day degrade.[47] Tissue regeneration scaffolds, drug delivery agents, and imaging agents are examples of natural polymers in biomedical applications. Natural polymer-based scaffolds are intriguing for skin repair and regeneration due to their resemblance to the extracellular matrix, mechanical tenability, high biocompatibility, and high water-holding capacity. Gelatin, starch, chitosan, alginate, and pectin are some of the most promising and widely used natural polymers in the biomedical and pharmaceutical fields. Gelatin is utilized in therapeutic settings as a plasma expander and stabilizer in various protein formulations, vaccinations, and gelatin sponges. Because gelatin is generated from collagen, it cannot form any toxic by-products during enzymatic degradation.[48] Due to its subcellular size, controlled and sustained release capabilities, and biocompatibility with a wide range of cells in the body's tissues,[49] biodegradable bone cement based on starch can provide rapid structural support and decompose at the site of application.

Furthermore, it can be coupled with bioactive particles, promoting new bone development at both the cement-bone contact and the volume left by polymer degradation.[50] For numerous medicinal drugs, including insulin, starch nasal bioadhesive microspheres with considerably longer half-lives have been identified.[51] Tablets made from modified starch are utilized to produce controlled-release medication delivery systems. Just one compound can represent many different medications from varied biochemical backgrounds. Their release rate remains stable over time.[47] Alginate has been utilized as a tablet disintegrant in compressed tablets meant for quick medication release, whereas sodium alginate has been employed as a tablet binding agent. In various wound dressings, alginates are engaged in a powder, a film, or a fiber. It is possible to use the hydrocolloid dressing on dry wounds by moistening them with saline first.[52] Chitosan-based contact lenses instead of typical injections allow for more precise drug release management. The former decreases the likelihood of an incorrectly injected substance and provides a more regulated medication release method.[53]

Biocompatible and noncytotoxic pectin regeneration applications revealed the adhesion and proliferation of human osteoblast cells on the surface of the scaffold.[54] A diverse range of biologically active compounds can be carried by it in pharmaceutical preparation and drug formulation, not only for sustained release applications but also as a carrier for medications that are targeted to the colon for local or systemic treatment. Pectin type, gelation circumstances, additional excipients, and coating agents can be adjusted to produce dosage forms with various morphologies and properties.[55] Pectin-based hydrogels, gels, tablets, beads, and aerogels are popularly used in medical treatment these days.[54]

4.5.2 NATURAL POLYMERS IN FOOD INDUSTRY

Natural polymers are also used in the food industry. Everything in food comprises macromolecules, and nearly every single one of these macromolecules is a polymer of some sort.[45] Natural polymers in food include proteins and polysaccharides, which are widely employed in various industrial applications, such as thickening and gelling aqueous solutions, emulsions, stabilizing foams and dispersions, inhibiting the formation of sugar and ice crystals, and controlling flavor release. At present, pullulan is widely used in the food business as it is a slow-digesting macromolecule with no flavor or smell, which is why it is employed as a low-calorie food additive to add bulk and texture. Pullulan holds moisture well as it contains oxygen barrier properties and inhibits microbial growth. These features make it a superior food preservation material extensively utilized in the industry.[56] Among the most often used starch-containing foods, breadfruit cookies, on the other hand, are made with wheat starch. The gelling characteristics of the starch were also discovered to be beneficial in meals. Corn starch aids in cake making by increasing the fiber present there. Corn has long been used to make syrup; exceptionally high fructose corn syrup is used in industry.[50] Starch is also applied as a thickening agent for filling blueberries in blueberry cheesecake. Gelatin has long been utilized in the food business as a clarifying agent, stabilizing agent, and protective coating substance to prevent bacterial growth. There have been reports that the food industry uses tons of gelatin each year, particularly in sweets, candies, bakery goods, jellied meat, ice cream, marshmallows, and dairy products.[48] According to certain reports, gelatin has been extensively employed to flavor syrups and canned soups.[57]

4.5.3 NATURAL POLYMERS IN COSMETIC SECTOR

Natural polymers have been employed in cosmetics more than synergic and semisynthetic ones. Their biocompatibility makes them ideal for various applications, including skin, haircare, and make-up modifiers and stabilizers. They are also safe and environmentally benign.[58] Owing to their wide range of properties like water-holding capacity, UV protection, film formation, greater skin penetration power, etc., natural polymers serve many applications in the cosmetic and personal care industries. They are commonly used as thickeners, delivery systems, moisturizers, skin peels, cleansers, etc. A total of 26 different collagen proteins have been found so far, all of which have a structure based on three helix-polypeptidic chains. Collagen is a major structural protein in humans, sustaining and maintaining tissues including tendons, skin, and teeth.[59] Type I collagen is the most widely employed fibril-forming collagen because of its outstanding biocompatibility with the human body in various cosmetic applications. While its moisturizing capabilities lend themselves to creating creams and gels with a strong moisturizing effect, other uses are also anticipated, including antiaging, anti-wrinkling, UV radiation shielding, and the healing of wounds, among others. Chitin is also intriguing in cosmetics because the skin tolerates it well. It's a hydrating agent that also happens to be a film-forming polymer, and it has two notable benefits: It keeps you hydrated by providing water when you require it. Chitosan's high molecular weight makes it challenging to permeate human skin. It is dissolved in an aqueous solution to overcome this constraint. Chitosan is a prominent ingredient in cosmetic formulations due to its compatibility with other substances like glucose, saccharose, starch, polyols, fats, oils, acids, and waxes. Chitosan is a water-soluble nonionic emulsifier. Chitosan is used in cosmetics (lipstick, eye shadow, and nail polish), deodorizing products, and microencapsulation of active agents. Shampoos and coloring agents, hairspray and setting lotion, and creams and lotions are all examples of haircare products.

4.5.4 NATURAL POLYMERS IN PACKAGING AND FOOD PACKAGING

A vital link in the food supply chain connects the food processing sector with consumers through food packaging. The typical approach to passive food packaging is to use a packaging system that acts as a physical barrier between the food and the environment. Some standard food storage containers provide robust support during transit, prevent physical damage, and protect the food from manipulation. In the food packaging industry, plastics are synthetic polymers that are particularly popular because of their lightweight properties, flexibility, durability, and chemical resistance. Plastics used in packaging, on the other hand, have the potential to harm the environment for decades or even centuries after they are utilized. Households are responsible for more than 60% of post-consumer plastic trash, the majority of which is single-use packaging. Typically biocompatible, nontoxic, biodegradable, and capable of undergoing chemical change, natural polymers are mostly made up of coupled monomeric units. They are also easily recyclable, and most of them undergo natural disintegration, which aids in the decomposition of organic matter. In order to have a more efficient influence on the environment and food system, natural polymers for food packaging should be used more frequently. Various biodegradable polymers show promise in terms of being utilized as packaging materials, and several are already being used in this capacity.[60]

4.5.5 NATURAL POLYMERS IN NUTRACEUTICAL INDUSTRY

There has been a boom in the nutraceutical sector as people become more health-conscious and aware of the poor quality of food due to rampant adulteration and the degradation of the environment. Biocompatibility, nonimmunogenicity, and nontoxicity have all been selling points for natural products. In collagen, significant research has been done to lower the fat content of meat, frozen desserts, dairy, and baked goods using fatty replacements or replacers and bulking agents. Chitosan and its derivatives also help people reduce their body fat, which adds the benefit of lowering blood pressure. As a beneficial prebiotic, chitosan can also help improve gastrointestinal health.[61] Figure 4.2 shows the biotechnological applications of natural polymers.

4.6 CONCLUSIONS AND FUTURE PERSPECTIVES OF NATURAL POLYMERS

The massive increase in the human population also leads to many environmental problems. This increase in population is also the reason for the depletion of natural resources. Reports say about 8–10% of our total oil supply makes plastic products from petroleum products. Though these polymers are cheap and easy to produce, their nondegradable nature is the major drawback of synthetic and petroleum-based polymers. Natural polymers tend to replace synthetic and petroleum-based polymers as they are biodegradable. Many

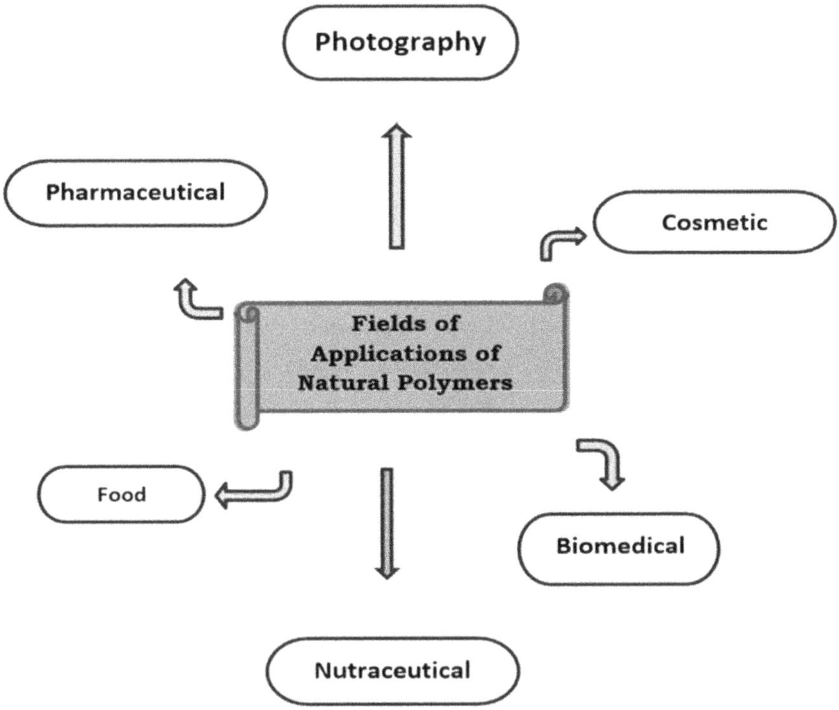

FIGURE 4.2 Biotechnological applications of natural polymers.

national and international government bodies, NGOs, and researchers from many institutes and industries are working hard to produce natural polymers at a low cost. Polysaccharides, starch, cellulose, protein, and PHA have been expected to be natural polymers extracted or isolated naturally, either from plant waste or animal sources or from the microbiota. These natural polymers are not restricted to packaging only. Still, these natural polymers also have many other biotechnological applications, such as in nutraceuticals, food and food packaging, pharmaceuticals, agriculture, and bioremediation. For example, collagen is widely used in the pharmaceutical industry to produce microneedles, scaffolds, wound dressings, etc. The same collagen has been applicable in cosmetics also. PHB, the class member family of PHA, shows the most promising properties to replace petroleum-based polymers for packaging and in different applications. Despite the significant gains in the past few years, there are still some limitations in the large-scale production of natural polymers. The cost of producing natural polymers is 10 times higher than that of synthetic polymers, which can be overcome by the enormous output of natural polymers at the industry level.

TABLE 4.2 Most Popularly Practiced Applications of Some Natural Polymers.

Natural polymer	Industry	Application	Reference
Collagen	Pharmaceutical	Microneedles	[62]
		Wound dressing	
		Cell migratory scaffolds	
		Hemostatic agent	
		DDS	
		Treatment of rheumatoid arthritis	
	Biomedical	Dental composites	[63]
		Regeneration template	
		Shields	
		Microcarrier in enzyme production	
		Artificial skin grafts	
		Treatment of rheumatoid arthritis	
	Cosmetic	Moisturizing	[59]
		UV protection	
		Antiaging	
		Hydration	
		Anti-wrinkling	
		Nutricosmetic drug	
	Food	Dietary supplements.	
		Drinks and beverages	
		Functional food	
		Confectionery and dessert	
		Clarification of alcohol	
		Edible films	
		Food additive	
Xanthium gum	Food and beverage	Stabilizer and thickener of low-fat products	[64]
		Improves texture	
		Precipitating alcohol	
	Cosmetic	Toothpaste	
		Stabilizer in lotions and creams	

TABLE 4.2 *(Continued)*

Natural polymer	Industry	Application	Reference
Pectin	Biomedical and pharmaceutical	Coating	[55]
		as an explicit	
		drug carrier	
		Antidote for cationic poisoning	
		Antihemorrhagic	
		hydrogels	
		tablets	
		Gels	
		Hydrogels	
		Beads	
		Aerogels	
		Colon-specific drug delivery vehicle	[65]
		Scaffolds	
		Wound dressings	
		Gelling agent	
	Food	Preservative	
		Stabilizer	
Gelatin	Pharmaceuticals	Hard and soft shells of capsules and microcapsules	
		Hydrogels for antioncogenic DDS	
		Vitamin and mineral gummies	
		Mouth-freshening tablets/chewing gums	
	Photography	Constituent of coating melts.	[49]
		Establish regular and stable coating flow	[66]
		Maintain mechanical stability during drying and reswelling	
	Food	Candies, toffees, gummies, marshmallows, ice creams, puddings, etc.	[67] [68]
		Shelf life increment	
		Fat-deficit cheese, yogurt, and margarines	[48]
		Coating food items	

TABLE 4.2 *(Continued)*

Natural polymer	Industry	Application	Reference
		Plasma volume expander (PVE)	
	Biomedical	Prosthetic heart valve and skin	
		Scaffolds	
		Gelling agent in facewash, moisturizers, lotions, bath soaps, shower gels, face, and body creams.	
Keratin	Animal husbandry	Feed supplement	
			[69]
		Scaffolds.	[70]
	Pharmaceutical	Sponges.	[71]
		Hydrogels	
Starch		DDS	
		Microparticles	
		Microcapsules	[47]
	Pharmaceutical		[50]
		Bone cement	
		Bone scaffolds	
	Biomedical	Cereals, breads, biscuits, cookies, pasta, noodles, pastries, etc.	
		Fiber content enhancement.	
	Food	Infant food	
		Syrups	
		Thickening agent in cheesecakes	
		Packaging	
		Longer shelf life of encased folic acid	
Alginate	Pharmaceutical	Encapsulating probiotics	
		Binding agent in tablets	
		Disintegrating agent in pills	

TABLE 4.2 *(Continued)*

Natural polymer	Industry	Application	Reference
	Biotechnology	Delivery of biomolecules	[52]
			[72]
	Biomedical	Rafts in treating esophagitis	[73]
		Wound dressing.	
	Food	Protective coating of food	
		Attractive coating	
		Gelling and emulsifying	
	Pollution control	Wastewater treatment	
Chitosan	Biomedical	Reduce ulcerogenic activity of drugs	
	Ophthalmology	Contact lenses for regulated DDS	
			[53]
		Scaffolds	
	Pharmaceutical	Capsule covers for controlled DDS	[61]
		Capsules	
		Microcapsules	
		Tablets	
		Hydrogels	
		Foam	
		Sponges	
		As a prebiotic	
	Nutraceutical	Antihypertension formulation	
		Antimicrobial chewing gums	
		Protective layer on enamel	
	Dentistry	In juices and beverage for:	
		Acidity regulation	
		Filtration	
		Clarification	
	Food	Color stabilization	
		Preservation	

TABLE 4.2 *(Continued)*

Natural polymer	Industry	Application	Reference
Pullulan	Pharmaceutical	Target drug delivery	
		Plasma expander	
		Scaffolds	[56]
	Packaging	Food packaging	
Cellulose	Pharmaceutical	Hydrogels	[66]
		Medical adhesive	
		Drug carrier	[74]
		Scaffolds	
		Transparent wound dressing	
	Nutraceutical	Encapsulation of probiotic cells and essential oils	
	Food	Packaging	
		Fat substitute	
		Bulking agent	
		Paper making	
	Cosmetic	Enhancing penetration of moisturizers and antiaging effect	
		Brightening masks	
Hyaluronic acid	Biomedical and aesthetic medicine	Osteoarthritic disease therapy.	[75]
		Artificial tear formula	
		Cosmetic filler	
		Reduction in free radicals count	
			[76]
	Ophthalmological surgery	Optimization of intraocular pressure	[77]
		Viscosupplementation	
	Dentistry	Treating chronic periodontitis	
		Cure mouth ulcers	

TABLE 4.2 *(Continued)*

Natural polymer	Industry	Application	Reference
Lignin	Industrial	Cement	[78]
		In batteries	
		Dispersant	
		Thickener in paint	
	Food	Additive	
	Packaging	High-resistant thermoplastic	
Plant seed mucilage	Food	Structuring	[79]
		Texturizing, emulsifying, thickening	
	Nutraceutical	Natural sources of dietary fiber, gut microbiota calorific density suppressor	
PHB	Pharmaceutical	Biodegradable plastic	
		Pills	[80]
		Microcapsules	[81]
		Swabs	
		Glove lubricant	
		Dressing	
		Scaffolds	
		Vascular grafts	
	Packaging	Biodegradable packaging	
	Miscellaneous	Aquaculture	
Guar gum	Miscellaneous	Hydraulic fracturing	
		textile and carpet printing	
		Gelling agent in explosives	
		Paper making	
		Stabilizer	
	Food	Emulsifier	
		thickener	

TABLE 4.2 *(Continued)*

Natural polymer	Industry	Application	Reference
		Beverages thickening and viscosity	[82]
		control	[83]
		Softens chapattis, breads, and sausages	
		Fat replacer in desserts and cheese	
		Improving texture and consistency of ketchups, cheese, and yogurt	
	Textile	Reduces breakage and	
		thickens the dye	
	Pharmaceutical	Binder, disintegrating agent in tablet	
		thickener, stabilizers, emulsifier, and suspending agent in syrups	
		Colon-specific, antihypertensive, protein, and transdermal hydrogel	
	DDS		
		Toothpastes	
	Cosmetic	Shaving cream	
		Shampoos cleansers hair colorants as thickener	
		Dry face mask as emulsifier	
	Bioremediation	Grafts and hydrogels for tapping heavy metals in aqueous solution	

KEYWORDS

- **natural polymers**
- **synthetic polymers gelatin**
- **PHB**
- **biotechnological applications**
- **thermoplastic elastomer**

REFERENCES

1. Zhong, Y.; Godwin, P.; Jin, Y.; Xiao, H. Biodegradable Polymers and Green-Based Antimicrobial Packaging Materials: A Mini-Review. *Adv. Ind. Eng. Polym. Res.* **2020,** *3*, 35.
2. Silva, A. C.; Silvestre, A. J.; Vilela, C.; Freire, C. S. Natural Polymers-Based Materials: A Contribution to a Greener Future. *Molecules* **2022,** *27*, 94.
3. Teixeira-Costa, B. E.; Andrade, C. T. Natural Polymers Used in Edible Food Packaging—History, Function and Application Trends as a Sustainable Alternative to Synthetic Plastic. *Polysaccharides* **2022,** *3*, 58.
4. Barouti, G.; Jaffredo, C. G.; Guillaume, S. M. Advances in Drug Delivery Systems Based on Synthetic Poly (hydroxybutyrate)(co) Polymers. *Prog. Polym. Sci.* **2017,** *73*, 31.
5. Mourey, T. H.; Schunk T. C. In *Synthetic Polymers*; Elsevier Inc, 1992.
6. Jain, R.; Tiwari, A. Biosynthesis of Planet Friendly Bioplastics Using Renewable Carbon Source. *J. Environ. Health Sci.* **2015,** *13*, 5.
7. Roohi; Bano, K.; Kuddus, M.; Zaheer, M. R.; Zia, Q.; Khan, M. F.; Ashraf, G. M.; Gupta, A.; Aliev, G. Microbial Enzymatic Degradation of Biodegradable Plastics. *Curr. Pharm. Biotechnol.* **2017,** *18*, 440.
8. Bano, K.; Kuddus, M.; Zaheer, M. R.; Roohi. A Novel, Thermotolerant, Extracellular PHB Depolymerase Producer Paenibacillus alvei PHB28 for Bioremediation of Biodegradable Plastics. *Turkish J. Biochem.* **2019,** *44*, 353.
9. Samadian, H.; Maleki, H.; Allahyari, Z.; Jaymand, M. Natural Polymers-Based Light-Induced Hydrogels: Promising Biomaterials for Biomedical Applications. *Coord. Chem. Rev.* **2020,** *420*, 213432.
10. Bhattarai, D. P.; Aguilar, L. E.; Park, C. H.; Kim, C. S. A Review on Properties of Natural and Synthetic Based Electrospun Fibrous Materials for Bone Tissue Engineering. *Membranes* **2018,** *8*, 62.
11. Tian, H.; Tang, Z.; Zhuang, X.; Chen, X.; Jing, X., Biodegradable Synthetic Polymers: Preparation, Functionalization and Biomedical Application. *Prog. Polym. Sci.* **2012,** *37*, 280.
12. Nurul Fazita, M. R.; Jayaraman, K.; Bhattacharyya, D.; Mohamad Haafiz, M. K.; Saurabh, C. K.; Hazwan Hussin, M.; Abdul Khalil, H. P. S. Green Composites Made of Bamboo Fabric and Poly (lactic) Acid for Packaging Applications-A Review. *Materials (Basel)* **2016,** *9*, 435.
13. Oliveira, F. C.; Dias, M. L.; Castilho, L. R.; Freire, D. M. Characterization of Poly (3-hydroxybutyrate) Produced by *Cupriavidus necator* in Solid-State Fermentation. *Bioresour. Technol.* **2007,** *98*, 638.
14. Piergiovanni, L.; Limbo, S. Oil Derived Polymers. *Food Packag. Princ. Pract.* **2016,** *38*, 41.
15. Youssef, A. M.; Abdel-Aziz, M. S. Preparation of Polystyrene Nanocomposites Based on Silver Nanoparticles Using Marine Bacterium for Packaging. *Polym. Plast. Technol. Eng.* **2013,** *52*, 613.
16. Wang, B.; Zhang, Q.; Xiong, G.; Ding, F.; He, Y.; Ren, B.; You, L.; Fan, X.; Hardacre, C.; Sun, Y. Bakelite-Type Anionic Microporous Organic Polymers with High Capacity for Selective Adsorption of Cationic Dyes from Water. *Chem. Eng. J.* **2019,** *366*–414.
17. Qian, S.; Ji, H.; Wu, X. X.; Li, N.; Yang, Y.; Bu, J.; Zhang, X.; Qiao, L.; Yu, H.; Xu, N.; Zhang, C. Detection and Quantification Analysis of Chemical Migrants in Plastic Food Contact Products. *PLoS One* **2018,** *13*, 11.

18. Bond, T.; Ferrandiz-Mas, V.; Felipe-Sotelo, M.; van Sebille, E. The Occurrence and Degradation of Aquatic Plastic Litter Based on Polymer Physicochemical Properties: A Review. *Crit. Rev. Environ. Sci. Technol.* **2018,** *48*, 722.
19. Miguel, M.; Leite, M.; Ribeiro, A. M. R.; Deus, A. M.; Reis, L.; Vaz, M. F. Failure of Polymer Coated Nylon Parts Produced by Additive Manufacturing. *Eng. Fail. Anal.* **2019,** *101*, 492.
20. Hashemi, M. J.; Jamshidi, M.; Aghdam, J. H. Investigating Fracture Mechanics and Flexural Properties of Unsaturated Polyester Polymer Concrete (UP-PC). *Constr. Build. Mater.* **2018,** *163*, 775.
21. Xiao, M.; Mohler, C.; Tucker, C.; Walther, B.; Lu, X.; Chen, Z. Structures and Adhesion Properties at Polyethylene/Silica and Polyethylene/Nylon Interfaces. *Langmuir* **2018,** *34*, 6204.
22. Herzberger, J.; Sirrine, J. M.; Williams, C. B.; Long, T. E. Polymer Design for 3D Printing Elastomers: Recent Advances in Structure, Properties, and Printing. *Prog. Polym. Sci.* **2019,** *97*, 101144.
23. Bokobza, L. Natural Rubber Nanocomposites: A Review. *Nanomaterials* **2019,** *9*, 12.
24. Zhang, X.; Cui, H.; Song, L.; Ren, H.; Wang, R.; He, A. Elastomer Nanocomposites with Superior Dynamic Mechanical Properties via Trans-1, 4-poly (butadiene-co-isoprene) Incorporation. *Compos. Sci. Technol.* **2018,** *158*, 156–163.
25. Öztürk, A.; Özçelik, N.; Bayrakçeken Yurtcan, A. Platinum/Graphene Nanoplatelets/ Silicone Rubber Composites as Polymer Electrolyte Membrane Fuel Cell Catalysts. *Mater. Chem. Phys.* **2021,** *260*, 124110.
26. Sharma, R. K.; Mohanty, S.; Gupta, V. Advances in Butyl Rubber Synthesis via Cationic Polymerization: An Overview. *Polym. Int.* **2021,** *70*, 1175.
27. Cichosz, S.; Masek, A. Thermal Behavior of Green Cellulose-Filled Thermoplastic Elastomer Polymer Blends. *Molecules* **2020,** *25*, 1279.
28. Asadi, N.; Rahmani, A.; Bakhshayesh, D.; Davaran, S. Common Biocompatible Polymeric Materials for Tissue Engineering and Regenerative Medicine. *Mater. Chem. Phys.* **2020,** *46*, 705.
29. Albuquerque, P. B. S.; Coelho, L. C. B. B.; Teixeira, J. A. Approaches in Biotechnological Applications of Natural Polymers. *AIMS Mol. Sci.* **2016,** *3* (3), 386–425.
30. Maraveas, C. Production of Sustainable and Biodegradable Polymers from Agricultural Waste. *Polymers* **2020,** *12*, 1127.
31. Xie, Y.; Niu, X.; Yang, J.; Fan, R.; Shi, J.; Ullah, N.; Feng, X.; Chen, L. Active Biodegradable Films Based on the Whole Potato Peel Incorporated with Bacterial Cellulose and Curcumin. *Int. J. Biol. Macromol.* **2020,** *150*, 480–491.
32. Rai, P.; Mehrotra, S.; Priya, S.; Gnansounou, E.; Sharma, S. K. Bioresource Technology Recent Advances in the Sustainable Design and Applications of Biodegradable Polymers. *Bioresour. Technol.* **2021,** *325*, 124739.
33. Åkesson, D.; Kuzhanthaivelu, G.; Bohlén, M. Effect of a Small Amount of Thermoplastic Starch Blend on the Mechanical Recycling of Conventional Plastics. *J. Polym. Environ.* **2021,** *29*, 985–991.
34. Rahmani, A.; Bakhshayesh, D.; Annabi, N.; Khalilov, R.; Samiei, M.; Alizadeh, E.; Alizadeh-, M.; Davaran, S.; Montaseri, A. Recent Advances on Biomedical Applications of Scaffolds in Wound Healing and Dermal Tissue Engineering. *Artif. Cells Nanomed. Biotechnol.* **2018,** *46* (4), 691–705.
35. Doppalapudi, S.; Katiyar, S.; Domb, A. J.; Khan W. In *Advanced Polymers in Medicine*; Springer, 2015; pp 33–66.

36. Olatunji, O. In *Natural Polymers: Industry Techniques and Applications*; Springer, 2015; pp 1–17.
37. Lee, H. Engineering and Functionalization of Gelatin Biomaterials : From Cell Culture to Medical Applications. **2019,** 1–48.
38. Shavandi, A.; Silva, T. H.; Bekhit, A. A., Bekhit, A. E. D. A. Keratin: Dissolution, Extraction and Biomedical Application. *Biomater. Sci.* **2017,** *5*, 1699–1735.
39. Jiang, T.; Duan, Q.; Zhu, J.; Liu, H.; Yu, L. Starch-Based Biodegradable Materials: Challenges and Opportunities. *Adv. Ind. Eng. Polym. Res.* **2020,** *3*, 8–18.
40. Wang, B.; Wan, Y.; Zheng, Y.; Lee, X.; Liu, T.; Yu, Z.; Huang, J.; Ok, Y. S.; Chen, J.; Gao, B. Alginate-Based Composites for Environmental Applications: A Critical Review. *Crit. Rev. Environ. Sci. Technol.* **2019,** *49*, 318–356.
41. Elieh-Ali-Komi, D.; Hamblin, M. R. Chitin and Chitosan: Production and Application of Versatile Biomedical Nanomaterials. *Int. J. Adv. Res.* **2016,** *4*, 411–427.
42. Ralph, J.; Lapierre, C.; Boerjan, W. Lignin Structure and its Engineering. *Curr. Opin. Biotechnol.* **2019,** *56*, 240–249.
43. Getachew, A.; Woldesenbet, F. Production of Biodegradable Plastic by Polyhydroxybutyrate (PHB) Accumulating Bacteria Using Low Cost Agricultural Waste Material. *BMC Res. Notes* **2016,** *9*, 9.
44. McAdam, B.; Fournet, M. B.; McDonald, P.; Mojicevic, M. Production of Polyhydroxybutyrate (PHB) and Factors Impacting its Chemical and Mechanical Characteristics. *Polymers (Basel)* **2020,** *12*, 2908.
45. Olatunji, O. *Natural Polymers: Industry Techniques and Applications*, 2015.
46. Abdullah, M. S. P.; Noordin, M. I.; Mohd Ismail, S. I.; Mustapha, N. M.; Jasamai, M.; Danik, M. F.; Wan Ismail, W. A.; Shamsuddin, A. F. Recent Advances in the use of Animal-Sourced Gelatine as Natural Polymers for Food, Cosmetics and Pharmaceutical Applications. *Sains Malaysiana* **2018,** *47*, 323–336.
47. Kulkarni, V.; Butte, K.; Rathod, S. Natural Polymers – A Comprehensive Review. *Int. J. Res. Pharm. Biomed. Sci.* **2012,** *3*, 1597–1613.
48. Djagny, K. B.; Wang, Z.; Xu, S. Gelatin: A Valuable Protein for Food and Pharmaceutical Industries: Review. *Crit. Rev. Food Sci. Nutr.* **2001,** *41*, 481–492.
49. Sahoo, N.; Sahoo, R. K.; Biswas, N.; Guha, A.; Kuotsu, K. Recent Advancement of Gelatin Nanoparticles in Drug and Vaccine Delivery. *Int. J. Biol. Macromol.* **2015,** *81*, 317–331.
50. Martins, E.; Rodrigues, A. Starch: From Food to Medicine. In *Scientific, Health and Social Aspects of the Food Industry*, 2012.
51. Rodrigues, A.; Emeje, M. Recent Applications of Starch Derivatives in Nanodrug Delivery. *Carbohydr. Polym.* **2012,** *87*, 987–994.
52. Tønnesen, H. H.; Karlsen, J. Alginate in Drug Delivery Systems. *Drug Dev. Ind. Pharm.* **2002,** *28* , 621–630.
53. Shariatinia, Z. Pharmaceutical Applications of Chitosan. *Adv. Colloid Interface Sci.* **2019,** *263*, 131–194.
54. Picot-Allain, M. C. N.; Ramasawmy, B.; Emmambux, M. N. Extraction, Characterisation, and Application of Pectin from Tropical and Sub-Tropical Fruits: A Review. *Food Rev. Int.* **2020,** *00*, 1–31.
55. Gałkowska, D.; Długosz, M.; Juszczak, L.; Takamine, K.; Abe, J.; Shimono, K.; Sameshima, Y.; Morimura, S.; Kida, K.; Sriamornsak, P. Chemistry of Pectin and Its Pharmaceutical Uses: A Review. *Silpakorn Univ. Open J. Syst.* **2003,** *3*, 206–228.

56. Diab, T.; Biliaderis, C. G.; Gerasopoulos, D.; Sfakiotakis, E. Physicochemical Properties and Application of Pullulan Edible Films and Coatings in Fruit Preservation. *J. Sci. Food Agric.* **2001,** *81*, 988–1000.
57. Baziwane, D.; He, Q. Gelatin: The Paramount Food Additive. *Food Rev. Int.* **2003,** *19*, 423–435.
58. Alves, T. F. R.; Morsink, M.; Batain, F.; Chaud, M. V.; Almeida, T.; Fernandes, D. A.; da Silva, C. F.; Souto, E. B.; Severino, P. Applications of Natural, Semi-Synthetic, and Synthetic Polymers in Cosmetic Formulations. *Cosmetics* **2020,** *7* (4), 75.
59. Rodríguez , M. I. A.; Barroso, L. G. R.; Sánchez, M. L. Collagen: A Review on its Sources and Potential Cosmetic Applications. J. Cosmet. Dermatol. **2018,** 17 (1), 20–26.
60. Charles, A. P. R.; Jin, T. Z.; Mu, R.; Wu, Y. Electrohydrodynamic Processing of Natural Polymers for Active Food Packaging: A Comprehensive Review. *Compr. Rev. Food Sci. Food Saf.* **2021,** *20*, 6027–6056.
61. Morin-Crini, N.; Lichtfouse, E.; Torri, G.; Crini, G. Applications of Chitosan in Food, Pharmaceuticals, Medicine, Cosmetics, Agriculture, Textiles, Pulp and Paper, Biotechnology, and Environmental Chemistry. *Environ. Chem. Lett.* **2019,** *17*, 1667–1692.
62. Olatunji, O.; Igwe, C. C.; Ahmed, A. S.; Alhassan, D. O. A.; Asieba, G.; Das, D. B. Microneedles from Fish Scale Biopolymer. *J. Appl. Polym. Sci.* **2014,** *131*, 1–10.
63. Meena, C.; Mengi, S. A.; Deshpande, S. G. Biomedical and Industrial Applications of Collagen. *Proc. Indian Acad. Sci. Chem. Sci.* **1999,** *111*, 319–329.
64. Kumar, A.; Rao, K. M.; Han, S. S. Application of Xanthan Gum as Polysaccharide in Tissue Engineering : A Review. *Carbohydr. Polym.* **2017**.
65. Freitas, C. M. P.; Coimbra, J. S. R.; Souza, V. G. L.; Sousa, R. C. S. Structure and Applications of Pectin in Food, Biomedical, and Pharmaceutical Industry : A Review. *Coatings* **2021,** *11* (8), 922.
66. Olatunji, O. Natural Polymers: Industry Techniques and Applications. *Nat. Polym. Ind. Tech. Appl.* **2015,** *1*, 370.
67. Sayyed, R. Z., Wani, S. J., Alarfaj, A. A., Syed, A.; El-Enshasy, H. A. Production, Purification and Evaluation of Biodegradation Potential of PHB Depolymerase of Stenotrophomonas sp. RZS7. *PLoS One* **2020,** *15*, e0220095.
68. Howe, A. M. Some Aspects of Colloids in Photography. *Curr. Opin. Colloid Interface Sci.* **2000,** *5*, 288–300.
69. Onifade, A. A.; Al-Sane, N. A.; Al-Musallam, A. A.; Al-Zarban, S. A Review: Potentials for Biotechnological Applications of Keratin-Degrading Microorganisms and their Enzymes for Nutritional Improvement of Feathers and Other Keratins as Livestock Feed Resources. *Bioresour. Technol.* **1998,** *66*, 1–11.
70. Rouse, J. G.; Van Dyke, M. E. A Review of Keratin-Based Biomaterials for Biomedical Applications. *Materials* **2010,** *3*, 999–1014.
71. Kakkar, P.; Madhan, B. Fabrication of Keratin-Silica Hydrogel for Biomedical Applications. *Mater. Sci. Eng. C* **2016,** *66*, 178–184.
72. Madziva, H.; Kailasapathy, K.; Phillips, M. Alginate-Pectin Microcapsules as a Potential for Folic Acid Delivery in Foods. *J. Microencapsul.* **2005,** *22*, 343–351.
73. Puscaselu, R. G.; Lobiuc, A.; Dimian, M.; Covasa, M. Alginate: From Food Industry to Biomedical Applications and Management of Metabolic Disorders. *Polymers* **2020,** *12*, 1–30.

74. Nsor-Atindana, J.; Chen, M.; Goff, H. D.; Zhong, F.; Sharif, H. R.; Li, Y. Functionality and Nutritional Aspects of Microcrystalline Cellulose in Food. *Carbohydr. Polym.* **2017,** *172*, 159–174.
75. Salwowska, N. M.; Bebenek, K. A.; Żądło, D. A.; Wcisło-Dziadecka, D. L. Physiochemical Properties and Application of Hyaluronic Acid: A Systematic Review. *J. Cosmet. Dermatol.* **2016,** *15*, 526.
76. Kogan, G.; Šoltés, L.; Stern, R.; Gemeiner, P. Hyaluronic Acid: A Natural Biopolymer with a Broad Range of Biomedical and Industrial Applications. *Biotechnol. Lett.* **2007,** *29*, 17–25.
77. Casale, M.; Moffa, A.; Vella, P.; Sabatino, L.; Capuano, F.; Salvinelli, B.; Lopez, M. A.; Carinci, F.'; Salvinelli, F. Hyaluronic Acid: Perspectives in Dentistry. A Systematic Review. *Int. J. Immunopathol. Pharmacol.* **2016,** *29*, 572–582.
78. Agrawal, A.; Kaushik, N.; Biswas, S. Derivatives and Applications of Lignin–An Insight. *SciTech J.* **2014,** *1*, 30–36.
79. Soukoulis, C.; Gaiani, C.; Hoffmann, L. Plant Seed Mucilage as Emerging Biopolymer in Food Industry Applications. *Curr. Opin. Food Sci.* **2018,** *22*, 28–42.
80. Bucci, D. Z.; Tavares, L. B. B.; Sell, I. PHB Packaging for the Storage of Food Products. *Polym. Test.* **2005,** *24*, 564–571.
81. Van Hung, N.; De Schryver, P.; Tam, T. T., Garcia-Gonzalez, L.; Bossier, P.; Nevejan, N. Application of Poly-β-hydroxybutyrate (PHB) in Mussel Larviculture. *Aquaculture* **2015,** *446*, 318–324.
82. Mudgil, D.; Barak, S.; Khatkar, B. S. Guar Gum: Processing, Properties and Food Applications - A Review. *J. Food Sci. Technol.* **2014,** 51, 409–418.
83. Thombare, N.; Jha, U.; Mishra, S.; Siddiqui, M. Z. Guar Gum as a Promising Starting Material for Diverse Applications: A Review. *Int. J. Biol. Macromol.* **2016,** *88*, 361–372.

CHAPTER 5

Pectin–Carbon Nanosphere: A Green Nanocomposite for the Sensing of Glucose

S. K. SUJA, J. SHAMILI, and S. GEETHANJALI

Department of Chemistry, Lady Doak College, Madurai, Tamil Nadu, India

ABSTRACT

Pectin was extracted from the peels of citrus fruits and carbon nanosphere (CNS) was synthesized from the shoots of gingelly oil. They were blended by the solvent casting method to get a nanocomposite. The formed composite was characterized by using UV-visible, IR spectroscopic methods. The surface morphological studies were performed using SEM and the crystallite size was determined by XRD studies and was found to be 20 nm and 9 nm for PEC and CNS, respectively, and which got reduced to 5 nm when CNS was incorporated into PEC. IR studies shows that glucose oxidase (GOx) interact with PEC through –OH group present in the ring of PEC and when CNS was introduced into PEC, the CNS interacted with PEC through –COOH group of PEC. The intensity of the voltammetric signal after incorporating CNS with PEC gets enhanced and this shows the presence of sp^2 hybridized carbons in CNS. In order to utilize the newly synthesized composite for sensing of glucose, it was fabricated into an electrode with carbon paste (CP). The electroactive surface area of the CPE/PEC–CNS modified electrode (0.3421 cm^2s^{-1}) was higher than the other electrodes studied—CPE (0.1242 cm^2s^{-1}), CPE/PEC (0.1530 cm^2s^{-1}), and CPE/CNS (0.2379 cm^2s^{-1}). For the GOx immobilized CPE/PEC–CNS electrode, the Michaelis-Menten constant, Km, value obtained was 0.09, and lower Km shows higher enzymatic activity of immobilized GOx. The CPE/PEC–CNS/GOx electrode showed high surface concentration of glucose, Γ (7.03×10^{-9} mol^{-1} cm^{-1}), high diffusion coefficient

Bio-Based Polymers and Composites: Properties, Durability, and Applications. Amadou Belal Gueye, Sabu Thomas, Nandakumar Kalarikkal, & Modou Fall (Eds.)

D0 (4.87×10^{-2} cm^2 s^{-1}), and the high electron transfer rate constant k_0 (0.397×10^2 s^{-1}). The limit of detection (LOD) and limit of quantification (LOQ) were very low, 0.78×10^{-15} M and 2.34×10^{-15} M, respectively, and this showed the high sensitivity of the fabricated electrode. The fabricated electrode was interference free and highly stable even after 120 days.

5.1 INTRODUCTION

Enzymes are the bio-recognition elements present in enzymatic biosensors and a limited number of enzymes are utilized for detecting clinical metabolites. These enzymes include glucose oxidase and glucose dehydrogenase for glucose assays. Enzymes confined on the surface of a material can easily recognize the metabolites[1,2] even though enzyme-modified electrodes have some disadvantages, such as instability, complicated immobilization procedures, and critical operating conditions.[3] The material used for surface confinement of enzymes should provide a conducive environment to the enzymes to retain their conformational stability and to be protected from denaturation. Also, the most intriguing property of enzymes is its specificity[4] and the ability to promote electron transfer between the enzyme and the metabolite.[5] Several carbohydrate-based materials like chitosan, cellulose, inulin, etc., served the purpose. Similar properties were expected from the same class of compound, pectin (PEC). Pectin is a naturally occurring polysaccharide and it is the methylated ester of poly-galacturonicacid.[6–10] It is commercially extracted from citrus peels and apple pomace under mild acidic conditions.[11] Because of the presence of large amount of –OH groups, this polymer is conducting in nature. To enhance the conductivity of polymers, carbon-based nanomaterials were widely used.

The ease of processing and forming of light materials may be one of the main reasons for the rapidly growing applications of carbon structures. Carbon nanosphere is one of the most widely used carbon materials. Compared to the studies of fullerenes, carbon nanotubes, and nanofibers, reports on carbon nanospheres are relatively low in number. Carbon spheres have similar properties to graphite or fullerene, also known as nanocapsules (CNC) with inner diameter 20–50 nm and shell thickness 10–15 nm.[12] It is having more applications which are related to aerospace structure, aircraft brakes, lubrication, and concrete structures. They are also known for their thermal application, environmental application, and biomedical application. The sp^2 and sp hybridized carbons are electrical conductors in the in-plane direction due to the delocalization of p_z electrons but in out of plane the conductivity is low due to the Vander waals bonding in this direction;

therefore, the electrical conductivity of the carbon material depends on the degree of orientation of carbon layers. The conductivity and alignment of carbon differs with respect to the processing procedures. Carbon nanosphere is more effective than 0.1 μm diameter of carbon filament and it has higher resistivity than the carbon filament. Carbon nanosphere is superior conductive additive because of their spreadability.[13] Carbon-based nanostructures have been widely used to immobilize enzymes due to their large surface area and fast electron transfer rate.[14] Carbon nanosphere (CNS), a novel kind of carbon nanomaterial possessing homogeneity of particle size, good chemical stability, and porous nanostructure of large loading guest molecules, has been employed as an immobilizing material for the fabrication of biosensors.[12]

Polymer composites are a new class of materials incorporating an ultrafine dispersion of nanomaterials in a polymeric matrix. The properties of polymer composite depend on the particles shape, size, loading interfacial bonding, and dispersion of the fillers.[15] Many polymers were used so far to synthesis polymer nanocomposite. CNS was synthesized and functionalized with ethylene glycol and dodecylamine and their thermal stability was studied.[16] MoS_2 on ordered mesoporous carbon nanospheres (MoS_2/MCNs) was synthesized and its properties were reported. The MoS_2/MCNs exhibit high catalytic activity for hydrogen evolution with a low overpotential and a very high current density.[17] Cu/carbon sphere composites were synthesized by a facile and effective polymer-assisted chemical reduction method. The composites showed good non-enzymatic electrocatalytic responses to glucose in alkaline media, and can be used for the development of enzyme-free glucose sensors.[18] Highly sensitive amperometric biosensor for glucose was developed using carbon nanosphere/sodium alginate composite matrix.[19] Limited numbers of biopolymer nanocomposite matrix with CNS were synthesized and reported for some applications. This is the first time that from the naturally available, cost-effective source both the PEC and CNS were synthesized. In this study, the synergistic effect was utilized to sense glucose by synthesizing the PEC–CNS bio-polymer nanocomposite.

5.2 EXPERIMENTAL

5.2.1 REAGENTS AND CHEMICALS

Glucose oxidase (GOx, from Aspergillus Niger, EC 61788(074040)[9001-37-0], 125 units/mg) was purchased from Sisco Research Laboratories Pvt. Ltd and D(+) glucose (98%) potassium dihydrogen phosphate (98%), dipotassium hydrogen phosphate (98%), ascorbic acid (99%), uric acid (99%), and graphite powder (98%) were purchased from Fischer Scientific, India. All other chemicals were of analytical grade reagent and used without further purification. All chemicals used were of analytical grade and were used as received. Deionized water was used throughout the study and all the experiments were carried out at room temperature.

Phosphate buffer solution (0.1 M, pH 7) was prepared by mixing equimolar solution of (0.1 M) potassium dihydrogen orthophosphate and (0.1 M) dipotassium hydrogen phosphate. GOx solution was prepared by dissolving 5 mg in 1 mL of deionized water and was stored at 2 °C in refrigerator. A stock solution of glucose (0.1 M) was prepared by dissolving 1.8016 g D-Glucose in 100 mL using deionized water and stored at 4 °C. Working solutions of glucose was prepared by diluting the stock solution using phosphate buffer (0.1 M, pH 7).

5.2.2 APPARATUS

UV-visible absorption spectra were recorded using Thermo Scientific Helios Alpha UV-Visible spectrophotometer. The Fourier Transform infrared (FT-IR) spectra of the samples in KBr pellets were recorded on an IR Affinity-I Fourier transform Infrared spectrophotometer, Shimadzu. X-Ray diffraction (XRD) patterns of PEC, CNS, and composite were carried out using Panalytical X "Pert Powder X" Celerator Diffractometer at 10–80° (2θ) equipped with Cu Kα1 radiation ($\lambda = 1.54$ A°). The surface morphology of the composite was observed by using scanning electron microscope (JOEL, JFM6390). All voltammetric measurements were performed on a CHI6063C electrochemical analyzer. All the experiments were carried out in a conventional electrochemical cell at room temperature. A three-electrode system, including Ag/AgCl (saturated KCl) as reference electrode, platinum wire as auxiliary electrode, and modified CPE as working electrode, was employed. The electrolyte solutions were purged with N2 for about 5 min to remove O_2 and kept under N_2 atmosphere throughout the measurements.

5.2.3 FABRICATION OF PEC–CNS COMPOSITE

5.2.3.1 EXTRACTION OF PEC

The citrus fruit peels were washed with plenty of water and rinsed with distilled water. 100 g of the washed peels were ground with 200 mL of deionized water using mixer grinder. To that mixture, 400 mL of deionized water and 200 mL of 0.1 N HNO_3 were added. This mixture was stirred continuously by magnetic stirrer for about 45 min followed by heating at 40 °C. Then, it was allowed to cool in refrigerator (4 °C) for 1 h and was filtered using muslin cloth. To the filtrate, equal amount of ethanol was added and stirred. A gel-like mass (pectin) was formed and was separated by using Buchner funnel.

5.2.3.2 SYNTHESIS OF CNS

CNS was prepared by flame vapor deposition method. Gingely oil was taken in a china dish and a bundle of cotton threads were placed in the oil. The cotton threads soaked in gingely oil was lighted. When the lamp glows brightly, a beaker was placed above the flame and the soot was allowed to get deposited on the sides of the beaker. The deposited soot (CNS) was collected.

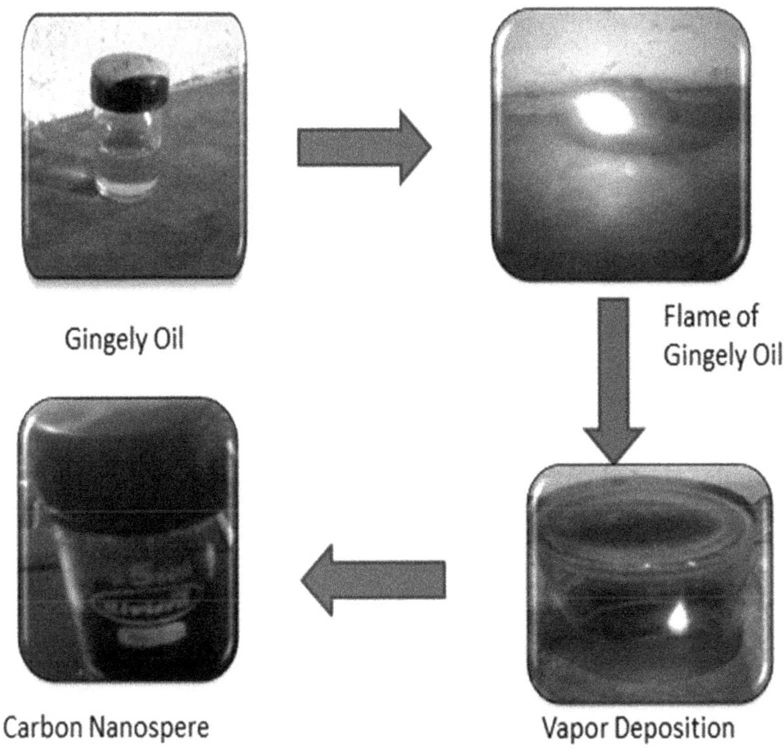

FIGURE 5.1 Synthesis of carbon nanosphere by flame vapor deposition method.

5.2.3.3 SYNTHESIS OF PEC–CNS COMPOSITE AND FABRICATION OF ELECTRODE

Pectin–CNS composite was synthesized by solvent casting method; 1 g of pectin was dissolved in 20 mL of deionized water and the polymer suspension was then magnetically stirred (10 min) for complete dissolution; 0.25 g of CNS was added to the polymer solution and magnetically stirred for 15 min. Subsequently, the mixture was sonicated for 15 min by ultra-waving in a water bath to improve the dispersion of CNS in the polymer matrix. To the homogenous mixture, equal volume of ethanol was added and the formed composite was separated by using Buchner funnel.

The bare CPE was prepared by hand mixing of 700 mg of graphite powder and three drops of paraffin oil in an agate mortar to produce a homogeneous carbon paste. The paste was packed into a 4 cm length hollow glass tube of 3 mm diameter and then the surface was smoothed on a filter paper. The electrical contact was provided by a copper wire. In order to prepare PEC–CNS/GOx/CPE, the base of the CPE was filled with 90 mg of PEC–CNS composite and a drop of the prepared GOx solution was added on the electrode surface by tilting the electrode upside down. The sensitivity was compared using the bare CPE, PEC/CPE, CNS/CPE, and PEC–CNS/ CPE and were prepared by the similar procedure as that of the PEC–CNS/GOx/CPE.

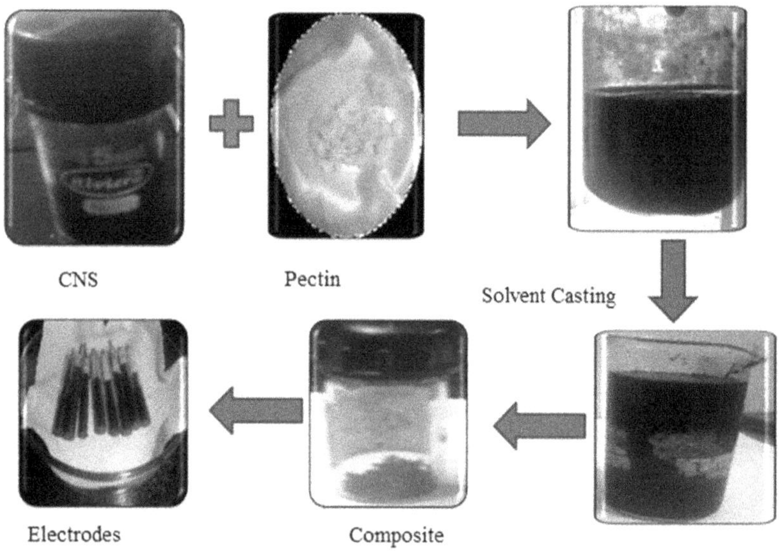

FIGURE 5.2 Synthesis of Pectin–CNS composite by solvent casting method and the fabricated electrode.

5.3 RESULTS AND DISCUSSION

5.3.1 CHARACTERIZATION OF PEC–CNS COMPOSITE

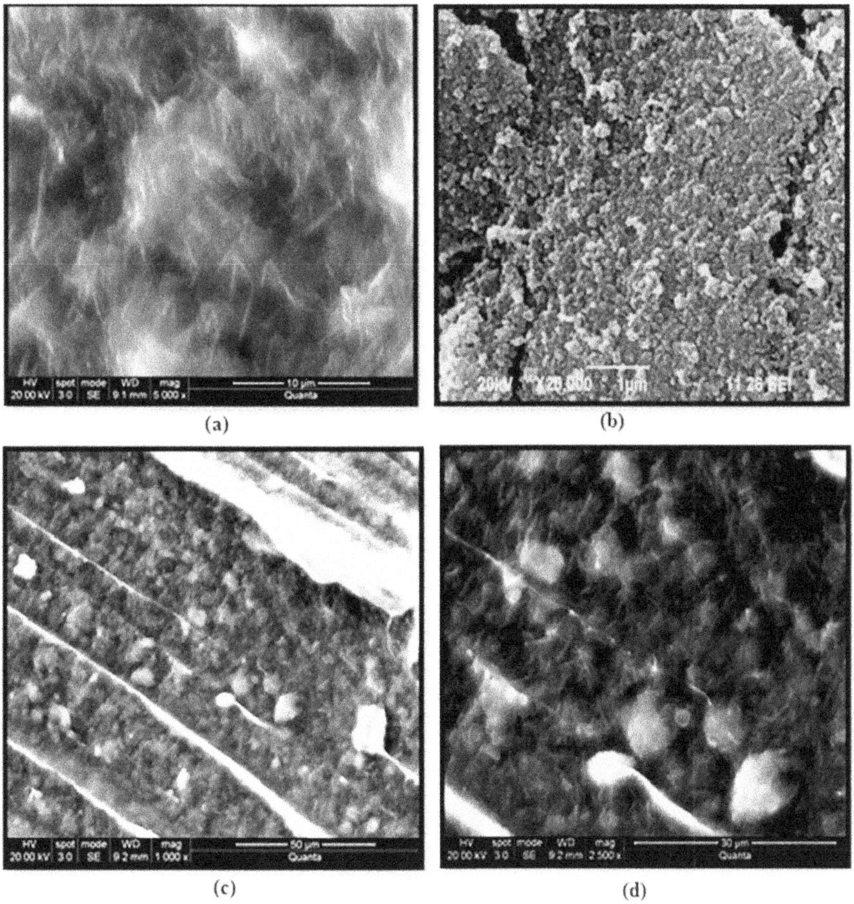

FIGURE 5.3 SEM images of (a) Pectin, (b) CNS, (c) and (d) Pectin–CNS composite and different dimensions.

The synthetic procedure of the carbon nanospheres does not involve any organic solvents, initiators, or surfactants. The bio-polymer PEC is used as a stabilizer. The morphology of pectin and pectin–CNS composite are observed using scanning electron micrographs (SEM). Figure 5.3 are the SEM images of (a) pectin, (b) CNS, and (c) and (d) pectin–CNS composite. The SEM image of pectin shows pores in it and are uniform in nature.

The spherical CNS nanoparticles are embedded into the pores of PEC and thereby prevents the agglomeration of CNS. Moreover, the pore size is suitable enough to accommodate the CNS particles. Also, the PEC matrix provided a layered orientation to the CNS and thereby the CNS becomes stable. The highly specific orientation of CNS enhances the specific surface area and more binding sites for biomolecules.[20] The incorporation of CNS in PEC was again confirmed by comparing the % carbon obtained from EDX spectra. The C % in PEC–CNS (85.36) was 27.66% higher than that of PEC (57.70). This confirms that the CNS is effectively embedded into PEC matrix in PEC–CNS composite.

The porosity is the parameter associated with pore size. The calculated quantity of PEC and PEC–CNS were taken in separate petri dishes and water was dropped into it slowly. This was allowed to stand for 24 h at room temperature. Swelling of PEC and PEC–CNS gels was observed. The weight of these two, both in dry and wet conditions, was calculated. The porosity was calculated using the formula,

$$Porosity = \frac{Ws - Wd}{Wd}$$

where Ws (g) is the weight of the component in swollen state and Wd (g) is the weight of the component in dry state. The porosity of PEC–CNS composite was found to be 62.5% and that of PEC was 84.76%. This shows that 22.26% of pores of PEC were entrapped with CNS.

5.3.2 SPECTRAL CHARACTERIZATION

UV–visible absorption spectroscopy has been performed to characterize the PEC–CNS composite and to establish the presence of GOx on PEC–CNS nanocomposite. UV–visible spectrum of PEC exhibited a strong band at 280 nm and that is due to $\pi - \pi^*$ transition. The peak around 324 nm in PEC is due to n - π^* transition, which disappeared when CNS was allowed to disperse in PEC. This shows that CNS was effectively incorporated into PEC matrix via –C=O group present in PEC. When GOx is incorporated into PEC, the absorption peak red shifted to 285 nm with the reduction in intensity of the signal. This shift represents the capping effect of GOx over PEC.

The FTIR spectra of pure PEC shows band at 3431 cm^{-1} due to the stretching vibration mode of –OH, 1750 cm^{-1} due to C=O stretching of ester or carboxylic acid, 1600 cm^{-1} due to amide group, C=O stretching along with NH-deformation mode, 1450 cm^{-1} is assigned to the symmetrical deformation of CH_2 group, 1250 cm^{-1} peak may be due to COO- group in ester, a sharp peak at 1383 cm^{-1} may be due to carboxylic acid, broad peak at 1071 cm^{-1} is assigned to the β (1-4) glycosidic linkage in polysaccharide unit, 1051 $cm^{-1}$is attributed to the stretching vibration mode of hydroxyl group, and 1000 cm^{-1} stretching vibration of C–O–C in glucose moiety. The –OH stretching band centered at 3431 cm^{-1} in PEC and 3412 cm^{-1} in CNS weakens in the PEC–CNS composite and shifted to lower wave number. When GOx was incorporated into the PEC, the peak at 1051 cm^{-1} shifted to 1020 cm^{-1} and this shows that GOx interacts with PEC through –OH group present in the ring. There is no change in the position of the peak at 1383 cm^{-1} when GOx was incorporated into PEC, but when CNS was introduced into PEC, the peak disappeared and this shows that the CNS interact with PEC through –COOH group in PEC.

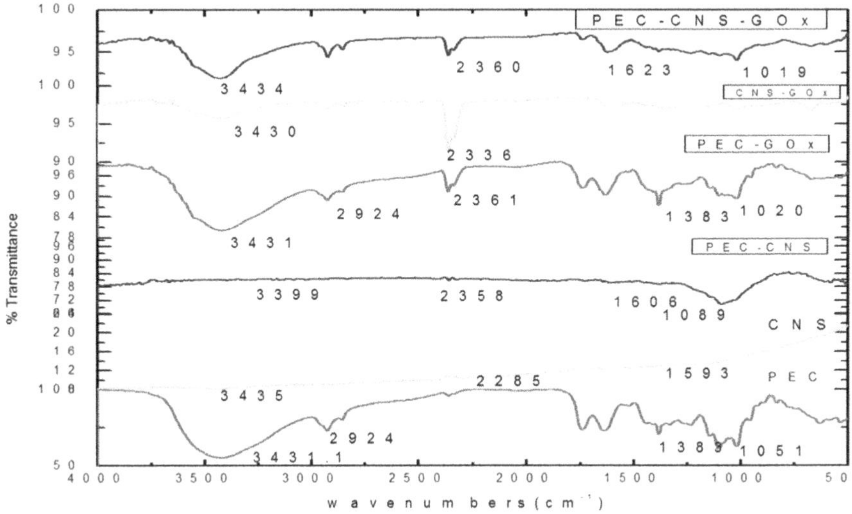

FIGURE 5.4 FTIR spectra of PEC, CNS, PEC–CNS, PEC-GOx, and PEC–CNS–GOx.

X-ray diffractogram of pectin indicated a single weak wide diffraction peak at 12.056° (2θ); it shows that pectin is amorphous material because there is no intense peak except a weak wide diffraction peak at 12.056°. X-ray diffractogram of CNS shows a sharp diffraction peak at 27.7^0 (2) and it shows that CNS is a crystalline material. When CNS was incorporated into PEC the intensity of the peak obtained for PEC gets diminished.[21] The crystallite size was calculated using the Debye–Sherrer equation. The crystallite size of PEC and CNS was found to be 20 nm and 9 nm, respectively. When CNS was incorporated into PEC, the crystallite size gets reduced to 5 nm.

5.3.3 DETERMINATION OF SURFACE PARAMETERS OF PEC–CNS COMPOSITE ELECTRODE

5.3.3.1 ESTIMATION OF ELECTROACTIVE SURFACE AREA

The electroactive surface area of the CPE, CPE/PEC, CPE/CNS, and CPE/PEC–CNS electrodes were studied. It is more important because the electrode is newly fabricated by the synthesized material. In order to estimate the surface area of the electrodes, the electrochemical behavior of the electrodes in which the electron transfer properties between redox system and the electrodes can be identified by using the cyclic voltammograms of $[Fe\ (CN)_6]^{4-/3-}$ redox couple.[22] The cyclic voltammogram of ferricyanide was chosen as a marker to investigate the changes of the electrode behavior after each modification.[23]

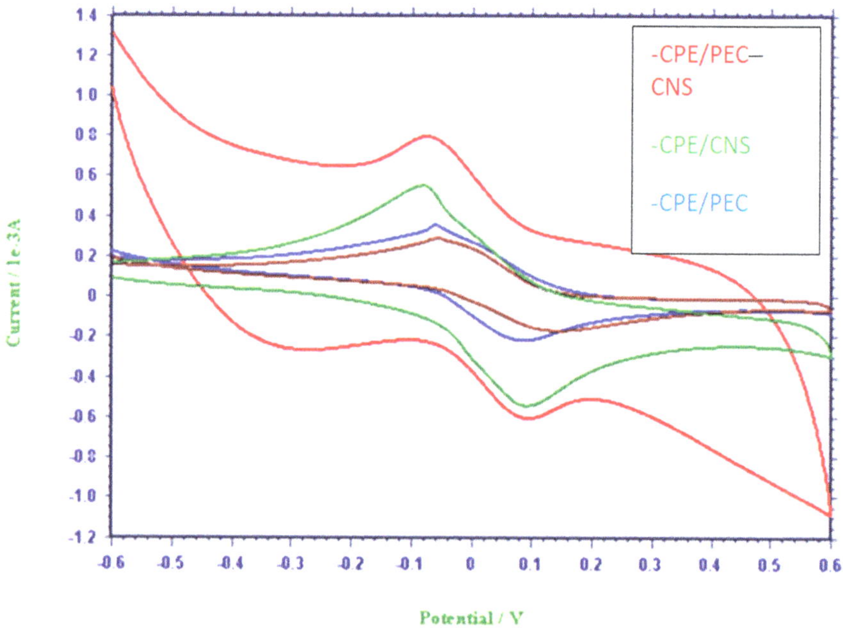

FIGURE 5.5 Cyclic voltammograms of 0.01 M potassium ferricyanide in 0.1 M KCl at a scan rate of 100 mVs^{-1}using CPE, CPE/PEC, CPE/CNS, and CPE/PEC-CNS electrodes.

Figure 5.5 shows the cyclic voltammograms of $[Fe(CN)_6]^{3-/4-}$ at the CPE, CPE/PEC, CPE/CNS, and CPE/PEC–CNS electrodes. The modification of the carbon paste electrode with CNS and PEC–CNS composite has resulted a positive electro-catalytic effect on the redox process of $[Fe(CN)_6]^{3-/4-}$. The redox peak current values increases when CNS-based electrodes were used as working electrodes; this indicate that the CNS-based CPE has high electrocatalytic activity. This enhanced peak current value indicates the enhanced reaction kinetics of the electrodes. As the modification of CNS was made with PEC, increase in peak current followed by decrease in peak-to-peak separation (ΔE_p) between the cathodic and anodic waves of the redox probe shows that the electron transfer kinetics of $[Fe(CN)_6]^{3-/4-}$ redox process was enhanced.[23] The ΔE_p values obtained during the redox process of $[Fe(CN)_6]^{3-/4-}$ using CPE, CPE/PEC, CPE/CNS, and CPE/PEC–CNS are 0.0881 V, 0.0267 V, 0.0116, and 0.019 V, respectively, and this shows that CNS-based electrodes provide perfect redox process ($\Delta E_p = 0.059$ V) than other electrodes for a single electron transfer process.

The electro active surface area for all the electrodes was calculated using Randles–Sevcik equation, $i_p = 2.69 \times 10^5\ n^{3/2}\ A\ Do^{1/2}\ \upsilon^{1/2}\ Co$, where Co is the concentration of electro active species (mol cm^{-3}), diffusion coefficient Do ($cm^2\ s^{-1}$), the projected area of the electrode A (cm^2), n is the number of transferred electrons, and υ, the scan rates (Vs^{-1}), and it is listed in Table 5.1.

This result shows the efficiency of the electrochemical functionalization of the electron transfer properties of CPE/PEC–CNS. This enhancement in the electron transfer rate provides a large number of electroactive defects created at the surface of CPE/PEC–CNS that leads to the efficient platform for enzyme immobilization.[24] The electroactive surface area of the CPE/PEC–CNS-modified electrode (0.3421 cm^2s^{-1}) is higher than the other electrodes studied, CPE (0.1242 cm^2s^{-1}), CPE/PEC (0.1530 cm^2s^{-1}), and CPE/CNS (0.2379 cm^2s^{-1}).

5.3.3.2 SURFACE CONFINEMENT OF GLUCOSE OXIDASE (GO_X) ON PEC–CNS COMPOSITE ELECTRODE

The electrochemical response of the immobilized GOx on the prepared electrodes were studied by comparing the electrochemical signals obtained with bare electrodes and GOx-immobilized electrodes in phosphate buffer solution at pH 7 and at the scan rate of 100 mVs^{-1}. No electrochemical signal was observed when a bare electrode was used as a working electrode and GOx-incorporated electrode gives corresponding redox peaks. This shows the presence of electrochemically active GOx on the electrode surface and the increased electron flow when GOx is present and the bare electrodes are electrochemically inert in phosphate buffer at pH 7. The peak current values get enhanced when PEC–CNS/GOx electrode is used and it seems that the PEC–CNS provides a bio-compatible environment to the GOx and CNS act as electron mediator, which accelerates electron transfer between PEC and GOx.

5.3.3.3 SURFACE ELECTRON TRANSFER KINETICS BETWEEN GLUCOSE AND PEC–CNS/GO$_X$ ELECTRODE

The cyclic voltamograms of all the enzyme-modified electrodes without glucose and upon addition of 0.1 M glucose solution were analyzed at a scan rate of 100 mVs^{-1} in phosphate buffer solution at pH 7. It was found that both the reduction and oxidation peak currents suppressed for all the enzyme-modified electrodes in absence of 0.1 M glucose. This observation shows that the glucose oxidase interact with glucose by intercalative mode. The most significant observation in this study was the oxidation potential of glucose by using CPE/CNS/GOx, CPE/PEC–CNS/GOx electrode. Glucose was oxidized at lower potential than other GOx immobilized electrodes and which can be described by the fact that less potential is needed to study the sensitivity of glucose by using the fabricated biosensor. When the concentration of glucose varies, the electron transfer rate at the surface of the electrode varies. From Figure 5.6, it was observed that the enzyme-loaded electrode response gradually deviates from the linear feature[25] as the glucose concentration reaches above 4.6 × 10^{-4} M. This is the characteristics of a typical Michelis-Menten Kinetics.

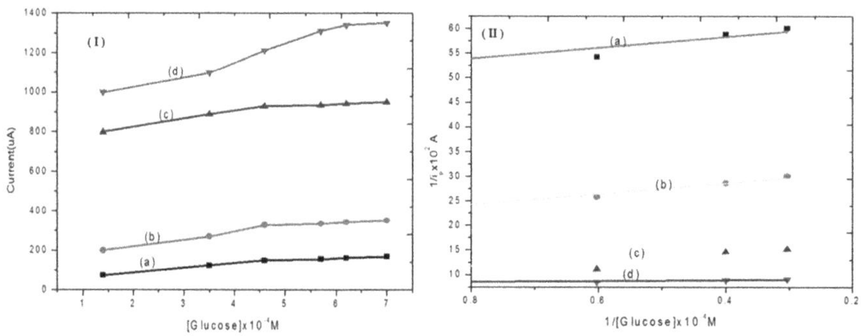

FIGURE 5.6 (I) Effect of concentration of glucose (× 10⁻⁴M) on the current responds; (II) Effect of 1/[glucose] (×10⁻⁴M) on the current responds of (a) CPE/PEC–CNS/GOx, (b) CPE/CNS/GOx, (c) CPE/PEC/GOx, and (d) CPE/GOx electrodes.

The apparent Michaelis–Menten constant (Km) gives an indication of the enzyme substrate affinity and the maximum current was observed for the fabricated biosensor. The K_m value was estimated using Lineweaver–Burk equation, $I^{-1} = (1+ K_m^{app} C_s^{-1}) I_{max}^{-1}$, where, I is the steady state current obtained at the glucose concentration C_s, K_m^{app} is the apparent Michaelis-Menten constant for the electrode, and I_{max} is the maximum current response. For the CPE/PEC–CNS /GOx, the K_m^{app} value obtained is 0.09, which is lower than the other electrodes studied, CPE/GOx (0.37), CPE/PEC/GOx (0.32), and CPE/CNS/GOx (0.33). The lower K_m^{app} means higher enzymatic activity of immobilized GOx; thus, the result further indicates that the CPE/PEC–CNS/GOx biosensor possesses a high affinity to glucose.[26]

5.3.3.4 ESTIMATION OF THE SURFACE EXCESS CONCENTRATION, DIFFUSION COEFFICIENT, AND ELECTRON TRANSFER RATE CONSTANT

The scan rate is probably the most important experimental parameter for evaluating the effects due to adsorbed reactant and/or due to reaction at the electrode via diffusion. The cyclic voltammograms of glucose solution using CPE/GOx, CPE/PEC/GOx, CPE/CNS/GOx, and CPE/PEC–CNS/GOx electrodes were recorded by varying the scan rate from 25 mV to 500 mV. The oxidation peak current increases linearly with increase in scan rate and this indicates that the electron transfer process occurring at the electrodes is a surface-confined process.[27]

The effect of υ (Vs^{-1}) on i_p (μA) for glucose solution (10^{-3} M) at CPE/GOx, CPE/PEC/GOx, CPE/CNS/GOx, and CPE/PEC–CNS/GOx electrodes were studied. Using the plots obtained the surface concentration (Γ) of all the electrodes was studied by the formula,[13] $i_p = n^2F^2A\Gamma\upsilon/4RT$, where R is the gas constant (8.314 $JK^{-1}mol^{-1}$), T is the Kelvin temperature, F is the Faraday constant (96487 $Cmol^{-1}$), n is the number of electrons transferred in the electrode reaction, A is the surface area of the working electrode, ν ($V s^{-1}$) is the scan rate, and i_p (μA) is the peak current of the reactant. The order of surface concentration was found to be,

CPE/PEC/GOx< CPE /GOx< CPE/CNS/GOx< CPE/PEC–CNS/GOx

The maximum Γ value was observed for CPE/PEC–CNS/GOx. From this result, it was concluded that the composite provides a matrix for the efficient immobilization of a large loading of glucose oxidase.[28]

In order to study the nature of electron transfer process at the electrode surface, the effect of $\upsilon^{1/2}$ (Vs^{-1}) on i_p (μA) for glucose solution (10^{-3}M) on CPE/GOx, CPE/PEC/GOx, CPE/CNS/GOx, and CPE/PEC–CNS/GOx electrodes was studied. A linear plot was obtained while plotting i_p (μA) Vs $\upsilon^{1/2}$ (Vs^{-1}) with regression coefficient 0.998 for CPE/PEC–CNS/GOx electrode. The regression coefficient values of i_p (μA) Vs $\upsilon^{1/2}$ (Vs^{-1}) are higher than that of i_p (μA) Vs υ (Vs^{-1}) for all the four electrodes studied. This shows that the electron transfer process between glucose and enzyme is a diffusion-controlled process. This was again confirmed by plot of log ip (μA) Vs log υ (Vs^{-1}); the slope value of which was between 0.5 and 0.7 for all the electrodes. From the plot of i_p (μA) Vs υ (Vs^{-1}), the diffusion coefficient (Do) was calculated and the order of diffusion coefficient was found to be,

CPE /GOx< CPE/PEC/GOx< CPE/CNS/GOx< CPE/PEC–CNS/GOx

The higher value of Do for CPE/PEC–CNS/GOx shows that a greater number of glucose molecules diffuse on the electrode surface.

The heterogeneous electron transfer process (rate constant k_o) from glucose to glucose oxidase and vice versa was calculated using the plot of ln $\upsilon^{1/2}$ (mVs^{-1}) on Ep (V) of glucose solution (10^{-3} M) on CPE/GOx, CPE/PEC/GOx, CPE/CNS/GOx, and CPE/PEC–CNS/GOx electrodes. It was calculated using the formula,

$$E_p = E_o - RT/\alpha F\ [0.780 + \ln (D_o^{1/2}/k_o) + \ln (\alpha F\ v/RT)^{1/2}]$$

where α is the transfer coefficient, Do is the diffusion coefficient, F is the Faradays constant, R is the gas constant (8.314 $JK^{-1}mol^{-1}$), T is the Kelvin temperature, and ν is the scan rate (Vs^{-1}).

It was observed that CPE/PEC–CNS/GOx electrode has greater k_o value compared to other electrodes, due to large surface to volume ratio.[29]

TABLE 5.1 Surface Properties of CPE/GOx, CPE/PEC/GOx, CPE/CNS/GOx, and CPE/PEC–CNS/GOx electrodes.

S.No	Electrode	Γ ($\times 10^{-9}$ mol^{-1} cm^{-1})	D_0 ($\times 10^{-2}$ cm^2 s^{-1})	k_0 (10^2 s^{-1})
1.	CPE/GOx	6.24	4.65	0.389
2.	CPE/PEC/GOx	6.23	4.66	0.347
3.	CPE/CNS/GOx	6.29	4.67	0.365
4.	CPE/PEC–CNS/GOx	7.03	4.87	0.397

From the results (Table 5.1), it was observed that CPE/CNS/GOx and CPE/PEC–CNS/GOx have very good affinity toward glucose molecules. The higher values of surface concentration (Γ), diffusion coefficient (Do), and k_0 for the CPE/PEC–CNS/GOx electrode show that this electrode attracts more glucose molecules toward itself and the electrode promotes the electron transfer process also in a better way than other electrodes. This shows that the pectin-based biosensors provide better platform for the redox process of glucose.

5.3.3.5 ANALYTICAL VALIDITY OF THE PEC–CNS COMPOSITE AS GLUCOSE SENSOR

Under the optimized experimental conditions, the change in oxidation peak currents for different concentrations of glucose solution was studied. The current responses increase linearly with the concentration of glucose in the range.

TABLE 5.2 Sensitivity of CPE/GOx, CPE/PEC/GOx, CPE/CNS/GOx, and CPE/PEC–CNS/GOx Electrodes.

S.No	Electrode	LOD (M)	LOQ (M)
1.	CPE/GOx	0.45×10^{-10}	1.35×10^{-10}
2.	CPE/PEC/GOx	1.52×10^{-12}	4.56×10^{-12}
3.	CPE/CNS/GOx	0.8×10^{-12}	2.4×10^{-12}
4.	CPE/PEC–CNS/GOx	0.78×10^{-15}	2.34×10^{-15}

The detection limit is very low for CPE/PEC–CNS/GOx electrode than the other electrodes. This shows the highest sensitivity of the fabricated electrode (Table 5.2).

The reproducibility of the electrodes was studied by doing ten repeated measurements by all the electrodes and the standard deviation was calculated.[30] It was found that the standard deviation was very low when CPE/PEC–CNS/GOx electrode was used to sense 1 mM solution of glucose. The stability of the CPE/PEC–CNS/GOx electrode were studied by recording the voltammograms 15 days once and no change in the peak current was noticed till 120 days and a slight decrease in peak current was noticed afterward. But very good signal was noticed even after 120 days also. This shows the stability of the bio-electrode.

5.3.3.6 INTERFERENCE STUDY

Interference tests were conducted by measuring the changes in the current and potential responses caused by the addition of 34 μM ascorbic acid[31,32] and 0.18 mM uric acid during glucose sensing. Glucose solution of 10^{-3} M was taken and ascorbic acid and uric acid were added and the cyclic voltamograms were recorded using CPE/PEC–CNS/GOx electrode. A slight decrease in peak current with no change in the peak potential was observed. This may be due to the dilution of the solution. So, ascorbic acid and uric acid has no interference in sensing glucose.

5.4 CONCLUSION

PEC–CNS composite was synthesized by solvent casting method and the structural effects are investigated by UV-visible, FT-IR spectroscopy, XRD, and SEM. CPE/PEC–CNS electrode has been employed as a best platform for GOx immobilization. This study provides a novel voltammetric GOx biosensor by immobilizing GOx on CPE/PEC–CNS/GOx electrode. PEC–CNS composite is a very good immobilizing matrix that enables to keep the enzyme with higher bioactivity and stability in the operational conditions. The newly developed glucose sensor presents a number of attractive features such as enhanced surface properties like high surface concentration of the analyte molecule, high diffusion coefficient and high electron transfer rate constant, high sensitivity, selectivity, and fast response. It exhibits a good linear concentration range for glucose sensing with a very low LOD and LOQ values. This new CPE/PEC–CNS/GOx electrode provides opportunity as an electrochemical sensing platform for glucose. Moreover, it exhibits a good electrocatalytic activity toward the analysis of glucose without having the common interferences under physiological conditions.

ACKNOWLEDGMENT

The authors would like to express their sincere gratitude to the Principal and Management for permitting us to do research and utilize the instruments available in the Science Instrumentation Centre, Lady Doak College, Madurai. We greatly acknowledge DST (DST-FIST Phase II grant, No. SR/ FST/ College-028/2010 (C)) and UGC (College with Potential for Excellence and Autonomy grants) for their support.

KEYWORDS

- **pectin**
- **carbon nanosphere**
- **enzymatic**
- **biosensor**
- **biopolymer matrix**

REFERENCES

1. Nguyen, H. H.; Lee, S. H.; Lee, U. J.; Fermin, C. D.; Kim, M. Immobilized Enzymes in Biosensor Applications. *Materials* **2019,** *12* (1), 121.
2. Wu, C. M.; Yu, S. A.; Lin, S. L. Graphene Modified Electrospun Poly(vinyl alcohol) Nanofibrous Membranes for Glucose Oxidase Immobilization. *eXPRESS Polym. Lett.* **2014,** *8* (8), 565–573.
3. Lin, K. C.; Lin, Y. C.; Chen, S. M. A Highly Sensitive Nonenzymatic Glucose Sensor Based on Multi-Walled Carbon Nanotubes Decorated with Nickel and Copper Nanoparticles. *Electrochim. Acta* **2013,** *96,* 164– 172.
4. Zhu, C.; Yang, G.; Li, H.; Du, D.; Lin, Y. Electrochemical Sensors and Biosensors Based on Nanomaterials and Nanostructures. *Anal. Chem.* **2015,** *87*, 230–249.
5. Kanga, X.; Wang, J.; Wu, H.; Aksay, I. A>; Liu, J.; Lin, Y. Glucose Oxidase–graphene–Chitosan Modified Electrode for Direct Electrochemistry and Glucose Sensing. *Biosens. Bioelectron.* **2009,** *25*, 901–905.
6. Jackcina Stobel Christy, E.; Suganya, K.; Jasmin Glory Kiruba, J.; Pavala Madhumitha, S.; Suja, S. K.; Jayanthi Kalaivani, G. Extraction of Pectin from Fruit wastes– An Effective Method of Municipal Solid Waste Management. *Int. J. Adv. Res.* **2014,** *2*, 936–944.
7. Sotanaphun, U.; Chaidedgumjorn, A.; Kitcharoen, N.; Satiraphan, M. Preparation of Pectin from Fruit Peel of Citrus Maxima. *Sci. Technol. Health Sci.* **2012,** *6* (1), 42–48.
8. Helene Canteri-Schemin, M. Extraction of Pectin From Apple Pomace. *Brazilian Arch. Biol. Technol.* **2005,** *48* (2), 259–266.
9. Burana-Osot, J. Partial Polymerization of Pectin by a Photochemical Reaction. *Carbohydr. Res.* **2010,** *345*, 1205–1210.
10. Priyaprasanth, S.; Sriamornsak, P. Flocculating and Suspending Properties of Commercial Citrus Pectin and Pectin Extracted from Pomelo (citrus maxima) Peel. *Carbohydr. Polym.* **2011,** *83*, 561–568.
11. Wong, T. W.; Colombo, G.; Sonvico, F. Pectin Matrix as Oral Drug Delivery Vehicle for Colon Cancer Treatment. *AAPS PharmaSciTech.* **2011,** *12* (1), 201–214.
12. Miao, J. Y.; Hwang, W. D.; Narasimhulu, V. K.; Lin, P. I.; Chen, Y. T.; Lin, S. H.; Hwang, L. P. Synthesis and Properties of Carbon Nanospheres Grown by CVD Using Kaolin Supported Transition Metal Catalysts. *Carbon* **2004,** *42* (4), 813–822.

13. Chung, D. D. L. Electrical Application of Carbon Material Science. *J. Mater. Sci.* **2004,** *39*, 2645–2661
14. Liang, W.; Zhuobin, Y. Direct Electrochemistry of Glucose Oxidase at a Gold Electrode Modified with Single-Wall Carbon Nanotubes. *Sensors* **2003,** *3* (12), 544–554.
15. Prakash, S.; Chakrabarty, T.; Singh, A. K.; Shahi, V. K. Polymer Thin Films Embedded with Metal Nanoparticles for Electrochemical Biosensors Applications. *Biosens. Bioelectron.* **2013,** *41*, 43–53.
16. Sobkowicz , M. J.; Dorgan, J. R.; Gneshin, K. W.; Herring, A. M.; McKinnon, J. T. Controlled Dispersion of Carbon Nanospheres Through Surface Functionalization. *Carbon* **2009,** *47*, 622–628.
17. Bian, X.; Zhu, J.; Liao, L.; Scanlon, M. D.; Ge, P.; Ji, C.; Girault, H. H.; Liu, B. Nanocomposite of MoS2 on Ordered Mesoporous Carbon Nanospheres: A Highly Active Catalyst for Electrochemical Hydrogen Evolution. *Electrochem. Commun.* **2012,** *22* (1), 128–132.
18. Liao, X.; Lu, S.; Huang, S. Preparation of Copper/Carbon Sphere Composites with Excellent Electrocatalytic Activity Toward Glucose. *Int. J. Electrochem. Sci.* **2012,** *7*, 11274–11280.
19. Han, E.; Li, X.; Cai, J. R.; Cui, H. Y.; Zhang, X. A. Development of Highly Sensitive Amperometric Biosensor for Glucose Using Carbon Nanosphere/Sodium Alginate Composite Matrix for Enzyme Immobilization. *Anal. Sci.* **2014,** *30*, 897–901.
20. Chau, J. L. H.; Lin, Y. M.; Li, A. K. Transparent High Refractive Index Nanocomposite Thin Films. *Mater. Lett.* **2007,** *6*, 2908–2910.
21. Mishra, R. K.; Anis, A.; Mondal, S.; Dutt, M.; Banthia, A. K. Preparation and Characterization of Amidated Pectin Based Polymer Electrolyte Membranes. *Chinese J. Polym. Sci.* **2009,** *27*, 639–646.
22. Derkus, B.; Emregul, E.; Emregul, K. C.; Yucesan, C. Alginate and Alginate-Titanium Dioxide Nanocomposite as Electrodematerials for Anti-myelin Basic Protein Immunosensing. *Sens. Actuators B* **2014,** *192*, 294–302.
23. Chen, X.; Zhang, K.; Zhou, J.; Yan, W.; Jiang, L.; Zhu, J. Electrochemical Immunosensor Based on Colloidal Carbon Sphere Array. *Biosens. Bioelectron.* **2010,** *25 (5)*, 1130–1136.
24. Moumene, M.; Rochefort, D.; Mohamedi, M. Electrochemical Functionalization as a Promising Avenue for Glucose Oxidase Immobilization at Carbon Nanotubes: Enhanced Direct Electron Transfer Process. *Int. J. Electrochem. Sci.* **2013,** *8*, 2009–2022.
25. Chu, X.; Zhu, X.; Dong, Y.; Chen, T.; Ye, M.; Sun, W. An Amperometric Glucose Biosensor Based on the Immobilization of Glucose Oxidase on the Platinum Electrode Modified with NiO Doped ZnO Nanorods. *J. Electroanal. Chem.* **2012,** *676*, 20–26.
26. Kaushik, A.; Khan, R.; Solanki, P. R.; Pondey, P.; Alam, J. Iron Oxide Nanoparticles – Chitosan Composite Based Glucose Biosensor. *Biosens. Bioelectron.* **2008,** *24*, 676–683.
27. Periasamy, A. P.; Chang, Y. J.; Chen, S. M. Amperometric Glucose Sensor Based on Glucose Oxidase Immobilized on Gelatin-Multiwalled Carbon Nanotube Modified Glassy Carbon Electrode. *Bioelectrochemistry* **2011,** *80*, 114–120.
28. Lyons, M. E. G.; Keeley, G. P. Carbon Nanotube Based Modified Electrode Biosensors. Part 1. Electrochemical Studies of the Flavin Group Redox Kinetics at SWCNT/Glucose Oxidase Composite Modified Electrodes. *Int. J. Electrochem. Sci.* **2008,** *3*, 819–853.
29. Kac, C.; Dalkiran, B.; Erden, P. E.; Kilic, E. An Amperometric Hydrogen Peroxide Biosensor Based on Co_3O_4 Nanoparticles and Multiwalled Carbon Nanotube Modified Glassycarbon Electrode. *Appl. Surf. Sci.* **2014,** *311*, 139–146.

30. Phongphut, A.; Sriprachuabwong, C.; Wisitsoraat, A.; Tuantranont, A.; Prichanont, S.; Sritongkham, B. A Disposal Amperometric Biosensor Based on Inkjet-Printed Au/PEDOT-PSS Nanocomposite for Triglyceride Determination. *Sens. Actuators B* **2013,** *178*, 501–507.
31. Wen, Y. X.; Liu, S, G.; Tao, B. X., Luo, H. Q.; Li, N. B. A Signal-Off Photocathode Biosensor Based on a Novel Metal-Organic Polymer for the Detection of Glucose. *Sens. Actuators B Chem.* **2020,** *304*, 127279.
32. Hassan, M. H.; Vyas, C.; Grieve, B.; Bartolo, P. Recent Advances in Enzymatic and Non-Enzymatic Electrochemical Glucose Sensing. *Sensors* **2021,** *21*, 4672.

CHAPTER 6

Structural and Functional Diversities of Lactic Acid Bacterial Polysaccharide

BINDHUMOL ISMAIL[1], SOUMYA M. P.[2,3], REEBA PARAMESWARAN[2], K. MADHAVAN NAMPOOTHIRI[2,3], A. JAYAKUMARAN NAIR[1], and A. GANGAPRASAD[1]

[1]*Department of Biotechnology, University of Kerala, Kariavattom, Trivandrum, Kerala, India*

[2]*Microbial Processes and Technology Division, CSIR- National Institute for Interdisciplinary Science and Technology (CSIR-NIIST), Trivandrum, Kerala, India*

[3]*Academy of Scientific and Innovative Research (AcSIR), Ghaziabad, India*

ABSTRACT

Microbial polysaccharides are bio-polymers that are bio-based, biodegradable or both. These exopolysaccharides are found on the surface of cells as bound form or as free form upon secretion into the extracellular space by the organism. Lactic acid bacteria (LAB) possess the ability to produce exopolysaccharides (EPS) with diverse structure and biological functions. Mainly they find applications in food industry as a rheology improving component and therapeutic applications. It is the structural diversity among LAB polysaccharides that make them a good candidate on usage in food industry. Both homo and heteropolysaccharides are being synthesized by LAB. The particular structural arrangement of exopolysaccharides contributes to their biological applications. Secondary and tertiary confirmation of the polysaccharide is strongly based on its primary structure and any change can

Bio-Based Polymers and Composites: Properties, Durability, and Applications. Amadou Belal Gueye, Sabu Thomas, Nandakumar Kalarikkal, & Modou Fall (Eds.)

give a tremendous flexibility in its application. This chapter deals with the structure-function aspect of LAB exopolysaccharides.

6.1 INTRODUCTION

Microbial exopolysaccharides find applications in the food industry by improving rheological and sensory properties of fermented food products and pharmaceutical industries as intelligent drug delivery systems (microsystems and nano systems for sustained delivery), interpenetrating polymer networks (IPNs), anticancer drug targets, recombinant macromolecular biopharmaceuticals, gene delivery, tissue engineering, and in diagnostics. Several microorganisms including bacteria, yeast, and fungi are good candidates for the production of these polysaccharides. *Xanthomonas campestris*, *Aureobasidium pullulans*, *Leucobacter* sp., *Leuconostoc mesenteroides*, *Pseudomonas elodea*, *Acaligenes* sp., *Alcaligenes faecalis* v. *myxogenes*, and *Bacilus megaterium* fall under this category which produces polysaccharides as xanthan, pullulan, dextran, cyclodextrins, gellan, welan, curdlan, and polyhydroxybutyrate, respectively.

Due to the generally regarded as safe (GRAS) status exopolysaccharides from LAB are given a special position in food industry. Almost all the LAB, especially *Fructilactobacillus*, *Lacticaseibacillus*, *Lactiplantibacillus*, *Lactobacillus*, *Lactococcus*, *Latilactobacillus*, *Lentilactobacillus*, *Leuconostoc*, *Limosilactobacillus*, *Pediococcus*, *Streptococcus*, and *Weissella* species synthesize variety of EPS.[1,2] However, the one produced by the *Weissella* sp. does not fall under GRAS status. These exopolysaccharides are long chain polysaccharides that are repeating units of carbohydrate or carbohydrate derivatives secreted as free form in to the environment or closely linked to the cell surface as capsular polysaccharides.[3]

EPS produced by bacterial strains are diverse and can be classified by diverse criteria. Based on the type of sugar units, they can be classified as homopolysaccharides and heteropolysaccharides. Classification can also be done on the basis of the glycosyl type, linkage variety and the position of carbon involved in the bond.[4] Homopolysaccharides present single type of monosaccharide (D-glucose or D-fructose) and the main ones are α-glucans, β-glucans, or fructans while heteropolysaccharides possess three to eight repeating units with two or more monosaccharides as rhamnose, fructose, galactose, glucose or mannose.

Heterogeneity of the polysaccharides is from the modifications of the repeating units. They may exhibit properties as different appearances or solubility (Table 6.1) distinct from their monosaccharide building units

TABLE 6.1 Basic Properties of LAB homopolysaccharides.

Polymer	Properties				
	Monomeric units	**Linkage**	**Solubility**	**Viscosity**	**Molecular weight**
Dextran	Glucose	[α-d-Glcp-(1→6)] associated with a low amount of (1→3) linkages and α (1→3;→6), α(1→2;→6), α (1→4;→6) branches	Water-soluble	Viscosity dependent on shear rate. Viscosity is decreased as shear rate is increased	40 and 70 kDa
Levan	Fructose	β-(2,6) linkages and β-(2,1) branching	Water and oil soluble, insoluble in organic solvents except DMSO	Low intrinsic viscosity	10^5–10^8 Da
Inulin fructans	Fructose	[β-d-Fruf-(2→1)] associated with β2→(6,→1) branches	Easily soluble in water	Viscosity is more compared to other polymers	-
Alternan	Glucose	α-(1,3) and α-(1,6) linkage	Highly soluble in water	Low viscosity	-
Pullulan	Maltotriose	α-(1,4)/α-(1,6) linkages	Soluble in water and insoluble in organic solvents	Low viscosity	4.5 × 104–6 × 10^5 Da

based on the structure and the linkage of the monosaccharides. This variability in their structure can greatly impart chemical and physical properties and in turn their biological properties (Table 6.2). In dairy and cereal-based products, they are being used because of their gelling, thickening, stabilizing, and viscosifying properties[5] They also display beneficial anticholesterolemic effect, antioxidant as well as anticancer property and immunomodulatory activities.

6.2 HOMOPOLYSACCHARIDES FROM LACTIC ACID BACTERIA

Homopolysaccharides are polymers of a single monosaccharide, either glucose or fructose. Several LAB genera, including *Lactobacillus*, *Streptococcus*, *Leuconostoc*, *Oenococcus*, and *Weissella* produce homopolysaccharides.[6] They are further classified based on the glucosidic or fructosidic linkage present in the polymer backbone. Dextran, mutans, alternan, and reuteran are α-glucan homopolysaccharides inulin and levan are fructan homopolysaccharides (Figure 6.1). Dextran possess α-(1,6) linkages between the glucosyl units while reuteran contains predominantly α-(1,4) linkages. Mutans possess α-(1,3) linkages whereas alternans possess α-(1,3) and α-(1,6) linkages. The fructan homopolysaccharides inulin and levan consist of β-(2,1) and β-(2,6) linkages, respectively[7] (Table 6.2).

6.2.1 *BIOSYNTHESIS OF HOMOPOLYSACCHARIDES*

Homopolysaccharides are mainly exopolysaccharides synthesized from sucrose by an extracellular enzyme among glycosyl hydrolases. Sucrose act as donor molecule as glucose or fructose. The glycosyl hydrolases that synthesize glucans and fructans are grouped into GH70 and GH68 families, respectively.[8]

Homopolysaccharide synthesis is relatively direct on comparison to heteropolysaccharides as it is mediated by glycan sucrase enzyme coded by a single gene.

Majority of homopolysaccharide producing LAB produce a single type of glycansucrase enzyme while some produce more than one type of homopolysaccharide as they possess more than one glycan sucrase enzyme. The biosynthesis of homopolysaccharides consists of two steps, including hydrolysis of a substrate the binding of the resulting monosaccharide residues to a glycan acceptor chain and direct release of polymerized chains extracellularly.[9]

TABLE 6.2 Physicochemical Properties and Functions of Homopolysaccharides.

Polymer	Bacterial strains	Linkages	Mol Wt	Solubility	Properties	Function
Dextran	*Lactobacillus reuteri*, *Lactobacillus casei*, *Weisella cibaria*, *L. mesenteroides* subsp. *Mesenteroides*, *Lactobacillus fermentum*	α-(1,6) and α-(1,3) glycosidic bonds	>40 kDa	High solubility in water	Nonionic, high stability, Newtonian, fluid behavior, water retention capacity	Good film forming moisturizer, thickener, anticoagulant, plasma lubricant
Alternan	*L. mesenteroides* NRRL B1355, NRRL B1501, NRRL B1498, and *Leuconostoc citreum*	α-(1,3) and α-(1,6) alternating glycosidic linkage	10^6–10^7 Da	High solubility in water and DMSO	Shows high stability at range of pH 3–9, low viscosity, low glycemic property	Act as bulking agent, extender, used as sweetener
Pullulan	*Lactobacillus pentosus* *Lactobacillus plantarum* L137	α-(1,4)/α-(1,6) glucopyranosyl linkage	4.5 × 104–6 × 10^5 Da	Soluble in water and insoluble in organic solvents except DMSO	Heat resistant and stable, hygroscopic in nature, shows remarkable hydrophilicity and permeability	Used in low calory foods, packing material, used in drug delivery
Mutan	*Streptococcus* spp., *L. mesenteroids*	α-(1,3) glycosidic linkage	8 MDa	Water insoluble	Adhesive	Slightly toxic
Reuteran	*L. reuterii,*	α-(1,4; 1,6)-glycosidic linkages	10–40 MDa	Water soluble	Low degree of branching	Thickener
B-glucans	*Pediococcus damnosus*, *Pediococcus parvulus*, *Lactobacillus suebicus* CUPV22, *Lactobacillus brevis* TMV 1.2112, *Lactobacillus rossiae*	β-(1,3)-D-glycosidic bond	10^7–10^4 Da	Water soluble	High water-binding capacity, viscosity, high inner cavity diameter	Act as fat replacer, used for immunomodulating scaffold for dressing

TABLE 6.2 *(Continued)*

Polymer	Bacterial strains	Linkages	Mol Wt	Solubility	Properties	Function
Levan	*L. reuteri*, *Lactobacillus sanfranciscensis*, *L. mesenteroides*, *Steptococcus salivarius*, *Steptococcus mutans*	β-(2,6) and β-(2,1) glycosidic linkage	10^5–10^8 Da	High solubility in oil and water, insoluble in organic solvents	Green adhesive, biocompatibility and film forming ability, water retention capacity, low intrinsic viscosity	Encapsulating agent, cryoprotector, flavor enhancer, moisturizer
Inulin-type	*Lactobacillus johnsonii* NCC 533, *S. mutans* strain JC2 *L. citreum* CW28, *L. plantarum L. reuteri*	β-(1,2) glycosidic bonds	6×10^7–9×10^7 Da	Soluble in water and organic solvents	Gelling property, short chain structures have high solubility, stable viscosity, hygroscopic in nature	Texturizer, helps to prolong shelf life, antioxidant

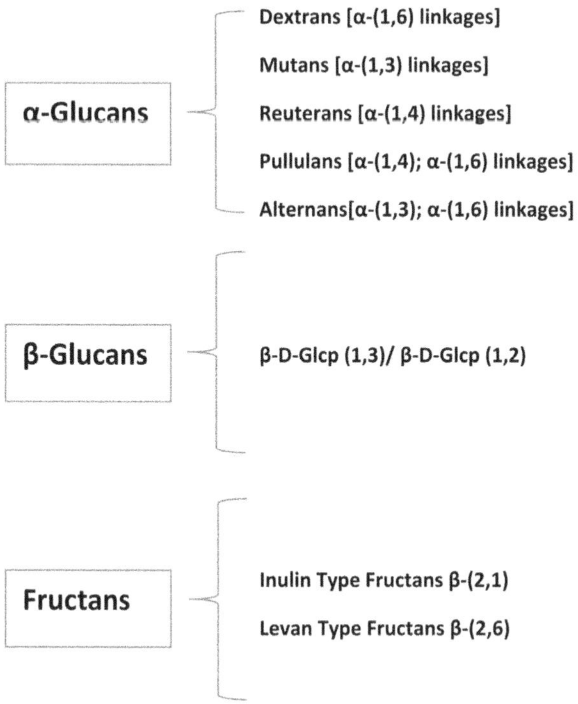

FIGURE 6.1 Classification of homopolysaccharides based on proportion of linkages.

The specific linkage and the carbon involved in the bond divides polymers mainly into three classes as α-D glucans (dextran, mutan, reuteran, and alternan), β-fructans (levan and inulin-types), and polygalactans. α-glucans and β-fructans are formed by glucansucrases, and the third group polygalactans, composed of mainly five repeating units of galactose. Glucansucrases are able to use the energy of the glycosidic linkages of sucrose catalyzing transfer of glycosyl moiety.[3]

6.2.2 ALPHA GLUCANS

6.2.2.1 DEXTRAN

6.2.2.1.1 Structure and Function

Dextran is a polymeric chain of glycosyl units categorized inside the large group of α–D glycans. Dextran synthesized mainly in sucrose rich medium[10] by dextransucrase enzyme by transferring D-glycosyl unit to acceptor molecules.[11,12] Potential application of dextran depends on the size and structure of the dextran which is decided by the type of enzyme produced by each strain.[12,13]

The genus *L. mesenteroides* generally synthesizes dextrans with a main chain linked by α-(1,6) bonds and branches with α-(1,3) bonds.[7,8,14] *L. mesenteroides* subsp. *mesenteroides* strain is mainly used for the large-scale production of dextran. As being produced by a LAB strain of GRAS status it is identified as safe to use dextran. Strains which can produce dextrans are *L. reuteri, L. casei, Lactobacillus sakei, L. fermentum, Lactobacillus parabuchneri, Lactobacillus curvatus,*[15] *L. mesenteroides* subsp. *mesenteroides, L. mesenteroides* subsp. *Dextranicum*[7,16,17] *S. mutans, Streptococcus downei, Streptococcus sobrinus, S. salivarius,* S*treptococcus gordonii,*[18–20] *W. cibaria.*[3] According to the molecular weight, dextrans are of two types dextrans and oligodextrans. The molecular weight greater than 40 kDa are called dextrans[21] and those having molecular weight less than 40 kDa are called oligodextrans.[21,22] The variations in the molecular weight and proportion of branches depends on the producing strain. Temperature is the next important factor which decides the structure. If the temperature of the environment greater than 25°C produces dextran with higher branching.[23] Linear dextrans have helical structure. According to the increase in branching units the structure and function differs.

Two major functionalities for dextran molecules are, water retention capacity and interactions with other components such as gluten network. These two functionalities benefit in dough stability and gas retention during bread-making processes. High Molecular weight dextrans increased wheat bread quality.[21] Linear structures and high molecular weight dextrans are preferred for good film forming abilities[24] utilized during baking. Dextrins with 3–5 degree of polymerization (DP) inhibit crumb hardening.[3] Dextrans, particularly the high Mw dextran,[22] allow more open structures by affecting the formation of gluten network through H-bonding and steric interactions. So, it is observed that high molecular weight dextrans (up to 2 × 106 247 g/mol) can retarding the aging of breads through hindering the recrystallization of amylopectin.[21,25,26]

In rheological properties, viscoelastic properties are essential to predict the quality of end product.[27] The highest Mw dextran (T2000) apparently promoted the stiff level of dough, reflected by higher G′ and G″ values. Low Mw dextran (T10) fitting well in the dough bulk may hinder associations of structure forming components such as starch and gluten, leading to a weakening of the dough network.[28] The solubility of the dextran depends on their ability to incorporate amount of water and hydrogels, this property reflects in pharmaceutical and chemical industries.[27] Viscosity of dextran is directly related to the concentration and the shear rate.[14,29] Being resistant to mammalian amylases and providing low calories, is used to produce low-calorie dietary products as starch replacers and functions as a prebiotic to promote the growth of beneficial bifidum bacteria. Moreover, its adhesive properties may be used as a binder and stabilizer in food pastes, and as an adhesive to stick nuts to cookies.

Water retention and film forming properties of dextran is an interesting feature to be used as a moisturizing and protective ingredient for cosmetic lotions, powders, facial packs, and protective agent for hair shampoos, hair dressings, and tooth powders. In the cosmetic industry, dextran is being used as a thickening and moisturizing agent, and its reducing property allows to be used as an antiaging agent.[14,30]

High molecular weight dextran (between 40 and 70 kDa) is used as an extender, anticoagulant, antithrombotic, osmotic agent, and intravenous plasma lubricant.[25] High molecular weight dextrans are also suitable for cryopreservation of vaccines and organs.[21,30]

6.2.2.2 MUTANS

6.2.2.2.1 Structure and Function

Mutan polysaccharides are alpha glucans having α-(1,3) glycosidic linkage, produced by strains *Streptococcus* spp., *L. mesenteroides* NRRL B-523, B-1149. It differs from dextran with a high proportion of α-(1,3) linkages. Mutan exopolysaccharide is synthesized by mutansucrase.

They are generally adhesive and insoluble in water,[7,31] which results from the proportions of different type of linkages. Mutan is used as an inducer of enzyme mutanase and also used as heavy metal adsorbent.[32] Their utility is mainly limited due to their low water solubility and invitro modifications preferred to enhance biological activities and pharmaceutical applications.

6.2.3 REUTERANS

6.2.3.1 STRUCTURE AND FUNCTION

Reuteran is a neutral polysaccharide and highly branched one synthesized mainly from sucrose as carbon source by enzyme glucansucrase. They are mainly produced by *L. reuterii* and are with α- (1,4; 1,6)- glycosidic linkages.

The mechanism of alternating linkage formation is carried out by glycoside hydrolase family 70 enzymes (GH70) reuteransucrases.

Reuteran is soluble in water and show degree of branching from 8 to 20% and have a molecular weight between 10 and 40 MDa.[33,34] They are used in food industries as stabilizers, gelling agents, and thickeners. Bacterial reuteran from *L. reuteri* TWM1.656 reduced the number of ETEC K88 colonizing piglet jejunum epithelial cells and also inhibited ETEC adhesion to IPEC-1 cells.[35] They also exhibited differential effect on wheat bread quality.

6.2.3.2 PULLULAN

6.2.3.2.1 Structure and Function

Pullulan is a homopolysaccharide produced by LAB and other kinds of bacteria. Pullulan is a linear polymer with no branching. They possess maltotriose repeating units linked by α-(1, 6) glycosidic bond.[36] The characteristic dimeric structure of pullulan has α-(1,4)/α-(1,6) glucopyranosyl linkage. It forms a highly ordered and branched structure with α-1, 4- to α-1, 6-glucano bonds present in a ratio of 2:1.[37] Pullulan exhibit a molecular weight range 4.5×10^4–6×10^5 Da. Pullulan is soluble in water and insoluble in organic solvents except DMSO[34] and is nonhygroscopic in nature. Pullulan is stable at a wide pH and temperature. It melts at 250°C. It is edible, odorless, tasteless, and have high concentration of hydroxyl groups with no toxicity and mutagenicity.

Pullulan producing strains are *L. pentosus*, *Bifidobacterium species*, *A. pullulans*, *L. plantarum* L137, *Bifidobacterium breve* UCC2003, *Streptomyces erumpens.*[37,38] Out of which the major producer of pullulan is *A. pullulans*. Pullulan is biosynthesized in the cytosol and then secreted into the extracellular environment.[39]

They find applications in the production of low-calorie foods, as viscosity filler in beverages and sauces and as adhesive agents in binding nuts to the cookies. Pullulan is highly oxygen impermeable and has GRAS status and is used as covering to prevent microbial spoilage. Pullulan can be a replacement for starch or fat and can be used to produce dietetic food by maintaining desired consistency and viscosity of the product.

Their permeability and retention effect helps in drug delivery systems. Its function as an insulin carrier helps in lowering excessive blood glucose level in diabetic patients. Temperature helps in internalizing pullulan-based drug nanogels in cancer cells. A coating of sugar solution with pullulan derivatives improves shelf-life and strength of capsules.[40] Extensive inter-molecular hydrogen bonding and remarkable hydrophilicity.

Pullulan forms fibers and films and is used as coating material. It is highly transparent and can be used as a packaging material. Pullulan thin films are edible odorless, fragrance free and flavor less. It is used as gene and drug carrier. Modified pullulan is used as a raw material in pharmaceutical applications,[41] such as in targeting, surface modification, nanoparticles, cancer therapy, bioimaging, plasma expander, and tissue engineering.[42] Furthermore, it has been used in oral care product. It is used as a sizing agent in paper industry. Pharmaceutical industry due to their higher stability, compatibility, and the ease of altering their physicochemical properties when mixed with other biopolymers.[42]

6.2.3.3 ALTERNANS

6.2.3.3.1 Structure and Function

Alternan is a homopolysaccharide belonging to the group of α-glucans. The major LAB strain which produces alternan as *L. mesenteroides* NRRL B1355, NRRL B1501, NRRL B1498, and *L. citreum.*[43] Alternan polysaccharides possess α-(1,3) and α-(1,6) alternating glycosidic linkage made by glucopyranosyl units.[43,44] Alternansucrase is the enzyme that synthesis alternan by transferring α-D glucosyl residue alternately from sucrose to the 6 position and the 3 position of the nonreducing terminal residue of an α-D glucan, thus producing a glucan having α-(1,6) and α-(1,3) linkages. Jeans et al 1954 were the first to report the production of two glucansucrase by *L. mesenteroides* NRRL B 1355.

Alternan solutions display high stability across abroad range of pH 3–9 and at temperature 4–70°C for a week. It is an exopolymer with high insolubility, but once dried to powder it is not a hydrophobic agent and forms a transparent film. Alternan has low viscosity so it is used as bulking agent and extender in food and cosmetics. It is used as sweetener in confectionaries utilizing its low glycemic property.[34] Compared to other glycopolymers, distinct properties of alternan are, it can self-assemble into nanoparticles at low concentration but becomes a gel-like material after solubilized at high concentration. Notably, three physical states of alternan, can be simply controlled by its concentration.[45]

6.2.4 β-GLUCANS

6.2.4.1 STRUCTURE AND FUNCTION

β-D-glucans are water-soluble polysaccharides synthesized in cereals, plants, mushrooms yeast, fungi, and bacteria. These are polysaccharides composed of glucose molecules. Glucans produced by bacteria and algae have β-(1,3)-D glycosidic bond with linear chain.[46] Main LAB strains known to produce β-D-glucans are *Lactobacillus spp.* G-77,[47] *L. suebicus, L. brevis* TMV 1.2112, *L. rossiae, Lactobacillus parabuchner* (Lactobacilli in breweries)[48] *and Lactobacillus paracasei* NFBC 338.[49] *Lactobacillus* spp. G-77 showed similarity to *P. damnosus.*[7] *L. suebicus* CUPV221 was able to synthesise (1,3)-β-D-glucan with masses of approximately 10^7 and 10^4 Da which can vary according to species.[50]

β-D-Glucans were recognized for their physical and chemical properties. β-D-glucans are well known for their water binding capacity and low molecular weight with high viscosity[51] and can provide a shiny and smooth appearance beneficial for yogurt production, dairy and candy industries. High water solubility and molecular weight can reduce the serum LDL cholesterol.[46] In medical field, β-glucans tried as wound dressing because of its inner cavity diameter.[52] Also glucans having linear chain and Mw less than 50×10^4 g/mol or larger than 110×10^4 g/mol can enhance immune system by stimulating macrophages, neutrophils, and natural killer cells as defense against foreign bodies.[53] According to the DP β-glucans can form single and triple helical structures which can be used for making resilient gels for drug delivery. Solubility reduces with increase in DP. The modified glucans can be used as substitute for other drugs. Linear (1→3)-β-D-glucans having degree of sulfation greater than 1.0 and with molecular weight between 18 k Da and 50 k Da can be used as alternative for heparin.[54] β-glucans are the most commonly consumed immunomodulator with strong anticancer, insulin resistance, antihypertension, and antiobesity effects.[55]

6.2.5 FRUCTANS

6.2.5.1 LEVAN TYPE FRUCTANS

6.2.5.1.1 Structure and Function

Levan is an exopolysaccharide made of β-(2,6) linkage forming a linear chain and β-(2,1) linkage,[56–58] while forming branching points by fructosyl units differentiated to the category of high molecular mass polymer known as Fructans.[59] Both plants and microorganisms are recorded in the production of levan exopolysaccharide. Levansucrase is an extracellular fructosyltransferase[38] which catalyzes the transfer of D-fructosyl residues[60] from fructose to yield Levan.[34] LAB strains producing levan are *S. salivarius*, *S. mutans*, *L. mesenteroides* NRRL B-512F, *L. sanfranciscensis* LTH 2590 and *L. reuteri* LB 121.[61,62]

Levan is synthesized mainly from sucrose but significant production of levan was also observed from wheat flour by *L. sanfranciscensis*. The polymer production is dependent on culture growth, pH, temperature, oxygen concentration, and media components.[63] Fructose units are released to the culture during synthesis, and these units are utilized by culture for further growth. The other Levan producing strains are *Zymomonas mobilis*, *S. salivarius*. *Acetobacter*, *Gluconobacter*, *Bacillus subtilis*, and *B. megaterium*.[57,63]

Levan is a homopolysaccharide with molecular weight ranges 10^5–10^8 Da[64,65] and have a high degree of polymerization (DP) > 100.[15,66] They possess distinguished properties as high solubility in oil and water,[67] insoluble in organic solvents such as methanol, acetone, ethanol except DMSO,[68] green adhesive, good biocompatibility with salts and surfactants and film-forming ability.[69] It have great potential as functional biopolymer in foods, feeds, cosmetics, pharmaceutical, and medical fields.[62]

Levan is accepted as nontoxic and safe and have cholesterol lowering, probiotic, and tri-acyl glycerol-lowering properties utilized in food industry.[70] It can act as emulsifier, stabilizer, and thickener agent in food.[70] Levan has been shown to exert excellent cell proliferation, skin-irritational levitating, and skin moisturizing effects utilized in cosmetic sectors.[15]

In medical field levan polymers have role in the healing of tissues that have been burned or mechanically damaged due to their metalloproteinase activation capability.[65] The low intrinsic viscosity of levan gives them antioxidant property.[65,71] Levan has attracted attention also for its antitumor properties. Levan find applications as encapsulating agent, osmoregulator and cryoprotector in addition to its uses in medicine as plasma substitute, prolongator of drug activity, radio protector, antitumor and antihyperlipidemic agent[15] and anti-inflammatory property.[57] They can be used as coating material in drug delivery systems,[62] color, and flavor enhancer in the manufacturing of tablets and capsules in the pharmaceutical side.[63]

6.2.5.2 INULIN TYPE FRUCTANS

6.2.5.2.1 Structure and Function

Inulin-type EPS are fructans or fructo-oligosaccharides containing β-(1,2) glycosidic bonds[72,73] in main chain and β-(2,6) raminification,[74] terminated generally by a glucose unit. Inulin type polysaccharides are formed in the similar way of levan production by different enzyme known as inulosucrase.[75] Inulin type fructans are widely distributed in numerous edible fruits and vegetables, such as chicory, onion, banana, garlic, and dahlia.[76] The various strains that produce insulin-type exopolysaccharides are *L. johnsonii* NCC 533, *S. mutans* strain JC2,[77] *L. citreum* CW28, *L. plantarum*,[78] and *L. reuteri* 121. *L. johnsonii* NCC 533 produces high molecular mass Inulin.[34]

These are nondigestible exopolysaccharides and they function as prebiotics in humans and animals. Inulin type fructo-oligosaccharides synthesize butyrate, which nourish the enterocytes, prevent pathogenic adherence,[3] and decrease pH of lumen.[34] Gelling tendency of inulin in aqueous solution while retaining the food texture and stability makes it a desirable ingredient in food products.[34,73,75] In food industry, depending on the inulin molecule's DP it is divided to three. Short chain inulin has more solubility[34] and mild sweetness. It is commercially used as sugar replacement as sweetener (low DP). Polymer having medium DP can act as texturizer and long chain inulin exhibits less solubility and stable viscosity which is used as a fat replacer (high DP) as it can provide smooth and creamy texture.[79]

Average DP of inulin-type exopolysaccharide is 2–60 monomeric units[75] with an average value of 10–12.[73] Average molecular weight varies between 5600 and 6300 g mol-1. When lower Mw inulin was more easily filled into the gluten network, higher Mw strongly impeded protein crosslinking, delaying the ability of the system to form an elastic structure. The solubility of inulin increases with rising temperature and decreases as its DP increases.

Inulin can act as an antioxidant[80] to scavenge ROS directly. Low DP inulin have high scavenging capacity which can reduce the inflammation cancer and aging.[78] Another promising application of inulin is in targeted drug delivery in colon cancer[74] and promising stabilizer for protein-based drugs (stabilizer for haemagglutinin, important component of influenza vaccine). The texture, stability, reactivity, and elasticity are influenced by molecular weight, added amount, processing conditions (temperature, moisture, pH, and salt), and flour components (gluten and starch).

While gelling lower temperatures served to slow molecular movements in the gel, which may have enhanced the stability of hydrogen bonding and hydrophobic interactions, causing storage modulus to rise rapidly with decreasing temperature. Compared to low Mw inulin, high Mw inulin features greater steric hindrance, rendering it more prone to macromolecular entanglement and insertion into the gluten protein matrix. In textural properties with increasing inulin content, the hardness of the mixed gels decreased.[79]

6.3 HETEROPOLYSACCHARIDES FROM LACTIC ACID BACTERIA

6.3.1 STRUCTURAL PATTERN

Heteropolysaccharides, as the name denotes, are made up of more than two types of sugar monomeric units, majority containing three to eight repeating units of two or more monosaccharides. The repeating units of heteropolysaccharides are either linear or branched and are mostly composed of different sugar moieties such as glucose, galactose, rhamnose, mannose, *N*-acetyl glucosamine, *N*-acetyl galactosamine, glucuronic acid, etc. These are produced by glycosyltransferases and then polymerized before secretion.[52]

Heteropolysaccharides are produced by a variety of mesophilic and thermophilic LAB. Some of the major heteropolysaccharides are produced by mesophilic and thermophilic LAB and *Streptococcus thermophilus* LAB strains.[81] The creamy, smooth texture of yogurt is improved by the ability of the yogurt bacteria to produce EPS, even in small amounts. EPS-producing LAB strains have been isolated from cheese, meat, and vegetables.[81]

The structures of heteropolysaccharide bonds, are diverse repeating units with diverse degrees of polymerization. The molecular weights of these polymers form the basis of viscosity, rigidity as well as strength of polysaccharides. The molecular structures of *Lactobacillus* polysaccharides can be linear to branched and can act as efficient candidates for protective barriers to probiotics. These candidates depend on viscosity, rigidity, and strength based on their molecular weight and functional groups present in polysaccharides.[82]

6.3.2 STRUCTURAL EVALUATION

There were many areas of research that focused on the structural analysis of HePS with the help of monomer and methylation analysis and NMR techniques (e.g., TOCSY, HMBC). These detailed structural studies have come up with structures of usually 3–8 types of monosaccharides or distinct linkage types or other decorative elements (e.g., acetate, phosphate).[83] Different chromatographic methods such as size exclusion chromatography, were employed to determine the molecular mass of the polysaccharides and usually yields magnitudes of 10^4–10^6 Da.[84–86]

There is a strong correlation between the molecular mass of the polysaccharides and their functional properties. However, the three-dimensional (3D) structure or conformation are the basis of physical and rheological properties of a polysaccharide in solution. Other factors such as the ability of the heteropolysaccharides to form intermolecular interactions and associations also plays an important role in these properties. Minor changes in the primary structure may result in tremendous effect on the confirmation and properties of the heteropolysaccharides. Urashima et al. 1999[87] first determined the structure of the repeating unit of a LAB heteropolysaccharides produced by *S. thermophilus.* Methods such as acid hydrolysis, methylation analysis, periodate oxidation, acetolysis, enzymatic digestion, Smith degradation and 1D and 2D proton nuclear magnetic resonance (^{1}H-NMR) spectroscopy, etc., were employed to elucidate the structural diversity of the LAB exopolysaccharides whose size ranges from disaccharides to heptasaccharides.

Compared with homopolysaccharides, the structure of the main chain of the heteropolysaccharide repeating unit and its branches is diverse, and a few heteropolysaccharide repeating units contain *N*-acetylglucosamine, *N*-acetylgalactosamine, etc. An exopolysaccharide containing a nonsugar residue such as a phosphate group, an acetyl group or glycerin may be found in a substituent such as uronic acid. There are reports supporting the evidence that the repeating units of *S. thermophilus* Sc136 and *S. thermophilus* Sfi20 heterotype exopolysaccharide contain *N*-acetyl groups. The repeating unit of *S. thermophilus* MR-1C contains an L-fructose residue. LAB produce a wide variety of extracellular polysaccharides, and their chemical structures are also significantly different. In general, the β-1,4 bond-linked exopolysaccharide is more viscous than the β-1,3 bond or β-1,2 bond-linked exopolysaccharide, and extracellularly linked by the α-binding bond. Polysaccharides are more elastic than β-binding bond-linked exopolysaccharides. For example, the main chain skeleton of the extracellular polysaccharide of *Lactococcus (Lc.) lactis* NIZOB40 is composed of β-1,4 bonds and has stronger rigid structure and specific volume than *L. Lactis* B891 which is composed of α-binding bonds.

It is a fact that the production levels of heteropolysaccharides are less when compared to homopolysaccharides, the structural diversity that it causes in food matrices are high. The creamy structure of yoghurt is a result of the heteropolysaccharide produced by LAB such as *S. thermophilus* or *Lc. lactis*, etc.[88] For example, HePS of the industrially exploited yoghurt starter culture *S. thermophilus* Sfi6 is a tetrameric repeating unit of galactose, glucose, and *N*-acetylgalactosamine (2:1:1). While *Lc. lactis* NIZO produces a phosphorylated and HePS of galactose, glucose, and rhamnose in the ratio 2:2:1.[89,90] Two of the well characterized heteropolysaccharides of *Lactobacillus* species are produced by *Lactobacillus delbrueckii* ssp. *bulgaricus* Lfi5 and *Lactobacillus rhamnosus* RW-9595M and both of them possess a heptameric repeating unit of galactose, rhamnose, and glucose in different ratios, whereas the HePS of *L. rhamnosus* RW-9595M possesses an additional pyruvate moiety.[91,92]

The structural data available for HePS produced by *L. plantarum* species shows that one type is glucomannan type EPS of *L. plantarum* BR2 with glucose and mannose as repeating units. They showed promising biological properties such as cholesterol lowering, antioxidant activities, etc.[93] While the HePS of *L. plantarum* MTCC9510 was a linear trimeric repeating unit of glucose and mannose residues, *L. plantarum* C88 produced a structurally complex HePS with pentameric repeating unit of glucose, galactose and an O-acetyl group in branching pattern.[16,78,94] *L. plantarum* KX041, produces a novel heteropolysaccharide of arabinose, mannose, glucose, and galactose with 38.67 kDa molecular weight for their monosaccharides. It exhibits higher free radical scavenging and also exhibits higher reducing power. A mechanism of chelation with a transition of metal ion catalysts exerted by the EPS was also proposed. The low molecular weight of EPS has significant effect in antioxidant activities.[95]

In a recent study, *L. delbrueckii* ssp. *bulgaricus* SRFM-1 produced a heteropolysaccharide of galactose and glucose with average molecular weight of EPS-1 and EPS-2 to be 3.97×10^5 Da and 3.86×10^5 Da (Table 6.3), respectively. The backbone of EPS-1 consisted of →6-β-D-Gal*p*-(1 → 4)-β-D-Glc*p*-(1 → 4)-α-D-Gal*p*-(1 → 4)- β-D-Gal*p*-(1 → 6)-β-D-Gal*p*-(1 → 4)-β-D-Glc*p*-(1 → 4)-α-D-Gal*p*-(1 → 4)-β-D-Gal*p*-(1 → 4)-α-D-Glc*p* (1→). They possess three branching points existing in terminal with D-Glc*p* residues with α/β-D-(1 → 6) linkages. In addition, EPS-2 had (1→6-β-D-Gal*p*-(1 → 4)-β-D-Glc*p*-(1 → 6)-α-DGal*p*-(1→) with a branching point attached in the terminal D-Glc*p* residue with β-(1 → 6) linkage. The strong scavenging activities on superoxide, hydroxyl, DPPH radicals, and ferrous ion chelation of EPS was contributed by the higher content of galactose which is promoted by the presence of glycosidic linkage and branching pattern by modulating their physiochemical properties. Glycosidic linkage is found to have a significant influence on the antioxidant capacity of EPS by modulating flexibility and water solubility.[96]

Most of the heteropolysaccharides are composed of D-galactose, D-glucose, and L-rhamnose as monomers, but in different ratios. The EPS from *Lactobacillus acidophilus* LMG 9433, *Lactobacillus helveticus* TYl-2, *L. helveticus* NCDO 766, *L. rhamnosus* C83, *S. thermophilus* Sfi20, *S. thermophilus* Sfi32, *S. thermophilus* LY03, *S. thermophilus* BTC, and *S. thermophilus* 480 lack rhamnose; *L. paracasei* 34-1 EPS only contains galactose; *S. thermophilus* OR 901 EPS contains galactose; and rhamnose and *L. sakei* 0-1 EPS consists of glucose and rhamnose.[97–100] Glucose and fructose, in the ratio 1:2, contributes to the EPS produced by *L. delbrueckii* subsp. *bulgaricus* CRL 420 and the octameric repeating unit of HePs produced by *S. thermophilus* MR-1C consists of D-galactose, L-rhamnose and L-fucose in a 5:2:1 ratio. Residues such as *sn*-glycerol- 3-phosphate, *N*-acetyl-amino sugars, phosphate, and acetyl groups can be present in the EPS backbone along with the above listed sugar moieties. As a drawback of the isolation and purification method chosen, especially when complex media are used, the pattern of secretion of more than one polysaccharide by a single strain was observed in several cases. There are reports of different compositions of exopolysaccharide material from the same strain when different fermentation conditions and isolation, purification techniques were used.[101] Researchers isolated two different EPS for a single strain in different fermentation medium with varying monosaccharide composition and EPS fractions of the same monomeric units but with low and high molecular weights were also observed in *L. delbrueckii* subsp. *bulgaricus* and *S. thermophilus* strain.[102] This variation in monomeric composition and glycosidic bonds are as a result of media and culture conditions.[103]

Apart from that, as mentioned above, molecular weight of the exopolysaccharides also has role in the properties of EPS. It predominantly effects the texture and viscosity of the EPS. There are reports of evaluating the structure as well as the molecular weight of the EPS using modern techniques such as multiangle laser light scattering (MALLS) coupled with the NMR and refractometer.[5]

6.4 STRUCTURE–FUNCTION RELATIONSHIPS OF EXOPOLYSACCHARIDES

LAB produces a wide variety of heteropolysaccharides at levels of 10–1000 mg/L, which is comparatively low to homopolysaccharide production, but with high molecular masses generally above 10^6 g/mol.[104]

TABLE 6.3 Lactic Acid Bacterial Heteropolysaccharide Composition with their Average Molecular Weights.

Organism	Monosaccharide composition	Functional groups involved	Mol. weight of EPS
Bacillus sp. S-1	Galactose, glucose, and mannose	Hydroxyl groups	1.765×10^4 Da
W. cibaria SJ14	Mannose, glucose, galactose, arabinose, xylose, and rhamnose	Hydroxyl, carboxyl, and acetyl groups	7.12×10^4 Da
Weissella confusa	Galactose, mannose, glucose, fructose, rhamnose, arabinose, xylose, and ribose	Hydroxyl and carbonyl groups	Not available
L. delbrueckii ssp. *bulgaricus* SRFM-1	Galactose and glucose	Hydroxyl, sulphate, and carboxyl groups	3.97×10^5 Da, 3.86×10^5 Da
L. plantarum KX041	Arabinose, mannose, glucose, and galactose	Hydroxyl, carbonyl, and amide groups	38.67 kDa
Bacillus amyloliquefaciens GSBa-1	Glucose	Carboxyl, carbonyl, and hydroxyl groups	54 kDa
S. thermophilus GST-6	Glucose and galactose	Sulfate groups	Not tested
L. plantarum ZDY2013	Galactose and xylose	-	51.7 kDa
Lactobacillus sp. *Ca*6	Glucose	. Carboxyl, hydroxyl, and amide groups	53.23 kDa
Enterobacter cloacae Z0206	Fucose, glucose, galactose, and glucuronic acid	-	1.1×10^6 Da
Bacillus velezensis MHM3	Glucose, fructose, rhamnose, and glucuronic acid	-	6.89×10^{20} kDa
L. plantarum TMW 1.1478	Glucose; rhamnose, galactose, and traces of the amino sugar glucosamine (GlcN)	-	2×10^6 Da
Lactobacillus kefiranofaciens	Glucose and galactose	-	7.6×10^5 Da
L. pentosus	Glucose and rhamnose	-	Not available
S. thermophilus S-3	*N*-acetyl-galactosamine, galactose, and glucose	-	5.7×10^5 Da

Some of the recent research studies on EPSs gives an idea that the structure of heteropolysaccharides is strain specific. They often differ by their monosaccharide composition, linkages between units, presence of repeated sidechains and substitutions, length and composition of branching, etc. Here the compactness of the EPS produced is affected thereby modifying the rheological properties.[105] EPS as good thickeners influence the rheology and texture of fermented products. However, EPS concentration in no way affects the apparent viscosities of fermented products, as the interaction of EPS with proteins may vary. The usual trends observed with viscosity are that high molar mass and chain stiffness contribute to high viscosity.[106] The presence of sidechains also affects chain stiffness and thereby thickening effect. It is also found that β-linkages contribute to stiffer chains than α-linkages.[107]

6.4.1 FUNCTIONAL PROPERTIES OF EXOPOLYSACCHARIDES

To achieve the industrial application and acceptance for commercialization, it is of great importance to study the relationship between chemical structures and physicochemical properties of EPS. These acquired knowledge is made useful to interpret the evidence of HePS from LAB on human health benefits including blood cholesterol-lowering, antioxidant, antitumor, antiulcer, and immunomodulating activity, in regard of their structure/function relationships. Several EPSs, especially the homopolysaccharides, have already been commercialized.

EPSs from LAB are also found to have beneficial physiological effects. The residence time of fermented milk, after ingestion, in the GI tract showed low levels of degradability due to the high viscosity of the EPS in those food products which will help the probiotic microflora in colonization.[108] Short-chain fatty acids (SCFAs) are obtained by colonic microflora from EPSs upon degradation in the gut. SCFAs help in prevention of colon cancer by providing energy to epithelial cells.[109] In vivo studies are crucial to demonstrate the health beneficial properties of EPSs as antitumor, antioxidant, antiulcer, immunomodulating or cholesterol-lowering activities. LAB also shows antioxidant activity due to some compounds it produces in its basic metabolic pathway. Major contribution for the antioxidant property of LAB is due to their exopolysaccharide production which itself has antioxidant properties and have been proved and reported in the past decades. Significant antioxidant and antitumor activities are observed in EPS from LAB. This potentiality of EPS from LAB is of great importance and a hot topic for exploitation and exploration. LAB have been known to contribute to the modulation of intestinal homeostasis by displaying antagonistic and antioxidant activities.[16] Therefore, their probiotic properties and functional capacities are crucial for their application as probiotics.[110]

Similarly, *L. plantarum* MA2 exhibited antioxidant effects via free radical scavenging and expression of antioxidant enzymes, such as SOD, NADH peroxidase, and glutathione reductase. Exogenous probiotic supplementation is recommended to modulate the oxidative stress, owing to the limited endogenous antioxidant capacity.[110] EPS of antioxidant activity was produced by *L. plantarum* C88 as they were found to have strong in vitro radical scavenging activity and antioxidant activity.[16]

Recently, Ref. [105] determined the structure of a polysaccharide from *Streptococcus macedonicus* Sc136 containing trisaccharide sequence of *β*-D-Glc*p*NAc- (1→3)-*β*-D-Gal*p*-(1→4)-*β*-D-Glc*p*. The same structure has been identified in human milk oligosaccharides for infant nutrition. Functional EPSs can be engineered by deletion and insertion of new genes coding for glycosyltransferases as seen in *S. macedonicus* Sc136. The same approach was observed in the case of LPS from *Escherichia coli* helping in in vivo neutralization of the Shiga toxin in mice. The display of in situ health beneficial polysaccharides of EPS-producing probiotics is a very promising approach as far as functional foods are considered. The quantity of EPS being produced by the LAB determine their potential in the use as food-grade ingredients. However, they are promising candidates offering health benefits specifically as consumer-specific nutrition.[52]

Biological activities of LAB EPS are the key factors offering beneficial properties.[111] It is a fact that most of the commercialized probiotics are known to produce EPSs of its kind. Adhesive properties and prebiotic properties were observed in trials with EPS producers.

6.4.2 COMMERCIAL POTENTIAL AND APPLICATIONS OF EXOPOLYSACCHARIDES

While exploring the functional properties of EPSs, a huge opportunity is there in front to exploit their properties for human benefits in different fields. LAB-derived EPSs physicochemical characteristics make them a natural alternative to commercial food additives. The use of EPS in food production to increase viscosity as well as syneresis reduction can have many advantages because they may impart desirable rheological changes in the food matrix in addition to their emulsifier and stabilizer effects. EPS can also be applied in the preparation of food packaging and also provides attractive body-texture or properties and mouthfeel.[112] They can be used as food additives and has to be mentioned in food packings so that the consumer is aware of the contents. However, when the in situ production of EPS by LAB rule out the labelling procedure. They are being widely used as replacement for biothickeners in food industries and also found to provide many other techno-functional properties. Polymer diversity can be better studied by understanding the mechanism of microbial biosynthesis of EPS for better functional exploitation.

It is particularly the structure of *Lactobacillus* polysaccharides that renders them a multitude of physicochemical characteristics with extensive roles in various industries. The relative physicochemical properties and functions are reported in Table 6.4. The use of polysaccharide from *L. delbrueckii* in improving the ripeness of yogurt was proved by Rawson.[113] The exopolysaccharides from *L. delbrueckii* and *S. thermophilus* were used as the starter culture of cheese. It was observed from literature that exopolysaccharide-producing starter cultures increased the moisture content in comparison to the one unable to produce exopolysaccharides. Furthermore, the polysaccharides derived from LAB also improve physical properties of fermented milk products. The polysaccharides produced from *Lactobacilli* exhibits high thermostability and mechanical stability according to Refs. [114,115]. This is in accordance to the plasticity of exopolysaccharides and their use in probiotic products.[116]

It was already stated that the function of EPS is greatly influenced by its structure. It is the presence of β- linkages in HePs from LABS which enables the EPS solution with increased consistency and high intrinsic viscosities. As well as the degree of branching directly affects the stiffness and the presence of functional group to the polysaccharide backbone gives EPSs the potentiality of promoting health benefits.

An average production of 60–150 mg/L and 30–890 mg/L of HePS was observed in *L. delbrueckii* subsp. *bulgaricus* and *S. thermophilus in situ* in. It is observed that these two bacteria when grown synergistically *L. delbrueckii* is the first to grow and secrete metabolites promoting the growth of *S. thermophilus* whereas formic acid and carbon dioxide produced by *S. thermophilus* stimulates the growth of *L. delbrueckii.* The EPS producing starter cultures not only improved the water retention in yoghurts thereby reducing the syneresis, but also, improved the rheological properties such as viscosity and elasticity.

In some European and American countries, LAB EPS has been used as food additives. The functional aspects of LAB EPS exploited in food sector are as follows: (1) texture improvement of the fermented product by syneresis

TABLE 6.4 Lactic Acid Bacteria Exopolysaccharide with their Functional Properties and Applications.

Organism	EPS composition	Functional properties and applications
L. delbrueckii subsp. *bulgaricus* DGCC291	Glucose and galactose	Increased viscosity, syneresis reduction in yogurt
L. helveticus MB2-1	Glucose, mannose, galactose, rhamnose, and arabinose	Antioxidant and prebiotics
Lactobacillus gasseri FR4	Glucose, mannose, galactose, rhamnose, and fucose	Antioxidant and viscosity-enhancing agent
Lactiplantibacillus (Lpb.) plantarum	Arabinose, mannose, glucose, and galactose	Improving texture and rheological properties
L. plantarum BR2	Glucose and mannose	Cholesterol lowering and antioxidant
Lpb. plantarum C88	Glucose and galactose	Antioxidant and prebiotics
Lpb. plantarum WLPL04	Xylose, glucose, and galactose	Functional food ingredient
S. thermophilus AR333	Galactose, glucose, and galactosamine	Improving quality of fermented milk
S. thermophilus CC30	Glucose and galactose	Emulsifier
S. thermophilus NIZO 2104	Galactose, ribose, glucose, and *N*-acetyl-galactosamine	Increased viscosity, syneresis reduction in yogurt
S. thermophilus S-3	*N*-acetyl-galactosamine, galactose, and glucose	Protein interaction and syneresis reduction in yogurt
Lpb. plantarum JLAU103	Arabinose, rhamnose, fucose, xylose, fructose, galactose, glucose, and mannose	Antioxidant and functional additive
L. delbrueckii ssp. *bulgaricus*	Glucose and galactose	Production of B-group vitamins, for example, folic acid

reduction and low-fat Mozzarella cheese. It was found that EPS produced by thermophilic LAB can prevent the precipitation of whey and improve the texture enhancing the food safety and nutritional value and prolonging the shelf life of the product (2) improve rheological properties as a thickener. If the concentration of LAB EPS is very low it will increase the viscosity but cannot form gel and a small shear force can increase its fluidity, and when the shear force is removed, the viscosity can be quickly recovered. It has an important improvement effect on the sensory characteristics and intrinsic quality of fermented dairy products.[117]

Another example is the low-fat cheddar cheese, which is a main food component in European countries are fermented using *Lc. lactis* that produces EPS taking into advantage of improving water desorption rate and cheese moisture content increased up to 9.5%.[5] A dairy beverage "Kefir" which is sour and slightly carbonated can be developed by some LAB and yeasts. EPS and some peptides are the by-products formed after fermentation of kefir grains developed with the help of some LAB and yeasts. EPS along with few peptides are formed after fermentation and the EPS is a HePs known as Kefiran containing six or seven repeating units (glucogalactan).[118,119]

Apart from all these applications in food industry, there are other areas where exopolysaccharides find significant role to play. That includes preparation of bio-based films, IPNs, anticancer drug targeting, tissue engineering to name a few. Their ability to form hydrogels and film forming units makes them eligible for drug delivery or as drug carriers. Some EPSs being insoluble, can hold large amounts of water inside the cross-linked 3D structural network. These gels can entrap drugs inside and also absorb/adsorb molecules outside the surface as well. In vivo studies are required to support EPS as drug delivery agents. EPS properties can be manipulated by different methods. One is by mixing it with other biopolymers. A mixture of galactomannan with xanthan is developed to form a gel structure and is an example of synergism. Chemical modification of EPS structural backbone by the addition of different functional groups can change the nature of the EPS. These chemical reactions with other constituents and biopolymers help to form cross links to achieve polymeric structures with enhanced functionality. Due to the non – Newtonian behavior and high viscosity in aqueous solutions, they find applications as thickening agents, stabilizers, and binding agents. Recent developments were seen in this field exploiting the ability of EPS to form novel structures such as micro/nanospheres, polymer beads and capsules that are edible so as to carry vitamins, antioxidant molecules, probiotics, and prebiotics, etc.[5]

Exopolysaccharides such as bacterial alginate, gellan gum, FucoPol, and GalactoPol have the capability to establish physical and chemical intermolecular interactions that can result in forming a cohesive polymer matrix thus forming a film that are edible and eco-friendly. These films in turn have many attracts as it can be used as food packaging material, drug delivery coatings as discussed earlier, including scaffolds, tissue engineering, wound dressing, and materials. This results in the development of bacterial EPS in pharmaceutical importance. Their inherent biological activity boosts this application of EPS being used as film forming units. FucoPol exhibits high anti-inflammatory, antitumor, and immune enhancer drugs as a result of high fucose contents. Xanthan, sulphated dextran, sulphated dextran shows antiviral and anticancer properties. Glucomannan EPS produced by *L. plantarum* has anticholesterolemic along with antioxidant property.[120]

All these applications and the commercial importance marks the future to flourish in this area of research. Enhancing the EPS productions with low cost, sustainable raw materials aids to the study. Fermentative production of EPS by genetic engineering the metabolic pathways of the EPS biosynthesis is yet another area to work on so as to increase the yield and diversity of EPS produced to get tailor-made EPS with varied applications and commercial relevance and significance.

6.5 CONCLUSION

Exopolysaccharides of LAB is a vast area of research as there is diversity in structure and functions of the EPS it produces. This diversity is observed in monomeric units, linkages, and the associated conjugates. Mainly they find applications in various fields as cosmetic industry, pharmaceutical industry, food and environmental industry. The major limiting factor is its production cost due to its low levels of yield. However, research leading to the exploitation of LAB in producing high quality and quantity EPS by bioprocess and rDNA methods exploring substrates available as by-products of industries will be helpful in improving the yield as well as the reduction of cost of production.

ACKNOWLEDGMENTS

BI, AJK, and AGP gratefully acknowledge University of Kerala for the financial support. SMP gratefully acknowledges the Council of Scientific and

Industrial Research (CSIR), India for the financial assistance and Academy of Scientific and Innovative Research (AcSIR) for the enrolment for doctoral program. KMN and RP acknowledges the grant from CSIR- FBR scheme, MLP0052.

KEYWORDS

- **exopolysaccharides**
- **structural diversity**
- **homopolysaccharides**
- **heteropolysaccharides**
- **therapeutics**
- **rheology**

REFERENCES

1. Angelin, J.; Kavitha, M. Exopolysaccharides from Probiotic Bacteria and their Health Potential. *Int. J. Biol. Macromol.* **2020,** *162*, 853–865. https://doi.org/10.1016/j.ijbiomac.2020.06.190
2. Zheng, J. S.; Wittouck, S.; Salvetti, E.; Franz, C. M. A. P.; Harris, H. M. B.; Mattarelli, P. A Taxonomic Note on the Genus Lactobacillus: Description of 23 Novel Genera, Emended Description of the Genus *Lactobacillus beijerinck* 1901, and Union of Lactobacillaceae and Leuconostocaceae. *Int. J. Syst. Evol. Microbiol.* **2020,** *70*, 2782–2858. https://doi.org/10.1099/ijsem.0.004107
3. Zannini, E.; Waters, D. M.; Coffey, A.; Arendt, E. K. Production, Properties, and Industrial Food Application of Lactic Acid Bacteria-Derived Exopolysaccharides. *Appl. Microbiol. Biotechnol.* **2016,** *100*, 1121–1135. 10.1007/s00253-015-7172-2
4. Prete, R.; Alam, M. K.; Perpetuini, G.; Perla, C.; Pittia, P.; Corsetti, A. Lactic Acid Bacteria Exopolysaccharides Producers: A Sustainable Tool for Functional. *Foods* **2021,** *10*, 1653. https://doi.org/10.3390/foods10071653 0
5. Daba, G. M.; Elnahas, M. O.; Elkhateeb, W. A. Contributions of Exopolysaccharides from Lactic Acid Bacteria as Biotechnological Tools in Food, Pharmaceutical and Medical Applications. *Int. J. Biol. Macromol.* **2021,** *173*, 79–89.
6. Dimopoulou, M.; Bardeau, T.; Ramonet, P. Y.; Miot-Certier, C.; Claisse, O. Exopolysaccharides Produced by *Oenococcus oeni*: From Genomic and Phenotypic Analysis to Technological Valorization. *Food Microbiol.* **2016,** *53*, 10–17.
7. Monsan, P.; Bozonnet, S.; Albenne, C.; Joucla, G.; Willemot, R. M.; Remaud-Simeon, M. Homopolysaccharides from Lactic Acid Bacteria. *Int. Dairy J.* **2001,** *11*, 675–685.

8. Leemhuis, H.; Dijkman, W. P.; Dobruchowska, J. M.; Pijning, T.; Grijpstra, P. 4,6-α-Glucanotransferase Activity Occurs More Widespread in Lactobacillus Strains and Constitutes a Separate GH70 Subfamily. *Appl. Microbiol. Biotechnol.* **2013,** *97*, 181–193.
9. Abarquero, D.; Renes, E.; Fresno, J. M.; Tornadijo, M. E. Study of Exopolysaccharides from Lactic Acid Bacteria and their Industrial Applications: A Review. *Int. J. Food Sci. Technol.* **2021,** 1–11.
10. Helmut, M.; et al. Guar, Locust Bean, Tara, and Fenugreek Gums. In *Industrial Gums: Polysaccharides and Their Derivatives*, Third Ed.; 2012; pp 181–226.
11. Jeanes, A.; et al. Characterization and Classification of Dextrans from Ninety-Six Strains of Bacteria1b. *J. Am. Chem. Soc.* **1954,** *76* (20), 5041–5052.
12. Purama, R. K.; Goswami, P.; Khan, A. T.; Goyal, A. Structural Analysis and Properties of Dextran Produced by *Leuconostoc mesenteroides*. *Carbohydr. Polym.* **2009,** *76* (1), 30–35. http://dx.doi.org/10.1016/j.carbpol.2008.09.018.
13. Seymour, F. R.; Knapp, R. D.; Chen, E. C. M.; Jeanes, A.; Stephen, H. B. Structural Analysis of Dextrans Containing 2-O-α-d-glucosylated α-d-glucopyranosyl Residues at the Branch Points, by Use of 13C-Nuclear Magnetic Resonance Spectroscopy and Gas-Liquid Chromatography-Mass Spectrometry. *Carbohydr. Res.* **1979,** *71* (3), 231–250.
14. Elsa, D.; et al. Dextran: Sources, Structures, and Properties. *Polysaccharides* **2021,** *2* (3), 554–565.
15. Tomulescu, C.; et al. Levan - A Mini Review. *Sci. Bull. Ser. F Biotechnol.* **2016,** *XX*, 309–320.
16. Zhan, X. B.; Lin, C. C.; Zhang, T. H. Recent Advances in Curdlan Biosynthesis, Biotechnological Production, and Applications. *Appl. Microbiol. Biotechnol.* **2012,** *93* (2), 525–531.
17. Ruas-Madiedo, P.; Hugenholtz, J.; Zoon, P. An Overview of the Functionality of Exopolysaccharides Produced by Lactic Acid Bacteria. *Int. Dairy J.* **2002,** *12* (2–3), 163–171.
18. Korakli, M.; Vogel, R. F. Structure/Function Relationship of Homopolysaccharide Producing Glycansucrases and Therapeutic Potential of Their Synthesised Glycans. *Appl. Microbiol. Biotechnol.* **2006,** *71* (6), 790–803.
19. Vasanthakumari, D. S.; et al. Physicochemical Characterization of an Exopolysaccharide Produced by a Newly Isolated *Weissella cibaria*. *Appl. Biochem. Biotechnol.* **2015,** *176* (2), 440–453.
20. Tieking, M.; Wolfgang, K.; Gänzle, M. G. Evidence for Formation of Heterooligosaccharides by *Lactobacillus sanfranciscensis* during Growth in Wheat Sourdough. *J. Agric. Food Chem.* **2005,** *53* (7), 2456–2461.
21. Klemm, D.; et al. Chapter 100, Functional Polymers Based on Dextran BT. In *Polysaccharides II*, 2006; vol 205, pp 199–291. http://link.springer.com/10.1007/12_100%0Apapers3://publication/doi/10.1007/12_100.
22. Gan, W.; Zhang, H.; Zhang, Y.; Hu, X. Biosynthesis of Oligodextrans with Different Mw by Synergistic Catalysis of Dextransucrase and Dextranase. *Carbohydr. Polym.* **2014,** *112*, 387–395. http://dx.doi.org/10.1016/j.carbpol.2014.06.018.
23. Sarwat, F.; Ul Qader, S. A.; Aman, A.; Ahmed, N. Production & Characterization of a Unique Dextran from an Indigenous *Leuconostoc mesenteroides* CMG713. *Int. J. Biol. Sci.* **2008,** *4* (6), 379–386.
24. Faucard, P.; et al. Macromolecular Structure and Film Properties of Enzymatically-Engineered High Molar Mass Dextrans. *Carbohydr. Polym.* **2018,** *181*, 337–344. http://dx.doi.org/10.1016/j.carbpol.2017.10.065.

25. Van De Velde, K.; Kiekens, P. Biopolymers: Overview of Several Properties and Consequences on their Applications. *Polym. Test.* **2002,** *21*, 433–442.
26. Aman, A.; Siddiqui, N. N.; Qader, S. A. U. Characterization and Potential Applications of High Molecular Weight Dextran Produced by *Leuconostoc* mesenteroides AA1. *Carbohydr. Polym.* **2012,** *87*, 910–915.
27. Suner, S. S.; et al. Responsive Biopolymer-Based Microgels/Nanogels for Drug Delivery Applications, In *Stimuli Responsive Polymeric Nanocarriers for Drug Delivery Applications*; Elsevier Ltd., 2018; vol 1 Types and Triggers, pp 453–500 https://doi.org/10.1016/B978-0-08-101997-9.00021-7.
28. Zhang, Y.; et al. Comparison of Dextran Molecular Weight on Wheat Bread Quality and Their Performance in Dough Rheology and Starch Retrogradation. *LWT* **2018,** *98*, 39–45. https://doi.org/10.1016/j.lwt.2018.08.021.
29. Zarour, K.; et al. Rheology and Bioactivity of High Molecular Weight Dextrans Synthesised by Lactic Acid Bacteria. *Carbohydr. Polym.* **2017,** *174*, 646–57.
30. Bhavani, A.; Nisha, J.; et al. Dextran - The Polysaccharide with Versatile Uses. *Int. J. Pharma Bio Sci.* **2010,** *1* (4), 569–573.
31. Bowen, W. H.; Koo, H. Biology of *Streptococcus mutans*-Derived Glucosyltransferases: Role in Extracellular Matrix Formation of Cariogenic Biofilms. *Caries Res.* **2011,** *45* (1), 69–86. DOI: 10.1159/000324598. Epub 2011 Feb 23.
32. Nowak, A.; Wiater, A.; Choma, A.; Wiącek, D.; Bieganowski, A.; Siwulski, M.; Waśko, A.; et al. Fungal (1 → 3)-α-d-glucans as a New Kind of Biosorbent for Heavy Metals. *Int. J. Biol. Macromol.* **2019,** *137*, 960–965. 10.1016/j.ijbiomac.2019.07.036
33. Van Leeuwen., Sander, S. Structural Characterization of Bioengineered α-D-Glucans Produced by Mutant Glucansucrase GTF180 Enzymes of *Lactobacillus reuteri* Strain 180. *Biomacromolecules* **2009,** *10* (3), 580–588.
34. Patel, S.; Majumder, A.; Goyal, A. Potentials of Exopolysaccharides from Lactic Acid Bacteria. *Indian J. Microbiol.* **2012,** *52* (1), 3–12.
35. Kšonžeková, P.; Bystrický, P.; Vlčková, S.; Pätoprstý, V.; Pulzová, L.; Mudroňová, D.; et al. Exopolysaccharides of *Lactobacillus reuteri*: Their Influence on Adherence of E. coli to Epithelial Cells and Inflammatory Response. *Carbohydr. Polym.* **2016,** *141*, 10–19.
36. Tabasum, S.; et al. A Review on Versatile Applications of Blends and Composites of Pullulan with Natural and Synthetic Polymers. *Int. J. Biol. Macromol.* **2018,** *120*, 603–632. https://doi.org/10.1016/j.ijbiomac.2018.07.154.
37. Nisha, M.; Satyanarayana, T. Characteristics, Protein Engineering and Applications of Microbial Thermostable Pullulanases and Pullulan Hydrolases. *Appl. Microbiol. Biotechnol.* **2016,** *100* (13), 5661–79. http://dx.doi.org/10.1007/s00253-016-7572-y.
38. İspirli, H.; Sagdic, O.; Dertli, E.; et al. Synthesis of Silver Nanoparticles Prepared with a Dextran-Type Exopolysaccharide from *Weissella cibaria* MED17 with Antimicrobial Functions. *Prep. Biochem. Biotechnol.* **2020,** *51* (2), 112–119. DOI: 10.1080/10826068.2020.1795673
39. Donot, F.; Fontana, A.; Baccou, J. C.; Schorr-galindo, S. Microbial Exopolysaccharides: Main Examples of Synthesis, Excretion, Genetics and Extraction. *Carbohydr. Polym.* **2012,** *87* (2), 951–962. http://dx.doi.org/10.1016/j.carbpol.2011.08.083.
40. Singh, R. S.; Kaur, N.; Rana, V.; Kennedy, J. F. Pullulan: A Novel Molecule for Biomedical Applications. *Carbohydr. Polym.* **2017,** *171*, 102–21. http://dx.doi.org/10.1016/j.carbpol.2017.04.089.

41. Akiyoshi, K.; et al. Self-Assembled Hydrogel Nanoparticle of Cholesterol-Bearing Pullulan as a Carrier of Protein Drugs: Complexation and Stabilization of Insulin. *J. Control. Release* **1998,** *54* (3), 313–320.
42. Vuddanda, P. R.; Montenegro-Nicolini, M.; Morales, J. O.; Velaga, S. Effect of Plasticizers on the Physico-Mechanical Properties of Pullulan Based Pharmaceutical Oral Films. *Eur. J. Pharm. Sci.* **2017,** *96*, 290–298. http://dx.doi.org/10.1016/j.ejps.2016.09.011.
43. Zikmanis, P.; Brants, K.; Kolesovs, S.; Semjonovs, P. Extracellular Polysaccharides Produced by Bacteria of the Leuconostoc Genus. *World J. Microbiol. Biotechnol.* **2020,** *9*, 1–18. https://doi.org/10.1007/s11274-020-02937-9
44. Srivastava, R. K.; Ish, P.; Safdarjung COVID-19 Vaccination Group. The Initial Experience of COVID-19 Vaccination from a Tertiary Care Centre of India. *Monaldi Arch. Chest Dis.* **2021,** *91* (4). DOI: 10.4081/monaldi.2021.1816.
45. Wangpaiboon, K.; et al. Modified Properties of Alternan Polymers Arising from Deletion of SH3-like Motifs in *Leuconostoc citreum* ABK-1 Alternansucrase. *Carbohydr. Polym.* **2019,** *220*, 103–109. https://doi.org/10.1016/j.carbpol.2019.05.002.
46. Theuwissen, E.; Mensink, R. P. Water-Soluble Dietary Fibers and Cardiovascular Disease. *Physiol. Behav.* **2008,** *94* (2), 285–292.
47. Dueñas-Chasco, M. T.; et al. Structural Analysis of the Exopolysaccharides Produced by *Lactobacillus spp*. G-77. *Carbohydr. Res.* **1998,** *307* (1–2), 125–133.
48. Fraunhofer, M. E.; et al. Characterization of β-Glucan Formation by *Lactobacillus brevis* TMW 1.2112 Isolated from Slimy Spoiled Beer. *Int. J. Biol. Macromol.* **2018,** *107* (PartA), 874–881. http://dx.doi.org/10.1016/j.ijbiomac.2017.09.063.
49. Kearney, N.; et al. *Lactobacillus paracasei* NFBC 338 Producing Recombinant Beta-Glucan Positively Influences the Functional Properties of Yoghurt. *Int. Dairy J.* **2011,** *21* (8), 561–567. http://dx.doi.org/10.1016/j.idairyj.2011.03.002.
50. Garai-Ibabe, G.; et al. Naturally Occurring 2-Substituted (1,3)-β-d-Glucan Producing *Lactobacillus suebicus* and *Pediococcus parvulus* strains with Potential Utility in the Production of Functional Foods. *Bioresour. Technol.* **2010,** *101* (23), 9254–9263. http://dx.doi.org/10.1016/j.biortech.2010.07.050.
51. Jolly, L.; Vincent, S. J. F.; Duboc, P.; Neeser, J. R. Exploiting Exopolysaccharides from Lactic Acid Bacteria. *Antonie van Leeuwenhoek.* **2002,** *82* (1–4), 367–374. DOI: 10.1023/a:1020668523541
52. Majtan, J.; Jesenak, M. β-Glucans: Multi-Functional Modulator of Wound Healing. *Molecules* **2018,** *23* (4), 806. DOI:10.3390/molecules23040806
53. Barsanti, L.; et al. Chemistry, Physico-Chemistry and Applications Linked to Biological Activities of β-Glucans. *Natl. Prod. Rep.* **2011,** *28* (3), 457–466.
54. Zeković, D. B.; et al. Natural and Modified (1→3)-β-D-Glucans in Health Promotion and Disease Alleviation. *Crit. Rev. Biotechnol.* **2005,** *25* (4), 205–230.
55. Liatis, S.; et al. The Consumption of Bread Enriched with Betaglucan Reduces LDL-Cholesterol and Improves Insulin Resistance in Patients with Type 2 Diabetes. *Diabetes Metab.* **2009,** *35* (2), 115–120.
56. Han, Y. W. Microbial levan. *Adv. Appl. Microbiol.* **1990,** *1* (35), 171–194.
57. Hövels, M.; et al. High Yield Production of Levan-Type Fructans by Gluconobacter Japonicus LMG 1417. *Int. J. Biol. Macromol.* **2020,** *164*, 295–303. https://doi.org/10.1016/j.ijbiomac.2020.07.105.

58. Nakapong, S.; Pichyangkura, R.; Ito, K.; Iizuka, M.; Pongsawasdi, P. High Expression Level of Levansucrase from Bacillus licheniformis RN-01 and Synthesis of Levan Nanoparticles. *Int. J. Biol. Macromol.* **2013,** *54*, 30–36.
59. Gupta, S. K.; et al. Microbial Levan, an Ideal Prebiotic and Immunonutrient in Aquaculture. *World Aquac.* **2011,** *42* (1), 61.
60. Ernandes, F.; Garcia-Cruz, C. Nutritional Requirements of *Zymomonas mobilis* CCT 4494 for Levan Production. *Brazilian Arch. Biol. Technol.* **2011,** *54* (3), 589–600.
61. Lorenzetti, M. F. S.; Moro, M. R. Alginate/PVA Beads for Levan Production by *Zymomonas mobilis*. *J. Food Process Eng.* **2015,** *38* (1), 31–36.
62. Sarilmiser, H. K.; et al. Effective Stimulating Factors for Microbial Levan Production by *Halomonas smyrnensis* AAD6 T. *J. Biosci. Bioeng.* **2015,** *119* (4), 455–463. http://dx.doi.org/10.1016/j.jbiosc.2014.09.019.
63. Öner, E. T.; Hernández, L.; Combie, J. Review of Levan Polysaccharide: From a Century of Past Experiences to Future Prospects. *Biotechnol. Adv.* **2016,** *34* (5), 827–844. http://dx.doi.org/10.1016/j.biotechadv.2016.05.002.
64. Gupta, S. K.; et al. Dietary Microbial Levan Ameliorates Stress and Augments Immunity in Cyprinus Carpio Fry (Linnaeus, 1758) Exposed to Sublethal Toxicity of Fipronil. *Aquac. Res.* **2014,** *45* (5), 893–906.
65. Barone, J. R.; Medynets, M. Thermally Processed Levan Polymers. *Carbohydr. Polym.* **2007,** *69* (3), 554–561.
66. Franken, J.; Bianca, A. B.; Siew, L. T.; Bauer, F. F. Biosynthesis of Levan, A Bacterial Extracellular Polysaccharide, in the Yeast *Saccharomyces cerevisiae*. *PLoS One* **2013,** *8* (10), 1–14.
67. Mantovan, J.; et al. Use of Microbial Levan in Edible Films Based on Cassava Starch. *Food Packag. Shelf Life* **2018,** *18*, 31–36. https://doi.org/10.1016/j.fpsl.2018.08.003.
68. Belghith, K. S.; et al. Hypolipidemic Effect of Diet Supplementation with Bacterial Levan in Cholesterol-Fed Rats. *Int. J. Biol. Macromol.* **2012,** *50* (4), 1070–1074. http://dx.doi.org/10.1016/j.ijbiomac.2012.02.024.
69. Erginer, M.; et al. Sulfated Levan from *Halomonas smyrnensis* as a Bioactive, Heparin-Mimetic Glycan for Cardiac Tissue Engineering Applications. *Carbohydr. Polym.* **2016,** *149*, 289–296. http://dx.doi.org/10.1016/j.carbpol.2016.04.092.
70. Srikanth, R.; et al. Review on Production, Characterization and Applications of Microbial Levan. *Carbohydr. Polym.* **2015,** *120*, 102–114. http://dx.doi.org/10.1016/j.carbpol.2014.12.003.
71. Dahech, I.; et al. Antidiabetic Activity of Levan Polysaccharide in Alloxan-Induced Diabetic Rats. *Int. J. Biol. Macromol.* **2011,** *49* (4), 742–746. http://dx.doi.org/10.1016/j.ijbiomac.2011.07.007.
72. Mudannayake, D. C.; Wimalasiri, K. M. S.; Silva, K. F. S. T.; Ajlouni, S. Selected Sri Lankan Food Plants and Other Herbs as Potential Sources of Inulin-Type Fructans. *J. Natl. Sci. Found. Sri Lanka* **2015,** *43* (1), 35–43.
73. Mensink, M. A.; Frijlink, H. W.; Maarschalk, K. V. D. V.; Hinrichs, W. L. J. Inulin, A Flexible Oligosaccharide I: Review of Its Physicochemical Characteristics. *Carbohydr. Polym.* **2015,** *130*, 405–419. http://dx.doi.org/10.1016/j.carbpol.2015.05.026.
74. Saadat, R. Y.; Khosroushahi, A. Y.; Gargari, B. P. A Comprehensive Review of Anticancer, Immunomodulatory and Health Beneficial effects of the Lactic Acid Bacteria Exopolysaccharides. *Carbohydr. Polym.* **2019,** *217*, 79–89. https://doi.org/10.1016/j.carbpol.2019.04.025.

75. Sun, Q.; et al. A Novel Inulin-Type Fructan from Asparagus cochinchinensis and Its Bene Fi Cial Impact on Human Intestinal Microbiota. *Carbohydr. Polym.* **2020,** *247*, 116761. https://doi.org/10.1016/j.carbpol.2020.116761.
76. Franck, A.; et al. Technological Functionality of Inulin and Oligofructose. *British J. Nutr.* **2002,** *87* (S2), S287–S291.
77. Matsuzaki, C.; et al. Immunomodulating Activity of Exopolysaccharide-Producing *Leuconostoc mesenteroides* Strain NTM048 from Green Peas. *J. Appl. Microbiol.* **2014,** *116* (4), 980–989.
78. Ismail, B.; Nampoothiri, K. M. Exposition of Antitumour Activity of a Chemically Characterized Exopolysaccharide from a Probiotic *Lactobacillus plantarum* MTCC 9510. *Biologia* **2013,** *68* (6), 1041–1047.
79. Liu, M.; et al. Heat-Induced Inulin-Gluten Gel: Insights into the Influences of Inulin Molecular Weight on the Rheological and Structural Properties of Gluten Gel to Molecular and Physicochemical Characteristics. *Food Hydrocoll.* **2021,** *111*, 106397. https://doi.org/10.1016/j.foodhyd.2020.106397.
80. Ismail, B.; Nampoothiri, K. M.. Production, Purification and Structural Characterization of an Exopolysaccharide Produced by a Probiotic *Lactobacillus plantarum* MTCC 9510. *Arch. Microbiol.* **2010,** *192* (12), 1049–1057.
81. De Vuyst, L.; Degeest, B. Heteropolysaccharides from Lactic Acid Bacteria. *FEMS Microbiol. Rev.* **1999**, *23* (2), 153–177.
82. Liao, N.; et al. Potential of Lactic Acid Bacteria Derived Polysaccharides for the Delivery and Controlled Release of Oral Probiotics. *J. Control. Release* **2020,** 323, 110–124. https://doi.org/10.1016/j.jconrel.2020.04.022.
83. Zeidan, A. A.; Poulsen, V. K.; Janzen, T.; Buldo, P.; Derkx, P. M. F.; Øregaard, G. Polysaccharide Production by Lactic Acid Bacteria: from Genes to Industrial Applications. *FEMS Microbiol. Rev.* **2017,** *41* (1), S168–S200. https://doi.org/10.1093/femsre/fux017
84. Pham, P. L.; Dupont, I.; Roy, D.; Lapointe, G.; Cerning, J. Production of Exopolysaccharide by *Lactobacillus rhamnosus* R and Analysis of its Enzymatic Degradation During Prolonged Fermentation. *Appl. Environ. Microbiol.* **2000,** *66* (6), 2302–2310. DOI: 10.1128/AEM.66.6.2302-2310.2000.
85. Remus, D. M.; van Kranenburg, R.; van Swam, II.; Taverne, N.; Bongers, R. S.; Wels, M. Impact of 4 *Lactobacillus plantarum* Capsular Polysaccharide Clusters on Surface Glycan Composition and Host Cell Signalling. *Microb. Cell Factories* **2012,** *11*, 149.
86. Ruijssenaars, H. J.; Stingele, F.; Hartmans, S. Biodegradability of Food-Associated Extracellular Polysaccharides. *Curr. Microbiol.* **2000,** *40*, 194–199, DOI: 10.1007/s002849910039
87. Degeest, B.; De Vuyst, L. Indication that the Nitrogen Source Influences Both Amount and Size of Exopolysaccharides Produced by *Streptococcus thermophilus* LY03 and Modelling of the Bacterial Growth and Exopolysaccharide Production in a Complex Medium. *Appl. Environ. Microbiol.* **1999,** *65* (7), 2863–2870.
88. Prechtl, R. M.; Wefers, D.; Jakob, F.; Vogel, R. F. Structural Characterization of the Surface-Associated Heteropolysaccharide of *Lactobacillus plantarum* TMW 1.1478 and Genetic Analysis of its Putative Biosynthesis Cluster. *Carbohydr. Polym.* **2018,** *202*, 236–245. https://doi.org/10.1016/j.carbpol.2018.08.115
89. Van Kranenburg, R.; Marugg, J. D.; Van Swam, II.; Willem, N. J.; De Vos, W. M. Molecular Characterization of the Plasmid-Encoded eps gene Cluster Essential for Exopolysaccharide Biosynthesis in *Lactococcus lactis*. *Mol. Microbiol.* **1997,** *24* (2), 387–97.

90. Stingele, F.; Neeser, J. R.; Mollet, B. Identification and Characterization of the eps (exopolysaccharide) Gene Cluster from *Streptococcus thermophilus* Sfi6. *J. Bacteriol.* **1996,** *178* (6), 1680–1690.
91. Van Calsteren, M. R.; Pau-Roblot, C.; Béginm A.; Roy, D. Structure Determination of the Exopolysaccharide Produced by *Lactobacillus rhamnosus* Strains RW-9595M and R. *Biochem. J.* **2002,** *363* (Pt 1), 7–17. DOI: 10.1042/0264-6021:3630007
92. Lamothe, G.; Jolly, L.; Mollet, B.; Stingele, F. Genetic and Biochemical Characterization of Exopolysaccharide Biosynthesis by *Lactobacillus delbrueckii* subsp. *bulgaricus*. *Arch. Microbiol.* **2002.**
93. Sasikumar, K.; Vaikkath, D. K.; Devendra, L.; Nampoothiri, K. M. An Exopolysaccharide (EPS) from a *Lactobacillus plantarum* BR2 with Potential Benefits for Making Functional Foods. *Bioresour. Technol.* **2017,** *241*, 1152–1156.
94. Wang, J.; Zhao, X.; Yang, Y.; Zhao, A.; Yang, Z. Characterization and Bioactivities of an Exopolysaccharide Produced by *Lactobacillus plantarum* YW32. *Int. J. Biol. Macromol.* **2015,** *74*, 119–126. http://dx.doi.org/10.1016/j.ijbiomac.2014.12.006.
95. Andrew, M.; Jayaraman, G. Structural Features of Microbial Exopolysaccharides in Relation to their Antioxidant Activity. *Carbohydr. Res.* **2020,** *487*, 107881. https://doi.org/10.1016/j.carres.2019.107881
96. Chen, X. Y.; Levy, C.; Gänzle, M. G.; Ubalua, A. O.; Liu, S.; Ren, F. Structural Features of Microbial Exopolysaccharides in Relation to their Antioxidant Activity. *Int. J. Food Microbiol.* **2016,** *15* (4), 107881. https://doi.org/10.1016/j.ijbiomac.2020.10.227
97. Qing, C.; et al. Probiotic Characteristics of Exopolysaccharides-Producing Lactobacillus Isolated from Some Traditional Malaysian Fermented Foods. *Food Hydrocoll.* **2019,** *137* (2), 123–130. https://doi.org/10.1007/s13205-020-02200-w
98. Grosu-Tudor, S.; Zamfir, M.; Van der Meulen, R.; Falony, G.; De Vuyst, L. Prebiotic Potential of some Exopolysaccharides Produced by Lactic Acid Bacteria. *Rom Biotechnol. Lett.* **2013,** *18* (5), 8666–8676.
99. Rodríguez-Carvajal, M. A.; et al. Structure of the High-Molecular Weight Exopolysaccharide Isolated from *Lactobacillus pentosus* LPS26. *Carbohydr. Res.* **2018,** *343* (18), 3066–3070. http://dx.doi.org/10.1016/j.carres.2008.08.028.
100. Torino, M. I.; Mozzi, F.; De Valdez., F. Exopolysaccharide Biosynthesis by *Lactobacillus helveticus* ATCC 15807. *Appl. Microbiol. Biotechnol.* **2005,** *68* (2), 259–265.
101. Petry, S.; Furlan, S.; Crepeau, M. J.; Cerning, J.; Desmazeaud, M. Factors Affecting Exocellular Polysaccharide Production by *Lactobacillus delbrueckii* subsp. *bulgaricus* Grown in a Chemically Defined Medium. *Appl. Environ. Microbiol.* **2000,** *66*.
102. Laws, A.; Gu, Y.; Marshall, V. Biosynthesis, Characterisation, and Design of Bacterial Exopolysaccharides from Lactic Acid Bacteria. *Biotechnol. Adv.* **2001,** *19*, 597–625.
103. Degeest, B.; Mozzi, F.; De Vuyst, L. Effect of Medium Composition and Temperature and pH Changes on Exopolysaccharide Yields and Stability During *Streptococcus thermophilus* LY03 Fermentations. *Int. J. Food Microbiol.* 2002, *79* (3), 161–174.
104. Parente, E.; Ricciardi, A. Production, Recovery and Purification of Bacteriocins from Lactic Acid Bacteria. *Appl. Microbiol. Biotechnol.* **1999,** *52* (5), 628–638.
105. Jacques-Edouard, G.; Michèle, D.; Nicola, D.; Vincent, S. J. F. Heterologous Expression and Characterization of the Exopolysaccharide from *Streptococcus thermophilus* Sfi39. *Eur. J. Biochem.* **2001,** *268* (19), 5149–5156.
106. Kumar, R.; Bansal, P.; Singh, J.; Dhanda, S. Purification, Partial Structural Characterization and Health Benefits of Exopolysaccharides from Potential Probiotic Pediococcus Acidilactici NCDC 252. *Process Biochem.* **2020,** *99*. 10.1016/j.procbio.2020.08.028.

107. Tuinier, R.; Van Casteren, W. H. M.; Looijesteijn, P. J.; Schols, H. A.; Voragen, A. G. J.; Zoon, P. Effects of Structural Modifications on Some Physical Characteristics of Exopolysaccharides from *Lactococcus lactis*. *Biopolymers* **2001,** *59* (3), 160–166.
108. Ruijssenaars, H. J.; Stingele, F.; Hartmans, S. Biodegradability of Food-Associated Extracellular Polysaccharides. *Curr. Microbiol.* **2000,** *40*, 194 199. DOI: 10.1007/s002849910039
109. Zhu, H.; Yang, T.; Xing, Y.; Chang, G. Advances in Nutritional Function of Extracellular Polysaccharides from Lactic Acid Bacteria. **2020,** *XXX*, 345–353.
110. Yu, Y. J.; Chen, Z.; Chen, P. T.; Ng, I. S. Production, Characterization and Antibacterial Activity of Exopolysaccharide from a Newly Isolated *Weissella cibaria* under Sucrose Effect. *J. Biosci. Bioeng.* **2018,** *126* (6), 769–777. https://linkinghub.elsevier.com/retrieve/pii/S1389172318302925
111. Hongpattarakere, T.; Cherntong, N.; Wichienchot, S.; Kolida, S.; Rastall, R. A. In-Vitro Prebiotic Evaluation of Exopolysaccharides Produced by Marine Isolated Lactic Acid Bacteria. *Carbohydr. Polym.* **2012,** *87* (1), 846–852.
112. Giles-Gómez, M.; García, J. G. S.; Matus, V.; Quintana, I. C.; Bolívar, F.; Escalante, A.. In Vitro and In Vivo Probiotic Assessment of *Leuconostoc mesenteroides* P45 Isolated from Pulque, a Mexican Traditional Alcoholic Beverage. *Springerplus* **2016,** *5*, 708 DOI: 10.1186/s40064-016-2370-7
113. Rawson, H. Rropy' Strains of *Lactobacillus delbrueckii* ssp. *bulgaricus* and *Streptococcus thermophilus* on Rheology of Stirred Yogurt. *Int. J. Food Sci. Technol.* **2003,** *32*, 213–220. 10.1046/j.1365-2621.1997.00395.x.
114. Abid, Y.; Casillo, A.; Gharsallah, H.; Joulak, I.; Lanzetta, R.; Corsaro, M.; Attia, H.; Azabou, S. Production and Structural Characterization of Exopolysaccharides from Newly Isolated Probiotic Lactic Acid Bacteria. *Int. J. Biol. Macromol.* **2018,** *108*, 719–728. DOI: 10.1016/j.ijbiomac.2017.10.155.
115. Saravanan, C.; Shetty, H. P. K. Isolation and Characterization of Exopolysaccharide from Leuconostoc lactis KC117496 Isolated from Idli Batter. *Int. J. Biol. Macromol.* **2015,** *90*, 100–106 DOI: 10.1016/j.ijbiomac.2015.02.007.
116. Korcz, E.; Varga, L. Exopolysaccharides from Lactic Acid Bacteria: Techno-Functional Application in the Food Industry. *Trends Food Sci. Technol.* **2021,** *110*, 375–384.
117. Bai, Y.; Dobruchowska, J. M.; Van Der Kaaij, R. M.; Gerwig, G. J.; Dijkhuizen, L. Structural Basis for the Roles of Starch and Sucrose in Homo-exopolysaccharide Formation by *Lactobacillus reuteri* 35-5. *Carbohydr. Polym.* **2016,** *151*, 29–39.
118. Cheirsilp, B.; Suksawang, S.; Yeesang, J.; Boonsawang, P. Co-Production of Functional Exopolysaccharides and Lactic Acid by *Lactobacillus kefiranofaciens* Originated from Fermented Milk, Kefir. *J. Food Sci. Technol.* **2018,** *55* (1), 331–340
119. Rehman, R.; Wang, Y.; Wang, J.; Geng, W. Physicochemical Analysis of Mozzarella Cheese Produced and Developed by the Novel EPS-Producing Strain *Lactobacillus kefiranofaciens* ZW3. *Int. J. Dairy Technol.* **2018,** *71* (1), 90–98.
120. Alves, V. D.; Ferreira, A. R.; Costa, N.; Freitas, F.; Reis, M. A.; Coelhoso, I. M. Characterization of Biodegradable Films from the Extracellular Polysaccharide Produced by Pseudomonas Oleovorans Grown on Glycerol Byproduct. *Carbohydr. Polym.* **2011,** *83* (4), 1582–1590.

CHAPTER 7

Comparison of Thermal Resistance by Insulation Board Made from Rice Husk and Bagasse Waste

SOWJANYA MADIREDDI, DEVA MOUKTHIKA K., ARCHANA C., and KAVYA CH.

CVR College of Engineering, Hyderabad, India

ABSTRACT

Agrowaste produced in India is of huge quantity (approximately 500 Mt/year) and most of it is being burnt leading to pollution. One of the interesting applications of the waste produced is the preparation of insulation material which can be used to resist heat from entering from outside the building to inside space. This will lead to a clean environment and extra income for the farmers. Preventing the heat transfer from outside to inside the building space will reduce the load on the air conditioning equipment. In the present work, two such biowaste materials—ice husk and bagasse waste—were used to make a thermal resistance board. The boards were tested for their thermal resistance to heat. The equipment used for this test simulates the real-time situations of inside and outside environmental conditions. Unground rice husk (RH)was observed to offer more resistance than ground/powdered samples. This is due to voids filled with air in the board made from unground RH.

7.1 INTRODUCTION

A hot environment is undesirable with respect to human comfort. Hence, the use of air conditioning equipment to maintain temperatures inside the

Bio-Based Polymers and Composites: Properties, Durability, and Applications. Amadou Belal Gueye, Sabu Thomas, Nandakumar Kalarikkal, & Modou Fall (Eds.)

building space. However, this leads to a load on the electrical costs and also to environmental issues such as global warming and those associated with chlorofluorocarbon (CFC) due to the refrigerants used for the purpose. Thermal insulation of the buildings is always a better option, having a scope for improvement. Presently materials such as rock wool, fiberglass, polystyrene, asbestos, etc. are being used as the insulating material. These materials have disadvantages because they can have adverse effects on people exposed to them in the event of damage to the packing. Hence, it is required to find safe alternatives. Moreover, the agrowaste is another cause of concern for the environment. This is due to the burning of waste to avoid the expenses for disposal after the crop. In the present work, the authors would like to bring both the issues together to give a common solution. Figure 7.1 shows few agrowaste materials. The agrowaste materials chosen in the present work are rice husk (RH) and bagasse waste.

FIGURE 7.1 Agrowaste materials.

Investigators have been trying to address both issues individually. Julieta et al.[1] developed a composite board using cork granules/recycled rubber granules and RH. This composite is tested for thermal conductivity and acoustic performance as a construction material. A lambda meter was used for testing the thermal conductivity. RH and EVA waste[2] were used as construction materials to reduce the noise. The composite was proposed for under-flooring construction material. It was observed that the new material reduced the building load and improved the acoustic performance. Nano-silica was extracted from RH waste by milder acid treatment and burning. Both nano and microparticles were extracted.[3] Silica has many applications in the polymer and electronic industries. Nano/ultrafine silica is useful as a thermal insulator as well.[4] Thermal insulator[5] was made by extrusion and pressing of RH ash with high silica content. Properties were found using the hot plate method. It was observed that the vitreous phase in the product leads to low thermal conductivity.

In the review of the application of RH composite, the properties of the composite were found to be influenced by the interaction of fiber-polymer. The main applications observed were as a construction material.[6] Secondary fillers were also found to influence the composite properties, which require further investigations. P.C. Kapur[7] suggested that the RH which is an agricultural waste can be used as a low to medium-cost thermal insulation material for dryers, furnaces, and in ceramic industries. The silica content in the agricultural waste was observed[8] to act as filler material and it reduced plasticity in the building bricks. However, the technological properties were not lowered due to the addition of these agrowaste (5 wt%) materials. Biocomposites developed from rice, wheat husk, and wood, textile fibers were tested for thermal conductivity.[9] The least thermal conductivity and minimum water absorption during immersion test was found in RH composite. The developed materials were suggested for indoor applications as an insulating material. The mechanical properties were observed to improve by adding fly ash. Thermal properties and water absorption rate were observed to increase with RH content.[10] 2–6% by mass of bagasse[11] was used as ash to prepare composites as insulating material. The composite was prepared to replace asbestos. When compared to commercial cement, these composites showed better insulating properties. Thermal insulation materials from RH, coconut coir, and bagasse showed better insulation properties such as low thermal conductivity.[12]

Hence, in the present work, rice unground husk, powdered RH, and bagasse powder are used to prepare the insulation board and tested for the thermal resistance offered when exposed to heating. A halogen lamp is employed as the heat source.

7.2 METHODOLOGY

7.2.1 MATERIALS

The materials used for the present study are ground and unground RH, and ground bagasse. Figure 7.2 shows the raw materials of RH and bagasse waste collected for the present study. In the present work, the RH was used as available unlike the extraction of fiber by a few investigators[12] for the preparation of insulation board. This shortened the process duration and cost. The RH was collected from a rice processing mill nearby. A part of it was ground to powder and the rest is used as unground raw material. Bagasse waste was collected from the juice extractor and dried. The dried waste was ground to powder. The present work is to find an economical way of producing the insulation boards. The procedure adopted can be repeated by a layman. Hence, this simple procedure is adopted by avoiding any chemicals for the processing and extraction of fiber from the RH or bagasse waste. Second, by avoiding any specific size of the ground powder. Powder is made with a simple kitchen grinder which in fact is easily available with most of the farmers. The raw materials used were collected free of cost. LY556 and HY951 were used as the resin and hardener for preparing the samples. The resin and hardener were taken in a ratio of 10:1. Samples were prepared for various quantities of the materials.

7.2.2 SAMPLE PREPARATION

RH collected from the rice mill nearby was ground to powder and some quantity was used in the unground form for the preparation of samples. Bagasse waste collected from the juice extractor is dried and ground to powder. No special care was taken to get uniform size of the powder etc. This was to make sure that the materials are used in their simplest form available. The resin and hardener are mixed in proportion. The raw material (RH/bagasse) were weighed and mixed with the resin and hardener. The mixture was placed in a mold made up of aluminum (10 cm × 10 cm × 1 cm). The material was allowed to set for nearly 12 h. Samples prepared are as shown in Figure 7.3. Weight of the sample before and after setting was measured. Thickness of the samples was measured after setting. Table 7.1 gives the details of the samples prepared for the experiments.

FIGURE 7.2 (a) RH; (b) RH powder; (c) Dried bagasse waste; (d) Bagasse powder.

FIGURE 7.3 Samples made from RH and bagasse.

7.3 EXPERIMENTAL PART

The setup for experimentation is shown in Figure 7.4. The design of the setup is detailed elsewhere by the author1[13,14] and hot and cold spots are identified. Based on the design, the chamber is divided into outer and inner space regions by a wooden board with a square hole at the center. The sample is placed at the central location of the board covering the hole. A halogen lamp at 400 W is used as a heat source to heat the upper chamber representing the outside environment. It is observed that it takes nearly 30 minutes to come to steady state.

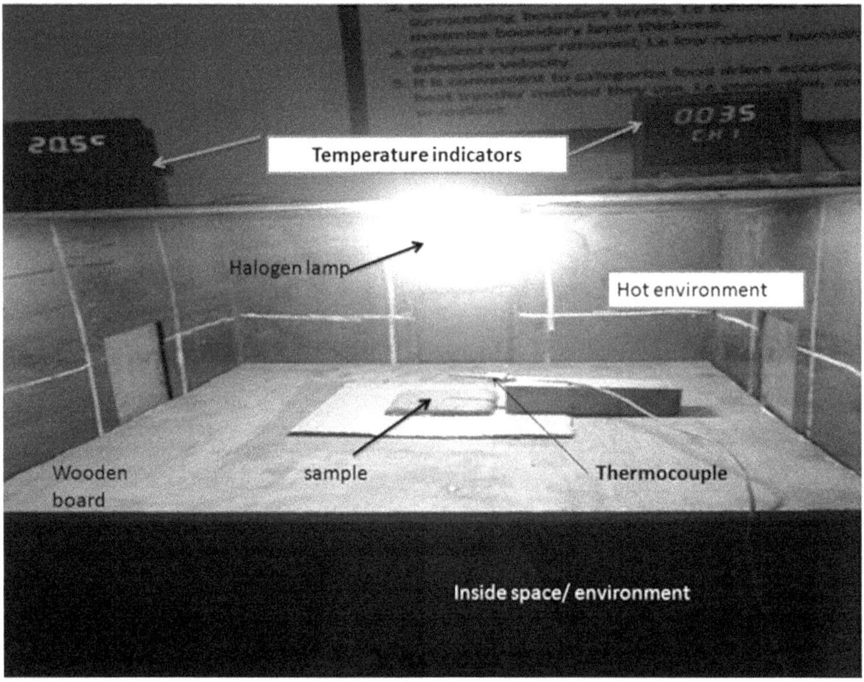

FIGURE 7.4 Testing chamber.

The temperatures of the (i) chamber air above and (ii) below the sample; surface temperatures of the sample on (iii) hot side and (iv) cold side; and (v) ambient temperatures are noted down for a period of 30 min. The temperatures are observed to come to steady state during this time. Hence, the time duration for all the experiments was fixed as 30 min to 1 h.

7.4 RESULTS AND DISCUSSION

The samples were placed centrally in the chamber below the heat source (Halogen lamp) as shown in the Figure 7.4. The power is set to 400 W. The readings of temperature are taken on the top and bottom surfaces of the sample and also above and below the sample. The space above the sample represents hot space (outside environment) and the space below the sample represents the cold space (inside environment). The difference in the temperatures after the steady state is reached was tabulated as given below in Table 7.1. The position of the temperatures recorded is as shown in Figure 7.5.

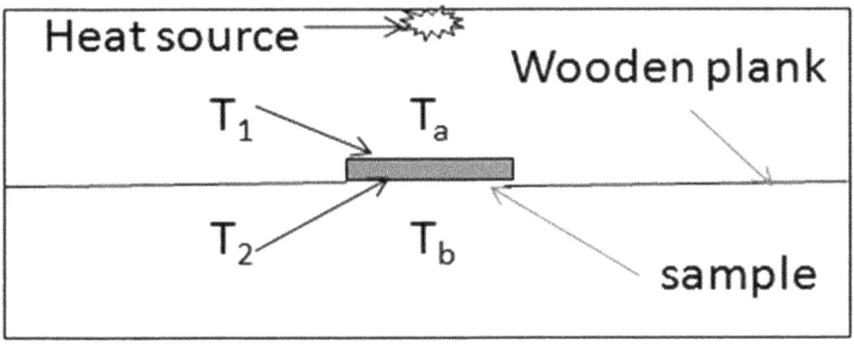

FIGURE 7.5 Position of temperature recording during the experiment.

TABLE 7.1 Details of Samples Prepared.

S. No.	Sample No.	(RH)/(BG)	RH (%)	R-Hd (%) 10:1	t (mm)	w (g/mm²)
1	2	RH powder	31.25	68.75	12	1.08
2	1	RH powder	30	70	6.5	0.53
3	3	RH powder	26	74	4.6	0.47
4	4	RH powder	26	74	5	0.47
5	11	RH powder	22.5	77.5	11	0.70
6	12	RH powder	19.52	80.48	9.2	0.75
7	5	RH (unground)	31.25	68.75	18.6	1.00
8	8	RH (unground)	26.66	73.34	12	0.47
9	9	RH (unground)	22.5	77.46	11	0.53
10	10	RH (unground)	19.52	80.48	10.4	0.69
11	6	BG Powder	26	74	5.6	0.73
12	7	BG Powder	19.51	80.48	7.2	0.85

RH, rice husk; BG, bagasse; R-Hd, resine-hardner; *t*, thickness; *w*, weight.

It is interesting to observe that powdered bagasse waste sample shows less difference (5.4°C) of temperature between upper and lower spaces. This difference is same for change in percentage of material in the mixture. For powdered RH samples, the difference of chamber upper and lower spaces is around 16°C. With increase in thickness due to increase in percentage of RH material the difference of upper and lower surface temperature of the sample is observed to increase. This is because of the absorption of heat by the material in a period of 30–60 min.

The difference of upper and lower space temperatures of the chamber representing outside and inside temperatures is observed as approximately 19°C in the case of unground RH material sample. The material percentage in the sample is same as in the case of powdered sample. However, due to more voids in the sample as RH being used as available, the difference in the temperatures is increased. These voids in the sample may be filled with air, which in fact acts as thermal barrier. Hence, unground RH sample offers more resistance to heat flow.

TABLE 7.2 Details of Temperature Measurements After Reaching the Steady State.

S. No.	(RH /(BG)	(RH/(BG) (%)	R-Hd (%)	Difference of upper and lower surface temperature of the sample ($\Delta T_{12} = T_1 - T_2$)	Difference of upper and lower chamber air temperatures ($\Delta T_{ab} = T_a - T_b$)
1	RH powder	31.25	68.75	6.4	16.2
2	RH powder	30	70	2.7	15.4
3	RH powder	26	74	0.4	14.1
4	RH powder	26	74	2	16
5	RH powder	22.5	77.5	5	17.2
6	RH powder	19.52	80.48	2.6	17
7	RH (unground)	31.25	68.75	16	15
8	RH (unground)	26.66	73.34	10	21.8
9	RH (unground)	22.5	77.46	9.7	18.3
10	RH (unground)	19.52	80.48	8.8	21.4
11	BG powder	26	74	5.4	14.1
12	BG powder	19.51	80.48	5.4	14.7

RH, rice husk; BG, bagasse; R-Hd, resine-hardner.

It is interesting to observe from the Figure 7.6 for the RH powdered sample 2 given in Table 7.1. that the temperature of the inside space is maintained less than 23°C. Steady temperatures of 54.4°C and 48°C are maintained on top and bottom surface of the sample. These higher temperatures are due to the heat retention by the sample material. For the air temperature of around 40°C, inside temperatures are maintained at around 23°C. These low temperatures reduce the usage of air conditioning equipment. Figure 7.7 shows the variation in temperature difference with percentage of RH on both sides of the sample and between inside and outside temperatures of air. The space temperature difference is nearly same for all samples of RH powder. However, the surface temperatures are minimum for sample with 26% RH powder.

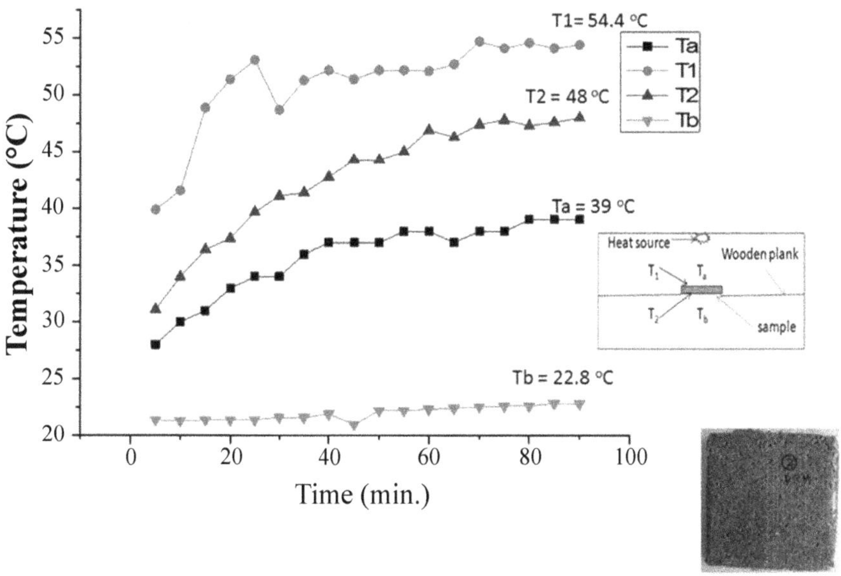

FIGURE 7.6 RH powdered sample 2 heat resistance to internal space.

Figure 7.8 shows the similar comparison for RH (unground) sample. The difference in space temperatures is same for all percentage of the material samples. However, the difference of temperatures is higher to that of RH powdered samples. However, the inconsistency may be due to the nonuniform distribution of the RH in the sample. Figure 7.9 shows the comparison between the RH powder and bagasse powder samples. Steady temperatures are observed after around 1 h. Bagasse sample is observed to absorb more heat compared to that of RH sample. However, the inside space is maintained at lower temperatures with RH material sample.

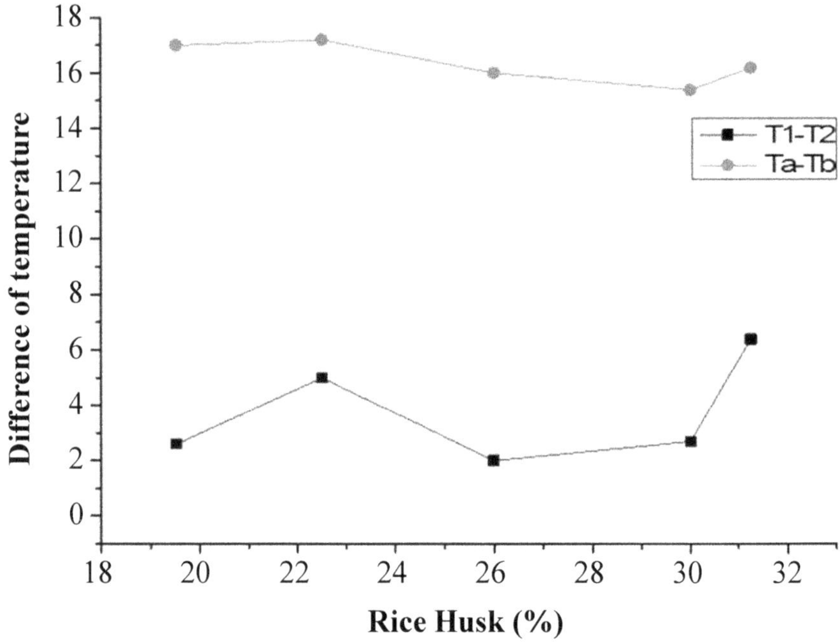

FIGURE 7.7 Difference in temperature with increase in percentage of RH (powder).

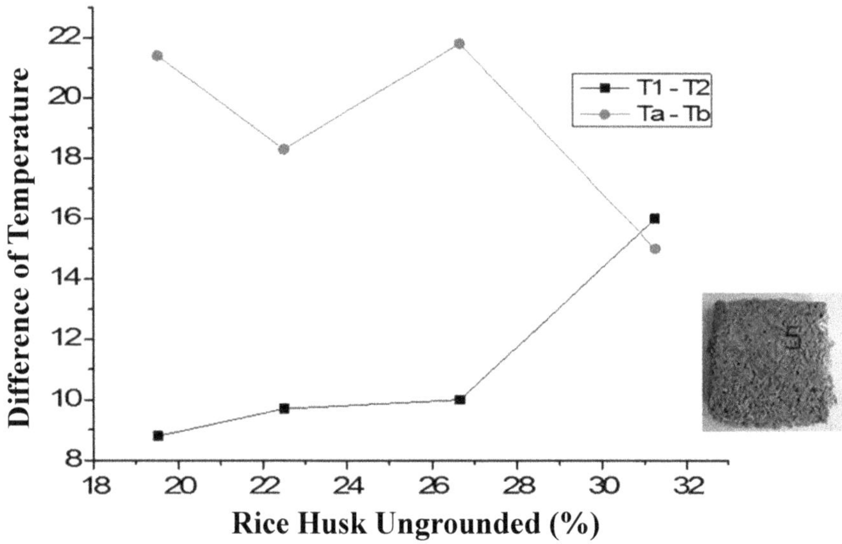

FIGURE 7.8 Difference in temperature with increase in percentage of RH (unground).

Table 7.3 shows the resistance offered by the material with increase in thickness. Thermal resistance is calculated using the following equation:

$$\text{Thermal resistance, } R_{\text{th}} = \frac{L}{kA} \tag{7.1}$$

where L (m) is the thickness of the sample, k is the thermal conductivity (W/mK), and A (m^2) is the surface area exposed to heating.

$$\text{Heat transfer rate, } Q = T_{12}/R_{\text{th}} \tag{7.2}$$

As thermal conductivity is constant for a material and sample area is same, with increase in thickness of the sample, the resistance offered by the material increases. Figure 7.10 shows the comparison of thermal resistance offered by the powdered RH sample and unground RH material sample. The resistance offered by the unground RH sample is higher for the same percentage of material in the sample. This may be due to the voids present in the nonuniform spread of the unground RH in the sample. This validates the experimentations conducted. The halogen lamp at 400 W generates 41 W of radiation heat transfer.[14] Heat transferred through the RH powdered sample is less than 1%. This is higher for unground RH sample but less than 2%. The rate of heat transferred through bagasse powdered sample is nearly 2%.

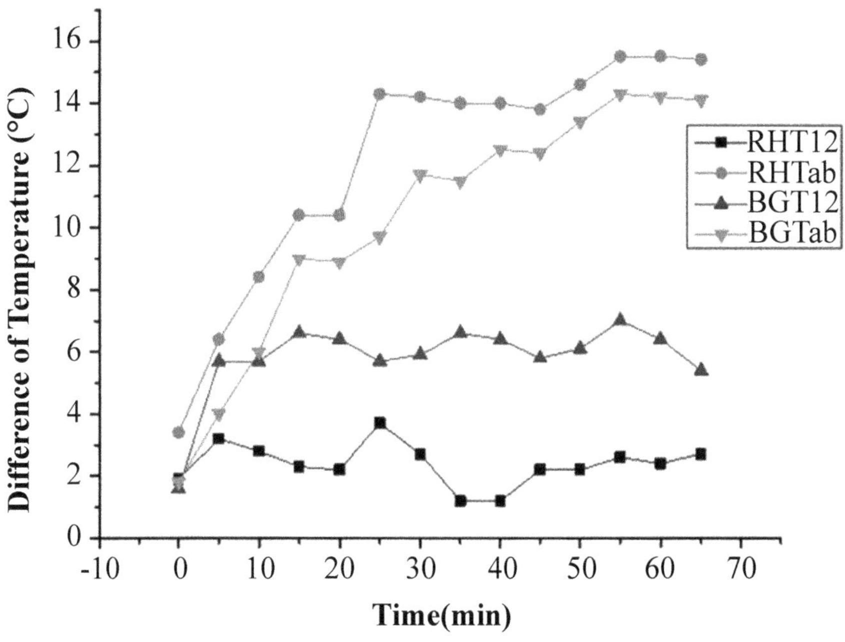

FIGURE 7.9 Bagasse waste vs powdered RH.

TABLE 7.3 Thermal Resistance of the Material Samples.

Sample No.	Material	T_{12} (°C)	L (mm)	K (W/mK)	A (m²)	R_{th} (K/W)	Q (W)	Q_p (%)
2	RH powder	6.4	12	0.056	0.01	21.43	0.30	0.73
1	RH powder	2.7	6.5	0.056	0.01	11.61	0.23	0.57
3	RH powder	0.4	4.6	0.056	0.01	8.21	0.05	0.12
4	RH powder	2	5	0.056	0.01	8.93	0.22	0.55
11	RH powder	5	11	0.056	0.01	19.64	0.25	0.62
12	RH powder	2.6	9.2	0.056	0.01	16.43	0.16	0.39
5	RH (unground)	16	18.6	0.056	0.01	33.21	0.48	1.17
8	RH (unground)	10	12	0.056	0.01	21.43	0.47	1.14
9	RH (unground)	9.7	11	0.056	0.01	19.64	0.49	1.20
10	RH (unground)	8.8	10.4	0.056	0.01	18.57	0.47	1.16
6	BG powder	5.4	5.6	0.085	0.01	6.59	0.82	2.00
7	BG powder	5.4	7.2	0.085	0.01	8.47	0.64	1.55

R_{th}, thermal resistance; Q, Heat transfer rate through the sample; Q_p, Percent heat transferred through the sample out of 41 W from heat source.

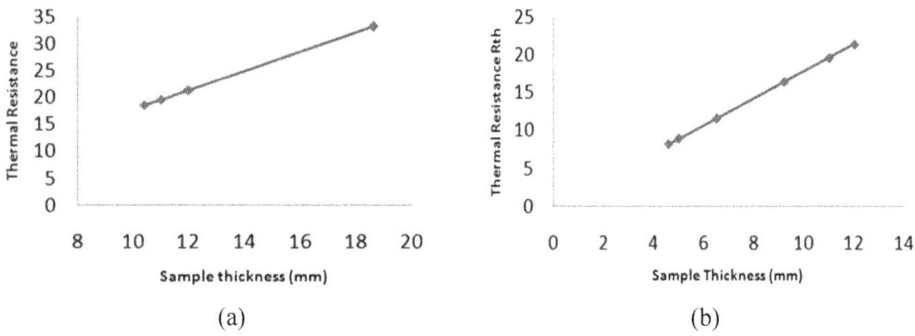

FIGURE 7.10 Variation of thermal resistance with thickness in RH sample made with (a) unground and (b) powdered material.

7.5 CONCLUSION

Agrowaste materials RH (unground), powdered RH, and bagasse powder are employed as raw materials for the preparation of insulation materials. The sample varied with percentage of raw material in the composite. With increase in thickness, the thermal resistance offered is increased. Samples made with bagasse powder offered less resistance. However, more samples with various percentage of bagasse are to be tested. Of the two varieties of samples powdered and unground RH, the latter one is observed to resist more heat. This is due to the air entrapped in the unevenly distributed RH in the sample. T_{ab} is the difference of temperature in the outside and inside spaces. T_{12} is the difference of temperature on both sides of the sample.

The inferences drawn are as follows:

- For powdered RH T_{ab}: 16°C and T_{12}: approx. 3°C only. This may be due to absorption of heat by RH. As the RH is powdered the granules are very small and closely packed in while mixed with the resin-hardener. Hence, the heat transferred like in a homogeneous solid material.
- For unground RH T_{ab}: 19°C and T_{12}: 10°C. This may be due to the air entrapped between the unground RH. As the thermal conductivity of the air is very less the resistance offered by the entrapped air augments the thermal resistance of the board.
- For powdered bagasse T_{ab}: 14.1°C and T_{12}: 5.4°C. This difference is same for change in percentage of material in the mixture and very near to that of RH powder board. This is because of the near-equal thermal conductivity values for RH and bagasse waste materials.
- The agrowaste like RH can directly be used for the preparation of insulation material board without going through the process of fiber extraction using chemicals which in turn is harmful to the environment.
- The process proposed will reduce the cost of processing and will make the preparation simpler and require less skill for making the insulation boards. This process can be adopted by the farmers and gain extra income and also reduce the pollution levels which is the need of the day.

ACKNOWLEDGMENTS

Authors thank the management of CVR College of Engineering for the support in the research activities in the Department of Mechanical Engineering.

KEYWORDS

- **rice husk**
- **bagasse waste**
- **heat resistance**
- **waste recycling**
- **agrowaste**

REFERENCES

1. Antonio, J.; Tadeu, A.; Marques, B.; Almeida, J. A. S.; Pinto, V. Application of Rice Husk in the Development of New Composite Boards. *Constr. Build. Mater.* **2018,** *176,* 432–439.
2. Borges, J. K.; Pacheco, F.; Tutikian, B.; de Oliveira, M. F. An Experimental Study on the Use of Waste Aggregate for Acoustic Attenuation: EVA and Rice Husk Composites for Impact Noise Reduction. *Constr. Build. Mater.* **2018,** *161,*501–508.
3. Carmona, V. B.; Oliveira, R. M.; Silva, W. T. L.; Mattosoa, L. H. C.; Marconcini, J. M. Nanosilica from Rice Husk: Extraction and Characterization. *Ind. Crops Prod.* **2013,** *43,* 291–296.
4. Liou, T.-H. Preparation and Characterization of Nano-Structured Silica from Rice Husk. *Mater. Sci. Eng.* **2004,** *A364,* 313–323.
5. Gonc¸alves, M. R. F.; Bergmann, C. P. Thermal Insulators Made with Rice Husk Ashes: Production and Correlation Between Properties and Microstructure. *Constr. Build. Mater.* **2007,** *21,* 2059–2065.
6. Arjmandi, R.; Hassan, A.; Majeed, K.; Zakaria, Z. Rice Husk Filled Polymer Composites. *Int. J. Polym. Sci.* **2015,** *501471,* 32.
7. Kapur, P. C. Thermal Insulations from Rice Husk Ash, an Agricultural Waste. *J. Ceram. Int.* **1980,** *6* (2), 75–78.
8. Barbieri, L.; Andreola, F.; Lancellotti, I.; Taurino, R. Management of Agricultural Biomass Wastes: Preliminary Study on Characterization and Valorisation in Clay Matrix Bricks. *Waste Manag.* **2013,** *33,* 2307–2315.
9. Muthuraj, R.; Lacoste, C.; Lacroix, P.; Bergeret, A. Sustainable Thermal Insulation Biocomposites from Rice Husk, Wheat Husk, Wood Fibers and Textile Waste Fibers: Elaboration and Performances Evaluation. *Ind. Crops Prod.* **2019,** *135,* 238–245.
10. Aravind, N. R.; Sathyan, D.; Mini, K. M. Rice Husk Incorporated Foam Concrete Wall Panels as a Thermal Insulating Material in Buildings. *Indoor Built Environ.* **2019,** *0* (0). 1–9.
11. Arsène, M. A.; Bilba, K.; Onésippe, C.; Rodier, L. Thermal and Flexural Properties of Bagasse/Cement Composites. *Green Mater.* **2016,** *3* (4), 132–143.
12. Usubharatana, P.; Phungrassami, H. Development of Thermal Insulation Materials for Packaging Made from Agricultural Wastes. *Acad. J. Sci.* **2015,** *4* (2), 133–142.
13. Madireddi, S.; Siddharth, V.; Amaan, M.; Adithya M. In *Heat Transfer in Food Crop Dryer Using Halogen Lamp*, International Conference on Emerging Trends in Engineering (ICETE). Learning and Analytics in Intelligent Systems, vol 2; Satapathy, S., Raju, K., Molugaram, K., Krishnaiah, A., Tsihrintzis, G., Eds.; Springer: Cham, 2020.
14. Madireddi, S.; Ashish, P.; Reddy, B. A.; Abhishek, D. Heat Transfer Analysis of A Multilayered Fabric-Material Assembly Exposed to Halogen Lamp Heating. *Mater. Today Proceed.* **2021,** *38* (Part 5), 3258–3263.

CHAPTER 8

Effect of Micro- and Nanofillers on Polarization, Depolarization Currents, and Dielectric Properties in Fiber-Reinforced Epoxy Composites

SUCHITRA M., GEETHASHREE A., and PANCHAMI S. V.

Vidyavardhaka College of Engineering, Mysuru, Karnataka, India

ABSTRACT

Polymer composites are preferred materials for different electrical and mechanical applications. However, the properties of polymers are seriously affected by hostile and extreme climatic conditions and environmental pollution. The performance of polymer insulators under such conditions is not fully investigated. In the present work, the glass fiber reinforced epoxy resin (G-E) based composites with nanosilica (SiO_2), alumina (Al_2O_3), and micro-alumina trihydrate (ATH) fillers were fabricated using the pultrusion technique to obtain a new hybrid composite insulator for high voltage applications. The relationship between dc conductivity and charge carrier mobility, the variations in polarization and depolarization current (PDC), with the micro- and nanofiller contents were investigated. The effect of micro- and nanofillers, their frequency dependence on electrical parameters, and the water aging effects of G-E hybrid nanocomposites were determined. For as-cast G-E hybrid nanocomposites, the results show a decrease in the dielectric constant and dissipation factor of G-EA, G-EAS, and G-EASA when compared to G-E composites. With water treatment, the dielectric constant and dielectric loss were slightly higher due to the formation of water layers around the fillers. A negligible effect on dielectric properties of G-EASA

Bio-Based Polymers and Composites: Properties, Durability, and Applications. Amadou Belal Gueye, Sabu Thomas, Nandakumar Kalarikkal, & Modou Fall (Eds.)

composite when aged in water is observed. This is due to reduced water absorption. The composites were exposed to water and maximum moisture content and apparent diffusivities were determined. From the obtained results there is a decrease in PDC, and conductivity with the incorporation of nano- and microfillers into the G-E composite. The best results were obtained from the G-E composite filled with micro- and nanofillers. The PDC and conductivity are reduced by 25%, 27%, and 77%, respectively, in G-EASA hybrid composite compared to the G-E composite. The Al_2O_3 nanofiller and ATH have different structures as compared to the SiO_2 filler and Al_2O_3 has more influence on the PDC of the nanocomposite. It was observed that the G-E composites with micro- and nanofiller had minimum water absorption.

8.1 INTRODUCTION

Polymer nanocomposite materials are emerging as a new material for insulation. The properties of the material can be enhanced by the addition of a few percentages of nanofiller. Fiber-reinforced polymer (FRP) composites have become a popular alternative to porcelain in the manufacture of high-voltage insulators. Many new technologies have been developed for HV equipment monitoring in recent years. The non-destructive method of testing, that is, polarization and depolarization current (PDC) testing is used to determine the conductivity of insulators.

Some authors have conducted significant research on the PDC technique for assessing the insulation conductivity in electrical equipment.[1] They have determined the conductivity of polyethylene and natural rubber compounds, with different quantities of TiO_2 and SiO_2 nanofiller by using the PDC measurement technique. The results have shown that the use of SiO_2 filler results in a reduction of the conductivity as compared to TiO_2 fillers. Few authors[2] have also discussed the interrelation of polarization, depolarization, absorption current, and charge carrier mobility of epoxy nanocomposites with different fillers. Alumina fillers have a better influence on the polarizability as compared to zinc oxide and silica fillers. A review-based study[3] is reported on the PDC method of determining the conductivity of the various types of insulators. It is established that the PDC technique can be successfully used for the estimation of the conductivity of insulating materials.

In the published research by some authors,[4] diffusion behavior and diffusion coefficient for glass fiber reinforced epoxy composites have been investigated with different solutions Benzene, HNO_3, and KOH. The results have shown that composites in distilled water had less mass gain than those immersed in kerosene solution while it was in between for the samples immersed in the other solutions. In a work published[5] on moisture diffusion by some authors, the behavior of glass fiber-reinforced epoxy fabric composite has been studied. The water absorption behavior at 60%, 96% R H, and a constant temperature of 60°C have been discussed. It is concluded that the maximum mass gain and diffusion coefficient of water depends both on the nature of the composite and hygrothermal conditions. Glass fiber is reinforced with different polymers such as epoxy, vinyl ester, and polyester laminates were investigated by some authors for the diffusion behavior.[6] The diffusion coefficient and water absorption in glass fiber with epoxy laminates were much less as compared to the other two laminates. In the work reported by some authors,[7] the dielectric properties of nano- and microepoxy composites under the effects of water absorption have been studied. The water uptake was much lower in microcomposites as compared to nanocomposites. This was reported as due to the blocking effect of microfillers and a slight increase in water absorption of nanocomposites was suggested due to an increase in the continuous path for percolation of water in the composites. The remedy reported to overcome this problem is by improvement of the bonding between matrix and inorganic filler. By using fillers with surface functionalization the surface of composites was rendered hydrophobic. The water absorption tests on E glass, low and high seed count ECR glass fiber with modified epoxy, polyester, and vinyl ester on samples fabricated by the pultrusion technique have been discussed in the literature.[8] The vinyl ester composites possess good moisture diffusion properties for insulator applications. The polyester-based E/ECR glass fiber composites have shown the highest rates of water absorption. The other two matrices had a lower rate of water absorption.

The effect of fillers on the dielectric properties of epoxy-based composites is reported in the literature.[9–12]

The objective of this paper is to find the best filler combination of G-E composite that can enhance the insulation property by experimental study of PDC, conductivity, absorption current, charge carrier mobility, and moisture absorption.

8.2 EXPERIMENTAL

8.2.1 MATERIALS

The matrix used was bisphenol-A epoxy resin (MY740). The color was pale yellow to yellow, with a density of 1.2 (g/cm^3) at 25°C and viscosity (mPa-s) at 25°C. A colorless or pale-yellow type liquid anhydride type hardener (HY918) with a density 0.9 (g/cm^3) and accelerator (DY062) were used. They were obtained from Huntsman, USA. Commercially available pure grades of uncoated particles of nanoalumina (Al_2O_3, average filler size 12 nm, density 4.0 g/cm^3), and nano-silica (12 nm, density 2.6 g/cm^3), were used as nanofillers. The coupling agent used was 3-glycidyloxypropyl trimethoxysilane. The nanofillers and silane procured by Sigma Aldrich were used. The solvent used to disperse the fillers was ethanol procured from Merck KGaA, Darmstadt, Germany. The alumina trihydrate (ATH), obtained from Akrochem Corporation, was used as a microfilter (2.6 μm). The surface-functionalized ECR-glass fiber (24 μm diameter) was used for reinforcement.

8.2.2 SPECIMEN FABRICATION

The epoxy resin and hardener were dried at 60°C for 1 h. Also, the SiO_2, Al_2O_3 nanofillers, and ATH microfiller were kept in a vacuum oven for 24 h at 110°C. In "as received" nanofillers, no hydroxyl group is available due to the absorption of water from the environment. Thus, preheating of fillers makes available the hydroxyl groups by evaporating the water and results in strong bonding between the fillers and epoxy. After preheating fillers were slowly introduced into ethanol with 1000 rpm of high shear mixing for 5 min, followed by 40 min of ultrasonication. These methods were used to break the agglomeration. This procedure was followed because directly introducing fillers into epoxy results in agglomerates due to their high viscosity. The next step was the introduction of the required amount of silane and then allowing for hydrolysis for up to 10 min. The solvent with fillers was then added to the resin along with sonication for 40 min. The next step is to remove the solvent from the mixture. For this, the mixture is kept in a vacuum oven at 85 C. The hardener (85 wt. %) and accelerator (DY062) of 2 wt% was added to the mixture of epoxy and fillers. A high shear mixing for 15 min at 3000 rpm was carried out. The addition of the hardener opens the epoxide rings and begins the hydrogen bond formation. Further, by pultrusion technique the ECR-glass fiber of 75 wt% were reinforced with the above mixture. The glass fiber-reinforced epoxy nanocomposites with and without fillers of 3 mm thickness were fabricated and identified as G-E, G-EA, G-EAS, and G-EASA.

8.2.3 MORPHOLOGICAL STUDIES

The morphology study of the G-E composites was conducted by ZEISS field emission scanning electron microscope (FESEM). The electron gun of the FESEM was energized in the range. The samples of size 5 mm × 5 mm × 3 mm were coated with gold particles with a thickness of 5 nm using a sputtering coater. Tecnai™ Spirit transmission electron micrograph (TEM) machine with optional scanning mode [scanning transmission electron microscope (STEM)] was used to capture the SEM photographs of the G-E composites.

8.2.4 PDC MEASUREMENTS

PDC of the fabricated G-E composites with and without fillers was measured in the time domain with a PDC analyzer (DIANA, Model EN 50110-1). The charging time was 300 s with 200 V DC and then shorted for some time of 1000 s and discharge current was recorded. 100 V/mm electric field strength was used. The inbuilt software records the process of charging and measurement of current. With PDC measurements, the dc conductivity, of the composites was estimated using equation 8.1:

$$\sigma = \frac{\varepsilon_0}{C_0 U_0}\left[i_p(t) - i_d(t)\right] \tag{8.1}$$

where σ is the composite conductivity, ε_0 is the relative permittivity, C_0 is the geometric capacitance, U_0 is the step voltage, $i_d(t)$ is the depolarization current and $IP_p(t)$ is the polarization current.

8.2.5 WATER ABSORPTION

A water absorption test was performed on fabricated composites as per ASTM D570 at room temperature in distilled water. The composites were cleaned with ethanol and then dried in a vacuum oven at 60°C. The weight gain was recorded with a digital balance (Citizen, CX228) with an accuracy of 0.1 mg by periodic removal of the composites from water. This process was repeated till the weight remained constant. The percentage gain at any time *t* (M) as a result of water absorption was calculated by using equation 8.2:

$$M\% = \frac{M_w - M_d}{M_d} \times 100 \tag{8.2}$$

Where M_w is the weight of water water-absorbed composite and M_d is the weight of the dried specimen.

8.2.6 FREQUENCY DOMAIN SPECTROSCOPY

Dielectric characterization was performed using a Precision high-frequency inductance, capacitance, and resistance (HF LCR) meter from 20 Hz to 10 MHz using Wayne Kerr Make, Model 6510P meter. The capacitance (C), dissipation factor (tan δ), and conductance (G) were measured over the frequency range, of 20 Hz to 10 MHz at an applied voltage of 1 V (RMS) as per ASTM D150-98.

The dielectric constant (ε') was calculated from the measured capacitance

$$C = \frac{\varepsilon_0 \varepsilon' A}{t} \tag{8.3}$$

where C is capacitance in F, ε_o is absolute permittivity ($\varepsilon_o = 8.854\times10^{-12}\,F/m$) A is an area of the electrode in (m), and is the thickness of the specimen (m).

The dielectric loss (ε'') was calculated from the measured tan δ, using the relation

$$\varepsilon'' = \varepsilon' \tan\delta \tag{8.4}$$

The *ac* conductivity (σ_{ac}) was computed using the relation

$$\sigma_{ac} = Gt / A \tag{8.5}$$

where G is the measured conductance in Siemens, t is the thickness of the sample (m), and is the area of the electrode (m^2).

8.3 RESULT AND DISCUSSION

8.3.1 MORPHOLOGICAL PROPERTIES

TEM of fabricated composites are shown in Figures 8.1(a-b) and 8.2(a-b). It is observed from TEM images that the unidirectional ECR glass fibers are uniformly embedded in the epoxy. An improved bonding interface between the fiber, filler, and epoxy is exhibited in this case due to the functionalization of nanofillers, the use of ECR glass fiber, and the fabrication method used.

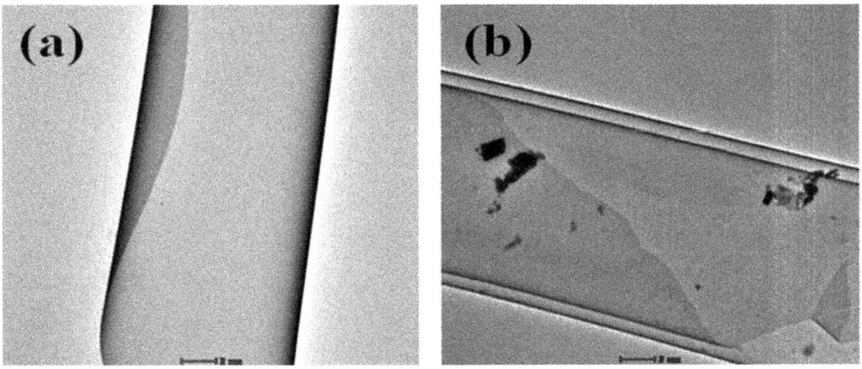

FIGURE 8.1 TEM images of: (a) G-E composite and (b) G-EA composite.

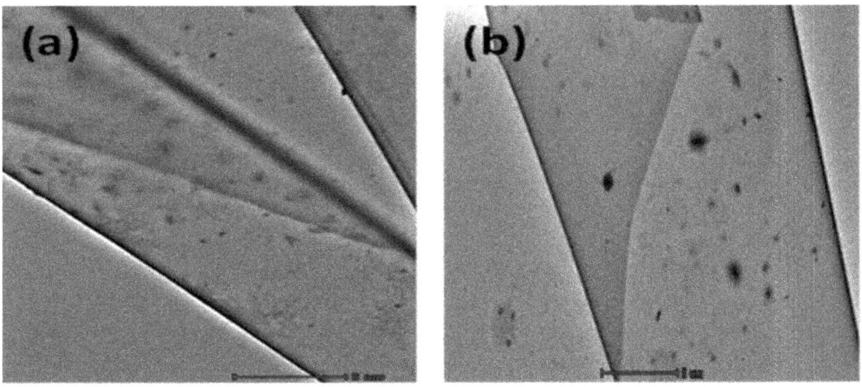

FIGURE 8.2 TEM images of (a) G-EAS composite and (b) G-EASA composite.

The SEM images of the G-EASA hybrid composite are shown in Figure 8.3. A uniform dispersion with fine agglomerate-free individual fillers of Al_2O_3, SiO_2, and microsized ATH particles is evident. The particle–particle contacts are unobserved and in addition, the transparency of the epoxy is observed with nanofiller and microfillers. Owing to the good dispersion, the scattering of light is minimized within the epoxy matrix. In hybrid composites, the interface between the fiber-filler and epoxy is the determining factor for the achievable properties.

8.3.2 PDC ANALYSIS

The results of PDC measured for fabricated samples are shown in Figures 8.4 and 8.5.

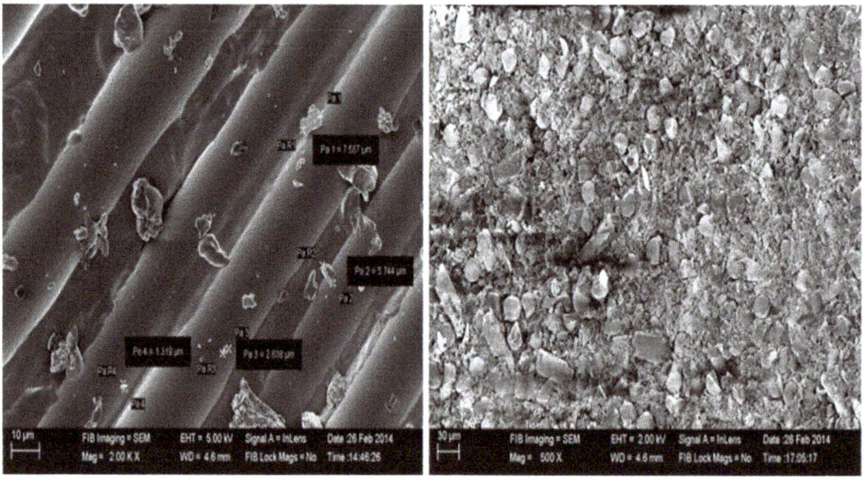

FIGURE 8.3 SEM images of G-EASA composite.

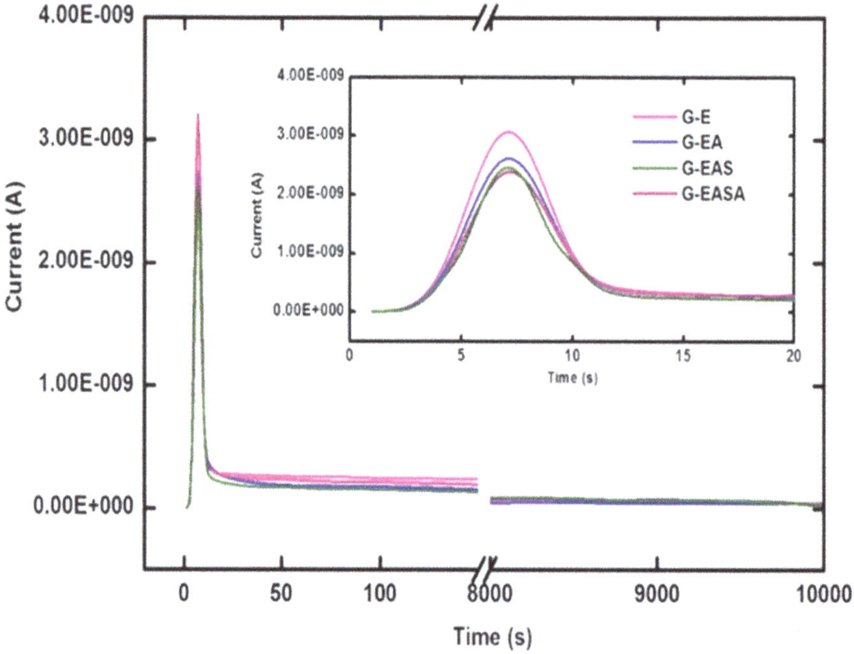

FIGURE 8.4 Polarization currents of G-E and hybrid composites.

In the G-E composite, the polarizing current initiates at 41.6 pA and attains a maximum current of 3.23 nA at ~8.5 s. The depolarizing current

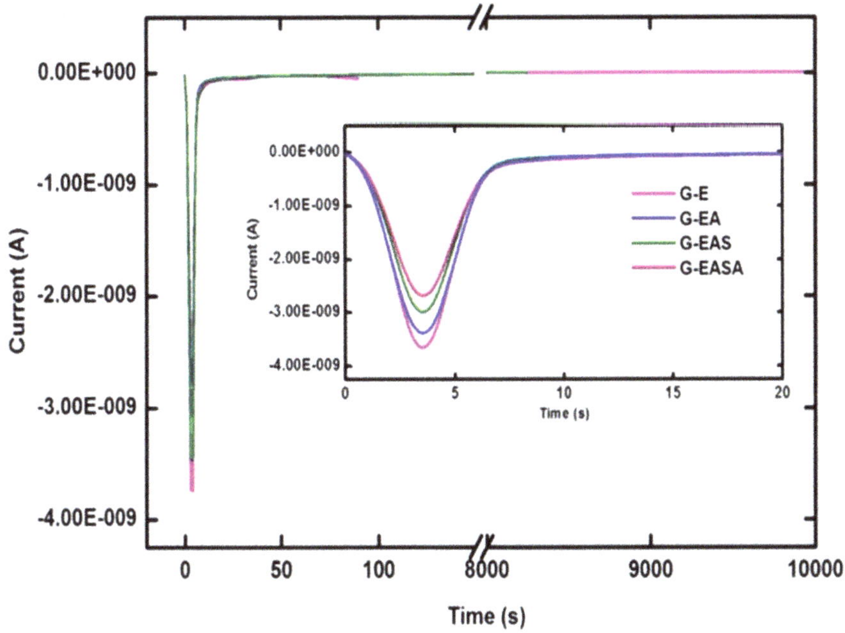

FIGURE 8.5 Depolarization currents of G-E and hybrid composites.

decreases from 0.12 pA to 3.73 nA after poling. With filler addition to the G-E composite, the peak current continues to decrease as compared to that of G-E. The filler addition, such as nano-silica, nanoalumina, and micro-ATH increases the resistivity of the composite system. G-EASA composite has low values of PDC compared to other composites. The PDC is reduced by 25% and 27% compared to the G-E hybrid composite. This shows that the G-EASA hybrid composite exhibits low conductivity as compared to the other composites and can be related to its higher resistivity. As observed from the SEM and TEM images, G-EA, and G-EASA composites due to a uniform distribution, strong interaction, and larger interfacial area have contributed towards good results. With good bonding between the matrix and fillers, the free volume has been reduced. As a result, moving charges have decreased and have contributed to reduced polarization current and conductivity. The conduction mechanisms within nanocomposites involve a complicated process. It involves charge injection from electrodes into epoxy; it depends on hopping and tunneling of charges between neighboring sites, traps and volumetric conduction, interaction mechanisms within the bulk-like covalent or ionic bonding, etc. The G-E composite without filler addition has the highest polarization current as compared to others. This may be due to less interfacial bonding between matrix and fibers. This leads to an increase in moving charges contributing to higher polarization current and conductivity.

8.3.3 DC CONDUCTIVITY

PDC measurement allows for an estimation of the conductivities of G-E hybrid composites. Conductivity variations of G-E hybrid composites areas are shown in Figure 8.6. It shows that the DC conductivity decreases with the filler addition to the G-E composite. The conductivity decreases by 69.3% in the G-EA hybrid composite and by 77.3% in the G-EASA composite compared to the G-E composite. Thus G-EASA composite has shown the best result with the least conductivity. This may be due to the uniform distribution of fillers in the matrix which is also evident from TEM micrographs. The nanofillers fill up the bonds in the matrix and thereby reduce the movement of charge carriers. Similar observations are reported by Jamail et al.[13] Also, the silica nanofiller tends to improve the dielectric properties.

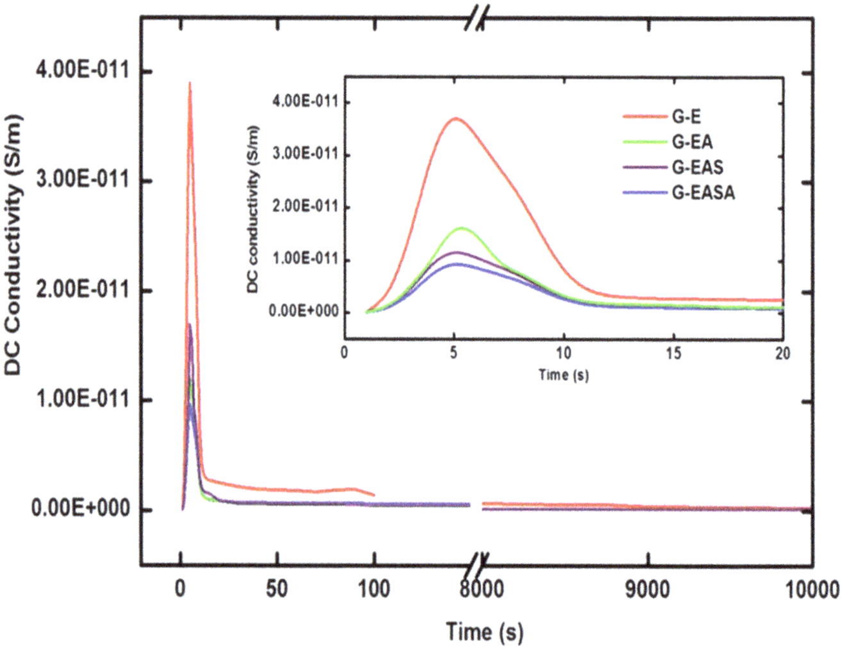

FIGURE 8.6 DC conductivity of G-E and hybrid composites.

It is observed from Figure 8.6 that in G-EA, G-EAS, and G-EASA the conductivity decreases. This may be due to preprocessing of the nanoparticles before use. Preprocessing includes either the heating of particles or the functionalization of particles with GPS after heating. As-received nanoparticles are known to have absorbed water on the surface. By heating the nanoparticles, the water from the surface is released and frees up the surface hydroxyl group, which may bond with the surrounding hydroxyl groups, epoxy via hydrogen bonding.

8.3.4 ABSORPTION CURRENT

The mobility of charge carriers, and absorption current measurements[14] were conducted on the G-E, G-EA, and G-EASA composites to evaluate the relative ease of charge propagation in composites. The mobility of the charge carriers is determined using equation 8.2.[15–19] t_p is the transient time at which absorption current is maximum, d is the dielectric thickness, and V is the applied voltage. All the measurements in the present work were carried out at room temperature.

$$\mu = \frac{0.786 \times d^2}{t\, pV} \tag{8.6}$$

From the observations of Table 8.1, the mobility of charge carriers slightly increases in G-EA, G-EAS, and G-EASA composites with the addition of fillers in comparison to the mobility of the base epoxy matrix.

TABLE 8.1 Charge Carrier Mobility (μ) at Room Temperature.

Samples	Mobility (μ) ±0.2 ($\times 10^{-9}$) m^2/Vs
G-E	6.83
G-EA	7.01
G-EAS	7.18
G-EASA	7.34

8.3.5 WATER ABSORPTION

The variation of water absorption with the square root of time for all G-E hybrid composites at 25°C is shown in Figure 8.7. An increase in time is noticed with the increase in water absorption up to 1700 h and thereafter it remains constant. The percentage of increase in water absorption is less in all G-E hybrid composites. However, after 1700 h it is noticed that water uptake is slightly higher in G-E and G-EA composites compared to G-EASA.

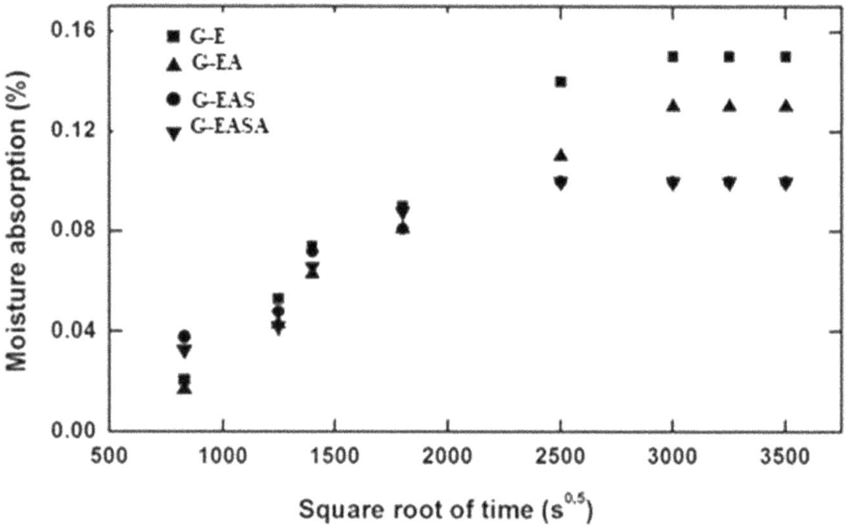

FIGURE 8.7 Variation of water absorption with the square root of time for all G-E hybrid composites at 25°C.

Two processes may be concerned with water diffusion. The first is the initial linear portion of the curve as in Figure 8.7. This diffusion follows Fick's law. Another slow process is described by the flat portion of the curves. The factors that are responsible for water absorption in composites are the hydrophilicity behavior of individual components and the structural arrangements of the filler within the epoxy. Water ingress can occur in the GRP rods of insulators through the end fittings, if not properly protected against moisture, water can then penetrate through the rubber shedding if the thickness is not sufficient or it is damaged due to cracks.[20] The rate at which water is absorbed by composites depends on the structure and chemical composition of polymers, temperature, the variation in water distribution within the composite, and the interaction between water and the matrix.

Water diffusion into composites takes place due to different mechanisms.[21] The main cause is the microvoids within voids present filler and matrix. The microcracks formed during the fabrication of composites.[22,23] This may induce effects such as chemical degradation, cracking, and de-bonding.[24]

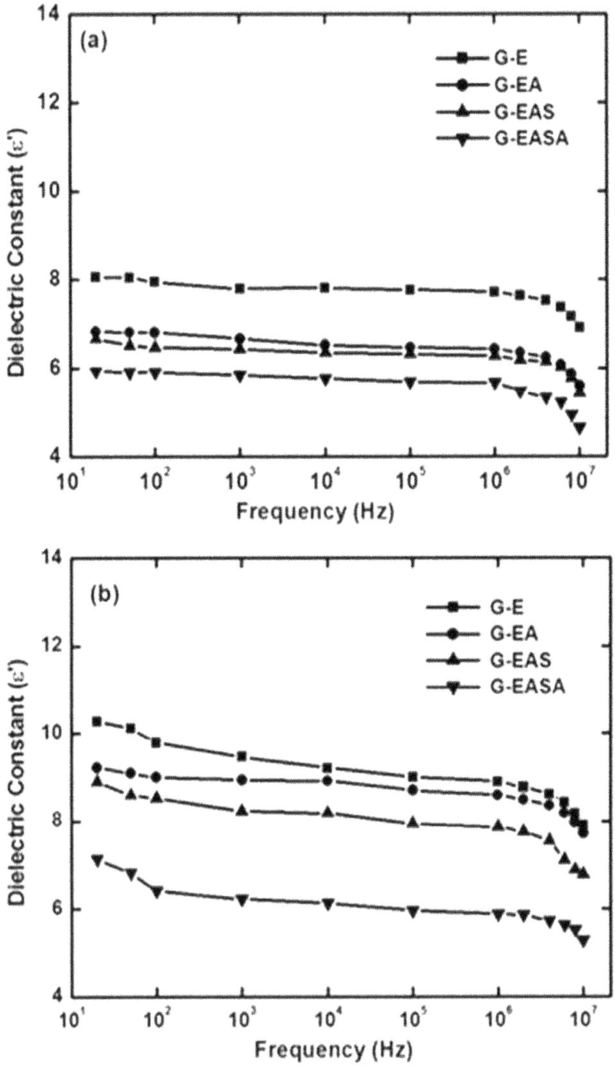

FIGURE 8.8 Variation of dielectric constant with frequency of G-E and hybrid composites: (a) as cast (b) water aged.

The ECR glass surface treated with silane has led to good interaction between epoxy and fiber and has the least water uptake. However, with the incorporation of fillers in G-E composites, the voids will be blocked and reduce the water absorption. The G-EASA sample showed less weight gain may be due to the addition of ATH and Al_2O_3 fillers.

8.3.6 DIELECTRIC CONSTANT

The plots of variation of the dielectric constant of 'as cast' and water-aged (500 h, 25 ± 3°C) samples with frequency are shown in Figures 8.8b and Figure 8.8a.

From Figure 8.8a it is observed that the dielectric constant of all the G-E and hybrid composites are almost constant and are slightly higher in the lower frequency range (up to 1kHz) and decrease at higher frequencies.

i) Effect of frequency

The dielectric constant depends on the dipoles that can orient themselves under the applied electric field. The dipoles of the epoxy-functional groups can orient themselves easily at low frequencies of the applied voltage resulting in a slight increase in insensitivity.

At higher frequencies, the rotational motions of the polar molecules of the dielectric are not sufficiently rapid to attain equilibrium with the field. As the frequency increases, some of the dipolar groups, e.g., larger molecules, cannot keep pace with the applied field. Therefore, the contribution of these inert dipolar groups to the dielectric constant disappears and this leads to a decrease in the dielectric constant of the G-E and hybrid composites. A similar observation is reported in the literature[25] on epoxy composites with E glass fiber.

ii) Effect of fillers

The effective dielectric constant of G-E and hybrid composites is governed by the interfacial polarization and electronic polarization of the composites. The number of dipoles that are oriented within the system determines its dielectric constant. With the addition of fillers into the epoxy, immobile nanolayers are formed due to the interaction between fillers and epoxy. This restricts the chain mobility and the orientation of the dipolar functional group is reduced, thereby reducing the dielectric constant.[26] The interfacial region between epoxy and fillers is responsible for the reduction in local density which in turn reduces the dielectric constant. Such a decrease in dielectric constant with the addition of nanofillers into epoxy was also observed in other studies[27] on epoxy with zinc oxide nanocomposites.

Apart from the phenomenon of chain immobilization, the inter-length scale between the epoxy chain and fillers is reduced due to the uniform distribution of fillers thereby reducing the wall-to-wall interface between the fillers. This results in the formation of a secondary epoxy chain of networks and causes entanglements.[28,29]

This further reduces the mobility of the epoxy chain leading to the reduction of the dielectric constant.

Thus, the reduction in the dielectric constant of G-EA, G-EAS, and G-EASA hybrid composites as compared to G-E composite can be summarized as due to i) micro/nanofillers' inherent dielectric constant, and ii) micro/nanofillers concentration iii) The other aspects is the density of fillers. The nanocomposites are also affected by drifting and ion migration which leads to the accumulation of hetero charges. Further, strong bonding and interaction between the micro/nanofillers and epoxy results in immobile nanolayers that restrict the movement of the end chain in the epoxy. Further, the silane treatment during fabrication would have resulted in a suitable interface with a lesser number of free ions and defects,[30,] and hence reduced interfacial polarization is observed. The OH radicals present due to the silane coupling agent restrict the molecular movement.

A similar comparison of dielectric permittivity of treated and untreated silica epoxy composites is reported.[31] It is suggested that treated nano-SiO_2 inhibits the molecular chain mobility and enhances the breakdown strength and also thereby improving the dielectric properties.

iii) Effect of water aging

From Figure 8.8b, it is observed that the dielectric constant has increased by 17% at 20 Hz and 14% at 10 MHz for water-aged samples compared to "as cast" samples. The reasons for the increase in dielectric constant with water aging are as follows:

(i) Presence of stronger dipole of water
(ii) Damage of epoxy main chains by absorbed water and increase in the segmental mobility by plasticization.[32]

The increase in dielectric constant is high in the case of water-aged G-E composite because of high void content; the water can sink within the voids and increase the dielectric constant.

8.3.7 DIELECTRIC LOSS

The variation of dielectric loss of "as cast" and water aged (500 h, 25 ± 3°C) samples with frequency are shown in Figures 8.9a and 8.9b.

i) Effect of frequency

As seen from Figure 8.9a, the dielectric loss values are slightly less in G-EA, G-EAS, and G-EASA composites as compared to G-E at all frequencies of measurements. The dielectric loss is reduced in the lower frequency range of 20–100 Hz and is almost constant in the frequency range of 100 Hz to 1 MHz and then starts increasing.

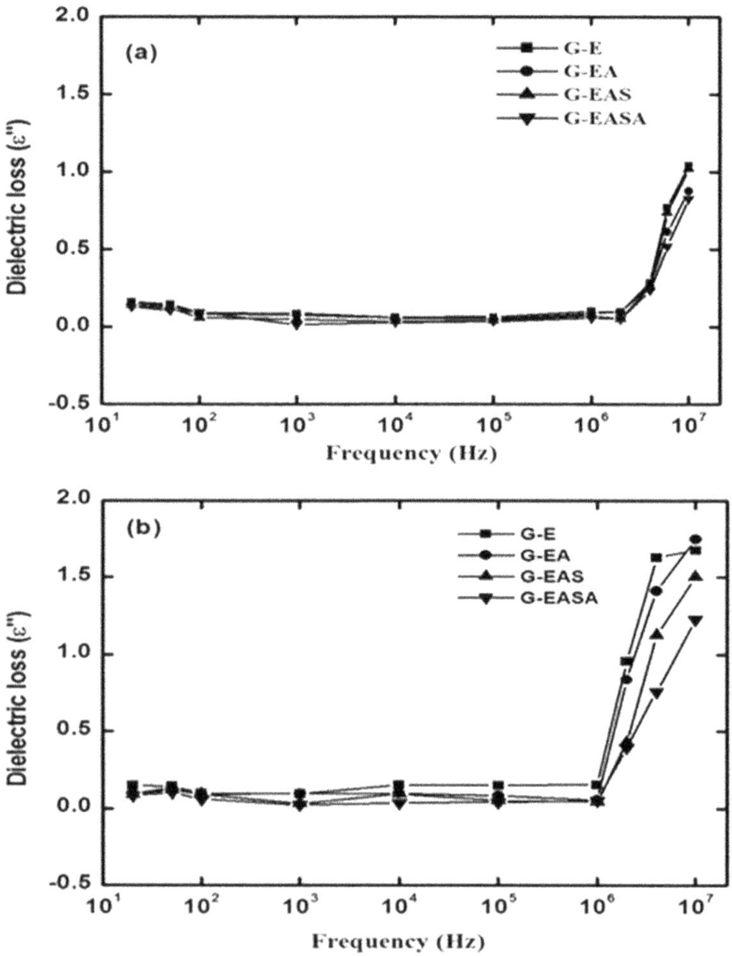

FIGURE 8.9 Variation of dielectric loss with frequency of G-E and hybrid composites: (a) as cast (b) water aged.

ii) Effect of fillers

The dielectric loss depends on the electrical conductivity of epoxy composites. This in turn depends on the number of charge carriers. The addition of nanofillers involves a large volume fraction at the interface and with the epoxy chain entanglements, the motion of the charge carriers is inhibited. This causes a reduction in the electrical conductivity.[33] The overlapping of the interfacial zone depends on the dispersion of micro/nanofillers in the G-E composite as well as on the size of the fillers. Uniform dispersion of fillers in the epoxy as evident from SEM images indicates no overlap of the interfacial area and hence there is no localized charge transport. This may be responsible for the reduction of dielectric loss of G-EA, and G-EAS.

The lower dielectric loss in G-EASA may be due to the inclusion of a microfilter, which changes the interaction zone and can manage the electrical charges better.

iii) Effect of water aging

In water-aged samples, an increase in dielectric loss is observed as depicted in Figure 8.9b. The variation of dielectric loss with frequency shows that it is not much affected up to 1 MHz. However, it is affected above 1 MHz in G-E and hybrid composites. The water molecule makes an angle at the tips with hydrogen atoms and the vertex with oxygen. Since oxygen has a higher electronegativity than hydrogen, the side of the molecule with the oxygen atom has a partial negative charge (dipole). The charge differences cause water molecules to be attracted to each other and another polar molecule,[34] the increase of percentage water absorption leads to an increase in the electrical conductivity.

A study on the effects of water treatment of alumina and aluminum nitride composites on dielectric properties is reported in the literature.[35] Experimental results have revealed that the aluminum nitride composites do not have significant water absorption and hence there are not many variations in DC conductivity. However, alumina composites have significant water absorption and the authors have studied that it results in broader dielectric relaxation and an increase in DC conductivity. A similar comparative study[36] on silicon nitride and silica composites has shown that silicon nitride exhibits improved properties as compared to silica composite.

The aging does not significantly affect the dielectric loss of the G-EASA composite due to the reduced water absorption of Al_2O_3 and ATH fillers. Further, the properties of SiO_2 nanofiller as a dielectric filler tend to improve the dielectric properties as a good insulator.

8.4 CONCLUSION

This study demonstrates the nano/microfiller combination of hybrid nanocomposites for insulating materials, the conclusions drawn are as follows:

i) Incorporation of micro/nanofillers reduces the conductivity thereby increasing the resistivity of the G-E composite.
ii) The PDC are reduced and charge carrier mobility is slightly increased with the incorporation of fillers compared to the G-E composite. Results also suggest that the spherical nanofiller particles will result in favorable interaction between fillers and epoxy matrices.
iii) Experiments carried out on water-aged samples have revealed reduced deterioration due to less water absorption when nano- and microfillers are included.
iv) Among all the fabricated composites the best results were obtained with G-E composite filled with micro/nanofiller (G-EASA) and hence considered as optimized composite.

ACKNOWLEDGMENT

The authors thank the Management and Principal of Vidyavardhaka College of Engineering, Mysuru, for their support and encouragement.

KEYWORDS

- **pultrusion**
- **charge carrier mobility**
- **polarization and depolarization current**
- **water absorption**

REFERENCES

1. Jamail, A. M.; Piah, M. A. M.; Muhamad, N. A.; Zainir, R. A.; bin Kasri, N. F.; Kamarudin, Q. E. DC Conductivity of Polymer Nanocomposites for Different Types and Amount of Nanofiller. *Int. J. Electr. Eng. Inform.* **2013,** *5,* 217–225.

2. Veena, M. G.; Renukappa, N. M.; Meghala, D.; Ranganathaiah, C.; Rajan, J. S. Influence of Nanopores on Molecular Polarizability and Polarization Currents in Epoxy Nanocomposites. *IEEE Trans. Dielec. Electr. Insul.* **2014,** *21,* 1166–1174.
3. Jamail, N. A. M.; Piah, M. A. M.; Muhamad, N. A. Comparative Study on Conductivity Using Polarization and Depolarization Current (PDC) Test. *J. Electr. Electron. Eng.* **2012,** *2* (4), 170–176.
4. Areef, S. R. The Study of Solution Absorption and Diffusion Coefficient in Epoxy Composite Reinforced with Glass Fibers. *Eng. Technol.* **2008,** *26,* 10.
5. Naceri, A. Moisture Diffusion Properties of Fabric Composite (Glass Fiber/Epoxy Resin). *Int. J. Eng. Trans. B: Appl.* **2008,** *22,* 205–210.
6. Sathyanarayanan, S. "Investigation of Moisture Diffusion Process in Fibre Reinforced Composites by Computing Diffusivity Factors Along 3-Perpendicular Directional Axes of Hot Wet Temperature Resin. *SSRG Int. J. Mech. Eng.* **2014,** *1* (2), 01–07.
7. Zou, C.; Fothergill, J. C.; Rowe, S. W. The Effect of Water Absorption on the Dielectric Properties of Epoxy Nanocomposites. *IEEE Trans. Dielec. Electr. Insul.* **2008,** *15,* 106–117.
8. Kumosa, L.; Benedikt, B.; Armentrout, D.; Kumosa, M. Moisture Absorption Properties of Unidirectional Glass/Polymer Composites Used in Composite (Non-Ceramic) Insulators. *Compos. A: Appl. Sci. Manuf.* **2004,** *35,* 1049–1063.
9. Tang, Y.; et al. Effect of Al_2O_3 with Different Nanostructures on the Insulating Properties of Epoxy-Based Composites. *Materials* **2020,** *13* (19), 4235.
10. Tang, Y.; Zhang, P.; Zhu, M.-X.; Li, J.; Li, Y.; Wang, Z.; Huang, L. Temperature Effects on the Dielectric Properties and Breakdown Performance of h-BN/Epoxy. *Compos. Mater.* **2019,** *12,* 4112.
11. Khattak, A.; et al. Effects of Compression and Silica Addition on the Dielectric Properties of Epoxy Composites. *Arab. J. Sci. Eng.* **2020,** *45* (8), 6741–6750.
12. Wang, Z.; et al. Simultaneously Enhanced Dielectric Properties and Through-Plane Thermal Conductivity of Epoxy Composites with Alumina and Boron Nitride Nanosheets. *Sci. Rep.* **2021,** *11* (1), 1–11.
13. Jamail, N. A. M.; Piah, M. A. M.; Muhamad, N. A.; Salam, Z.; Kasri, N. F.; Zainir, R. A.; Kamarudin, Q. E. Effect of Nanofillers on the Polarization and Depolarization Current Characteristics of New LLDPE-NR Compound for High Voltage Application. *Adv. Mater. Sci. Eng.* **2014,** 416–420.
14. Das-Guptha, D. K. Conduction Mechanisms and High High-Field Effects in Synthetic Polymers. *IEEE Trans. Dielec. Electr. Insul.* **1997,** *4,* 149–156.
15. Malec, D.; Truong, H.; Essolbi, R.; Hoand, T. G. Carrier Mobility in LDPE at High Temperature and Pressure. *IEEE Trans. Dielectr. Electr. Insul.* **1998,** *5,* 301–303.
16. Malec, D.; Truong, H.; Essolbi, R.; Hoand, T. G. Carrier Mobility in LDPE at High Temperature and Pressure. *IEEE Trans. Dielectr. Electr. Insul.* **1998,** *5,* 301–303.
17. Smith, R. C.; Liang, C.; Landry, M.; Nelson, J. K.; Schadler, L. S. The Mechanisms Leading to the Useful Electrical Properties of Polymer Anodielectrics. *IEEE Trans. Dielectr. Electr. Insul.* **2008,** *15,* 187–196.
18. Lau, K. Y.; Vaughan, A. S.; Chen, G.; Hosier, L.; Holt, A. F. Absorption Current Behavior of Polyethylene/Silica Nanocomposites. *J. Phys: Dielectr. Conf. Series* **2013, 472,** 1–6.
19. Smith, R. C.; Liang, C.; Landry, M.; Nelson, J. K.; Schadler, L. S. In *Studies to Unravel Some Underlying Mechanisms in Nanodielectrics*, IEEE Conference. Electr. Insul. Dielectr. Phenomena (CEIDP), **2007,** 286–289.

20. Kumosa, M.; Narayan, H. S.; Qiu, Q.; Bansal, A. Brittle Fracture of Non-Ceramic Suspension Insulators with Epoxy Cone End-Fittings. *Compos. Sci. Technol.* **1997,** *57,* 739–751.
21. Espert, A.; Vilaplana, F.; Karlsson, S. Comparison of Water Absorption in Natural Cellulosic Fibers from Wood and One Year Crops in Polypropylene Composites and Its Influence on Mechanical Properties. *Compos. A: Appl. Sci. Manuf.* **2004,** *35,* 1267–1276.
22. Lin, Q.; Zhou, X.; Dai, G. Effect of Hydrothermal Environment on Moisture Absorption and Mechanical Properties of Wood Flour Filled Polypropylene Composites. *J. Appl. Polym. Sci.* **2002,** *85,* 2824–2832.
23. Barsberg, S.; Thygesen, L. G. Nonequilibrium Phenomena Influencing the Wetting Behavior of Plant Fibers. *J. Colloid Interface Sci.* **2001,** *234,* 59–67.
24. Pavlidou, S.; Papaspyrides, C. D. The Effect of Hydrothermal History on Water Sorption an Interlaminar Shear Strength of Glass/Polyester Composites with Different Interfacial Strength. *Compos. A Appl. Sci. Manuf.* **2003,** *34,* 1117–1126.
25. Ali, N. A.; Noori, F. T. M.; Hussin, S. I. Dielectric Properties of Industrial Polymer Composite Materials. *J.. Al Univ.* **2010,** *13,* 73–77.
26. Wang, Q.; Chen, G.; Alghamdi, A. S. Influence of Nanofillers on Electrical Characteristics of Epoxy Resins Insulation. *IEEE Intl. Conf. Solid Dielectr.* **2010,** 1–4.
27. Singha, S.; Thomas, M. J. Influence of Filler Loading on Dielectric Properties of Epoxy-ZnO Nanocomposites. *IEEE Trans. Dielectr. Electr. Insul.* **2009,** *16,* 531–542.
28. Picu, R. C.; Ozmusul, M. S. "Structure of Linear Polymeric Chains Confined Between Impenetrable Spherical Walls. *J. Chem. Phys.* **2003,** *118,* 11239–11248.
29. Sternstein, S. S.; Zhu, A. J. Reinforcement Mechanism of Nanofilled Polymer Melts as Elucidated by Nonlinear Viscoelastic Behavior. *Macromolecules* **2002,** *35,* 7262–7273.
30. Roy, M.; Nelson, J. K.; MacCrone, R. K.; Schadler, L. S.; Reed, C. W.; Keefe, R. Polymer Nanocomposite Dielectrics—the Role of the Interface. *IEEE Trans. Dielectr. Electr. Insul.* **2005,** *12,* 629–643.
31. Wang, Q.; Chen, G. Effect of Pre-treatment Nanofillers on the Dielectric Properties of Epoxy Nanocomposites. *IEEE Trans. Dielectr. Electr. Insul.* **2014,** *21,* 1809–1816.
32. Reid, J. D.; Lawrence, W. H.; Buck, R. P. Dielctric-Properties of an Epoxy Resin and its Composite 1, Moisture Effects on Dipole Relaxation. *J. Appl. Polym. Sci.* **1986,** *31,* 1771–1784.
33. Kattayam, J.; Ohki, Y.; Fuse, N.; Kozako, M.; Tanaka, T. Effect of Nanofiller Materials on Dielectric Properties of Epoxy Nanocomposites. *IEEE Trans. Dielectr. Electr. Insul.* **2013,** *20,* 157–165.
34. Singha, S.; Thomas, M. J. Permittivity and Tan Delta Characteristics of Epoxy Nanocomposites in the Frequency Range of 1 MHz-1 GHz. *IEEE Trans. Dielectr. Electr. Insul.* **2008,** *15,* 2–11.
35. Hosier, I. L.; Praeger, M.; Vaughan, A. S.; Swingler, S. G. The Effects of Water on the Dielectric Properties of Aluminium based Nanocomposites. *IEEE Trans. Nanotechnol.* **2017,** *16* (4), 667–676.
36. Hosier, I. L.; Praeger, M.; Vaughan, A. S.; Swingler, S. G. The Effects of Water on the Dielectric Properties of Silicon based Nanocomposites. *IEEE Trans. Nanotechnol.* **2017,** *16* (2), 169–179.

CHAPTER 9

Ni/Co/Cu and TiO_2 Incorporate Ragi Husk-Derived Biosilica Nanocomposite as Photocatalyst and Adsorbent for the Degradation of Methyl Blue Dye

A. ANANTHI[1], D. GEETHA[1], and P. S. RAMESH[2]

[1]*Department of Physics, Annamalai University, Tamil Nadu, India*

[2]*Thiru Kolanjiappar Government Arts College, Virudhachalam, Tamil Nadu, India*

ABSTRACT

Transition metals (Ni/Co/Cu) and TiO_2 incorporate ragi husk derived silica have been synthesized by sol–gel route using $NaNo_3$ as the precipitant. The samples were named as RaHA silica; RaHA–TiO_2 and Ni/Co/Cu doped RaHA–TiO_2 are the prepared nanocomposites. Calcinations of RaHA–TiO_2 at 700 °C gave spherical SiO_2–TiO_2 nano structure clusters. The structure identification is made from the X-ray diffraction pattern analysis and comparing it with the internationally recognized database containing the reference pattern (JCPDS). Strong diffraction peaks at 25° and 48° indicate TiO_2 in the anatase structure. Silica can exhibit the formation of anatase by impeding direct contact between TiO_2 particles forming the Ti–O–Si binding. The SEM images of RHS/TiO_2/Ni nanocomposites are mono dispersive and highly crystalline. The EDX spectrum for RHS/TiO_2/Co nanocomposites reveals that Ti, O, Si, and Co are the only elementary species in the sample. Transition metals (Ni/Co/Cu) doped SiO_2–TiO_2 sample shows the highest photocatalytic degradation of MB. The adsorption of MB on these catalysts was found to be a second-order kinetic model.

Bio-Based Polymers and Composites: Properties, Durability, and Applications. Amadou Belal Gueye, Sabu Thomas, Nandakumar Kalarikkal, & Modou Fall (Eds.)

9.1 INTRODUCTION

Waste water treatment is a severe problem in most of the countries. The dying industry is the main source of polluters, discharging organic pollutants causing environmental disturbances.[1,2] The different wellsprings of contaminations are the material, kicking the bucket, printing, restorative, food clothing, paramedical, engineered, cleansers, tanneries, and paper enterprises. The modern materials utilize not just influence the climate change but also cause the color of water environment.[3] The bacterial debasement of these colors is harmful; many colors are cancer-causing, mutagenic, and tetragonal on both sea-going and human existence.

Severe standards and guidelines have been implemented in numerous nations on squandering water releases to limit the obliteration of the climate. A number of wastewater treatments are currently practically speaking viz., physical, synthetic, and organic techniques. Among these strategies, the conventional actual methods of adsorption are by and large effective on the evacuation of MB. The compound oxidation techniques are viable for many organics, and they can bring about complete mineralization of natural contaminations. Among the numerous photocatalyses, TiO_2 acquired extensive consideration because of its photograph soundness, non-poisonousness, minimal expense, synthetic and organic latency, and warm dependability. Because of its wide band hole anatase type of TiO_2 is more interesting.

The little band hole uses just a tiny part of the sun-powered range. A few strategies have been created to build the photocatalytic movement of TiO_2 by expanding the surface region and the planning blended materials. Hardly any examination bunches attempted that blended metals oxide impetus utilizing titanium and tin oxide might enhance the photocatalytic movement.

This work is expected to create a new photocatalyst utilizing SiO_2 and TiO_2 nanocomposites that have improved photocatalytic properties than unadulterated TiO_2 and SiO_2 nanostructures. It has been accounted for that surface bond formed SiO_2/TiO_2 photocatalyst has higher synergist conduct than TiO_2 in the expulsion of azo dyes. The joining of TiO_2 with bio SiO_2 is a powerful way to deal with work on the photocatalytic movement in light of the fact that the stage change from anatine to retile is restrained because of the upgraded warm steadiness of TiO_2/SiO_2 blended oxide framework. In this work, the photocatalytic removal of MB in fluid arrangement over titanium and change metals (Ni/Co/Cu) integrated silica as the impetus.

Silica from ragi husk offers a modest option in contrast to business silica. Simultaneously, the use of ragi ball contributes to cleaner environment, also many investigations have revealed the use of agrobio-derived silica as a boost medium. Since the impetuses depicted having contained silica the chance of expulsion of MB by the peculiarity of adsorption cannot be precluded. Ragi husk has been accounted for as modest adsorbed for the expulsion of weighty metals containing viz. Cd, Ni, Zn, and Pb and furthermore for the evacuation of poisonous colors congo red, indigo carmine, and green. This exploration portrays the utilization of Cu/Ni/CO on SiO_2:TiO_2 as a photocatalyst as well as an adsorbent for the expulsion of MB from fluid media.

9.2 EXPERIMENTAL DETAIL

9.2.1 *MATERIALS*

Ragi husks were collected from the Faculty of Agriculture, Botanical Garden of Horticulture Department, Annamalai University, India. Titanium isopropoxide ($Ti(OC_3H_7)_4$ 97%) was used as a precursor of titania, nickel nitride ($Ni(NO_3)_2.6H_2O$), cobalt (II) nitride ($Co(NO_3)_2.6H_2O$), and copper (II) nitride ($Cu(NO_3)_2.6H_2O$) were used as dopant sources. Other materials used were nitric acid, sodium hydroxide pellets and methylene blue (MB). They all were purchased from sd fine Mumbai and used directly without further purification Methanol and distilled water were used as solvents.

9.2.2 *PREPARATION OF BIO ASH*

The bio debris of chosen bio deposits is cleaned and washed with refined water. The cleaned 100 gm husk was stunned in 250 mL of 3M HNO_3 and the blend was mixed for 3 hours at 800 °C temperature. After the reflux, the solids are washed with deionized water until unbiased (acid-free) and dried at a temperature of 800 °C in short term. Husk was dried and afterward calcined at a temperature of 7000 °C for 5 h.

9.2.3 *PREPARATION OF SILICA GEL*

0.5 g bio debris was blended with 250 mL of 5 N sodium hydroxide fluid arrangement. The combination was refluxed for 5 h at 800 °C. The arrangement was sifted and put away in a cooler for additional use. A 5 mL example of sodium silicate arrangement was taken from the stock arrangement and filled with 15 mL refined water at once for a particular time frame. The pH of the arrangement was acclimated to 10–12 by the drop-wise expansion of hydrochloric corrosive under blending velocity to frame silica gel. The time needed for gelation is not less than five minutes after the sol arrangement was killed. The tumult was gone on for another hour to get wet gel nanoparticles. The prearranged gel was matured at 7000 °C for 5 h under settled conditions. Refined water, NaCl, and different pollutions were isolated from the gel by dissolvable trade with weakened and unadulterated methanol 4–5 times. The weak methanol (75%) was utilized to keep away from any primary shrinkage during the dissolvable trade. In total, methanol trade was completed multiple times and matured at 7000 °C. The complete season of maturing was 52 h. The wet gels were then dried. The end result is named bio-silica.

$$SiO_2\ (Ash) + 2NaOH\ Na_2SiO_3 + H_2O \tag{3.1}$$

$$Na_2SiO_3 + H_2SO_4\ SiO_2 + Na_2SO_4 + H_2O \tag{3.2}$$

During the addition of sulfuric acid, Si-OH formed and condensed to form siloxane bond as shown below:

$$Si - OH + HO - Si \longrightarrow Si - O - Si + H_2O \tag{3.3}$$

9.2.4 SYNTHESIS OF TITANIA-DOPED BIO SILICA

The synthesis of titania incorporated on silica nanocomposites was prepared by the sol–gel method. The molar ratio of silica to titania sols was 2.5:1. The stock solution of silica from bio ash was added to titania solution while stirring for 5 h until a gel was formed. The gel was aged for 24 h and then rinsed with water followed by heating at 80 °C for 5 h. Finally, the solution was filtered at Whatmann filter paper and dried at an oven. The final product was labeled as $SiO_2 - TiO_2$.

9.2.5 SYNTHESIS OF (Ni/Co/Cu) DOPED SiO_2/TiO_2 NANOCOMPOSITES

The base was prepared (2.5:1) using bio silica and TiO_2 and stirred for 30 minutes, 1.5 mM of nickel, cobalt, and copper were added onto the base immediately, and the gel was formed. The wet gel was dried at hot air oven.

9.2.6 CHARACTERIZATION

The crystalline structure delivered was X-Ray diffraction tests of rice husk debris were performed with a Rigaku mini flex diffract meter for 2 qualities from 10o to 90o utilizing Cuk focus at a frequency of 1.5406. Our examples were described by checking electron microscopy (SEM) (Jeol-JSM-6360LV) at a voltage of 20 kV and recently covered with gold under vacuum by faltering utilizing a BAL-TEC'S device. Auxiliary electron pictures were acquired and basic examination was concentrated on utilizing EDX. Substance holding states were researched by X-beam photoelectron spectroscopy XPS (Kratos Analytical). The absorbance spectra of impetuses were estimated by UV-Visible Spectrophotometer (Hitachi, UV3501) at 200–800 nm frequency and within the sight of distilled water as a standard material.

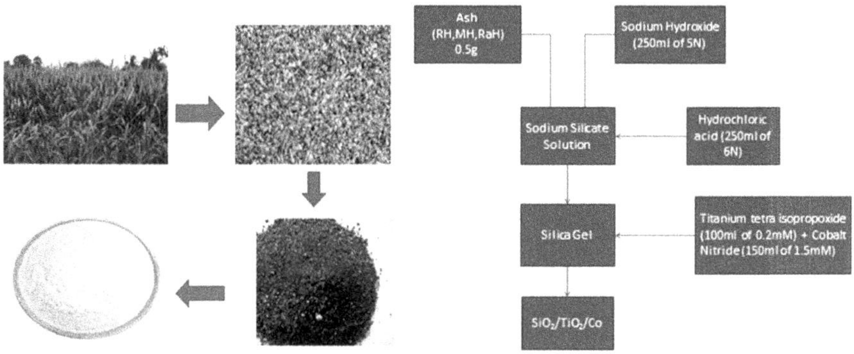

Flow chart for the synthesis of Ni/Co/Cu doped on bio silica: TiO_2.

9.3 RESULTS AND DISCUSSION

9.3.1 STRUCTURED ANALYSIS (XRD)

Figure 9.1(a-d) shows the XRD pattern of ragi husk silica incorporated with TiO_2 and doped with transition metal viz., Ni, Co, and Cu measured at a 2Θ of 10°–80°. It is seen that the nanocomposites have the characteristic of weak anatase crystalline phase. Figure 9.1(a) shows that the diffraction pattern of nickel oxide is presented. In fact, only the Ni-doped RaHS/TiO_2 displays intense and well-resolved NiO particles. The XRD pattern for cobalt-doped RaHS/TiO_2 is shown in Figure 9.1(b). The broad peak at low angles might show the presence of SiO_2. By examination with the JCPDS information base for SiO_2, the normal outcomes are gotten. The tops at 310 contain a hint of SiO_2 alongside the top at 350. Be that as it may, both these pinnacles are exceptionally veiled by the cobalt oxide tops in a similar region. Figure 9.1(c) shows RaHS-TiO_2 doped copper (1.5 mM) nanoparticles.

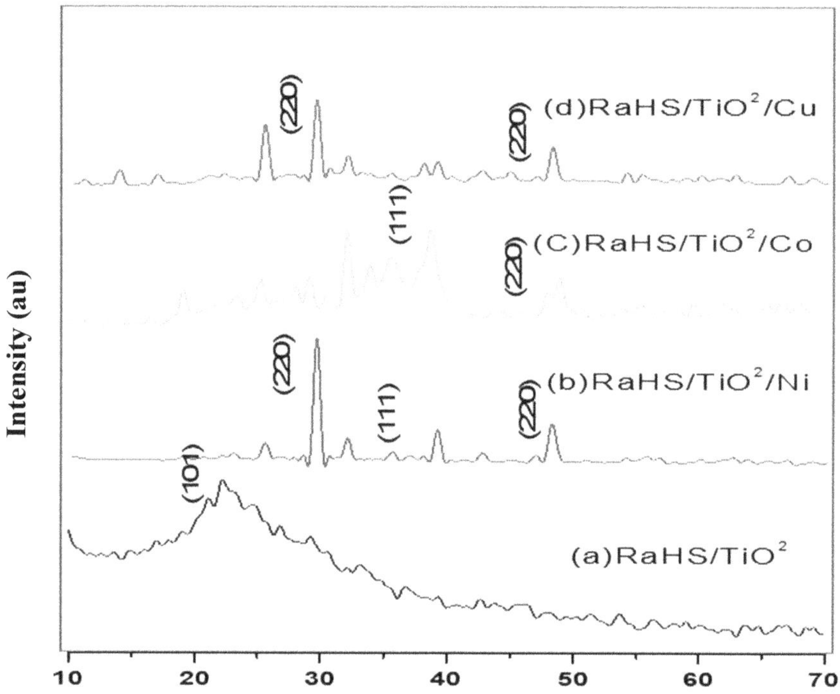

FIGURE 9.1 XRD pattern of a (a) RaHS/TiO_2, (b) RaHS/TiO_2/Ni, (c) RaHS/TiO_2/Co, (d) RaHS/TiO_2/Cu.

The peak in diffraction pattern exhibits the cubic structure of copper oxide plane (111) and the angle $2\theta = 43.41°$ (JCPDS 04 -0836) and also determined the crystalline size as 25 nm. Therefore, the extent of structure defection increased with increasing Cu concentration. The intensity of XRD peaks of the sample replicates that the formed nanoparticles are crystalline, and the broad diffraction peaks indicate very small size crystallite.[4] A small peak is also obtained at around 29° specifying that a small amount of copper was oxidized and converted into copper oxide.

TABLE 9.1 Particle Size and Dislocation Density of the Transition Metals (Ni/Co/Cu) doped RaHS/TiO_2 Nanocomposites.

Sample Name	Particle size nm	Strain	Dislocation Density (m^2)	Morphology Index
RaHS/TiO_2/Ni	38.30	0.008	0.743×10^{-15}	0.29 varies from 0.09
RaHS/TiO_2/Co	28.33	0.009	0.755×10^{-15}	0.52 varies from 0.35
RaHS/TiO_2 /Cu	25.26	0.005	0.380×10^{-15}	0.79 varies from 0.89

9.3.2 MORPHOLOGICAL ANALYSIS

The SEM images confirm the presence of particles of microsized diameter, high homogeneity, almost spherical shape and show a small tendency to form agglomerate structures in Figure 9.2(e). Figure 9.2(f) shows the elemental signature presence of Ti, O, and Si.

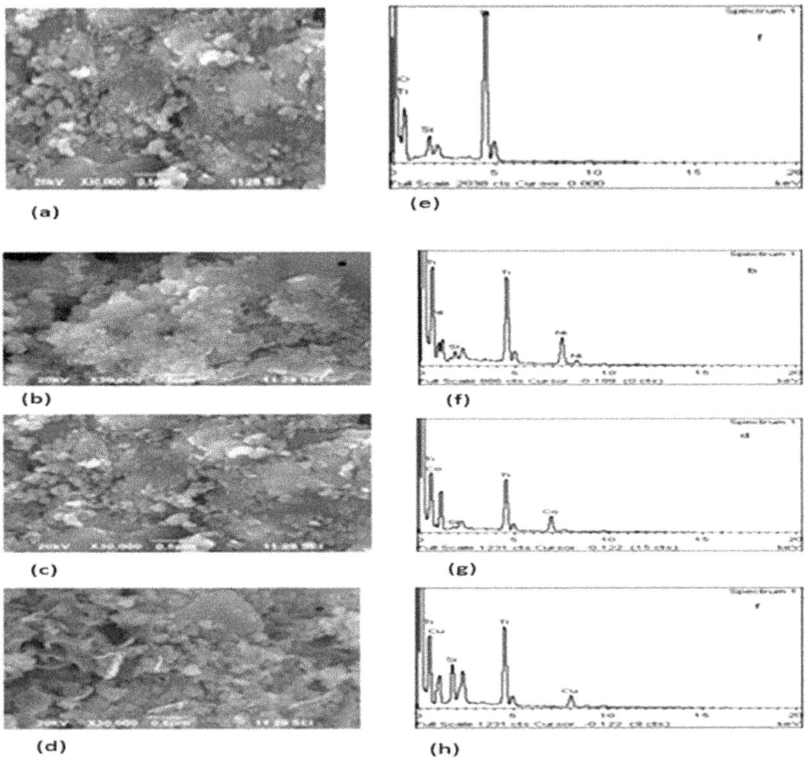

FIGURE 9.2 SEM image of (a) RaHS/TiO_2, (b) Ni-RaHS/TiO_2, (c) Co-RaHS/TiO_2, (d) Cu-RaHS/TiO_2 and their respective EDS spectrum in (e)–(h).

Figure 9.2(a,c,e) displays ragi husk silica with TiO_2-doped transition metals such as Ni, Co, and Cu. The SEM image shows that the particles are spherical in shape and are polycrystalline. All the samples show nearly the same morphological features. This infers that the addition of transition metal does not affect the crystal structure which confirms the incorporation of transition metals on the RaHS/TiO_2 lattice site.[5] The EDX spectra for RaHS/TiO_2/Ni, RaHS/TiO_2/Co, and RaHS/TiO_2/Cu nanocomposites respectively are shown in Figure 9.2(b,d,f). EDX spectra reveal that Si, Ti, Ni, Co, and Cu are the only elements in the sample. Moreover, no additional peaks corresponding to any other element are observed. These results also provide evidence that the transition metals Ni/Co/Cu doping suppressed the formation of the SiO_2/TiO_2 which inhibited the agglomeration of SiO_2/TiO_2.[6]

9.3.3 SURFACE ANALYSIS (XPS)

Figure 9.3(a) shows the XPS survey scan spectrum of nickel-doped RaHS/TiO_2. In order to check the valence states of Ni ions XPS measurements are carried out. The core level binding energy of $Ti2p_{3/2}$ is of about 458.4 eV and that of $Ti2p_{1/2}$ is about 464.16 eV. Additionally, Ni doping may also result in oxygen vacancy content due to the replacement of Ti^{4+} by Ni^{2+} ions.[7] The main peak moved to a higher binding energy of 900.9 and 904.5 eV due to the Ni-O bond formed, leading to the oxidation state of Ni^{2+} and Ni^{3+}. Figure 9.3(a) shows XPS spectra of Ti2p survey scan spectrum of copper-doped RaHS/TiO_2. Two pairs of peaks for silicon (Si 2s and 2p) and only one main peak for titanium (Ti 2p doublet) were detected in the spectra. The binding energy of the Si 2p peak arranged in 104.96 eV in Figure 9.3(c) may testify the formation of the Ti-O-Si bonding.

It can be seen that there are two peaks, showing that there are two types of copper species on the surface. Cu2O has a binding energy of 941.12 eV, whereas CuO has a lower binding energy of 936.10 eV.[8] The key difference between Cu^+ and Cu^{2+} species was the prominent satellite peak present on the high binding energy sides.

9.3.4 PHOTOCATALYTIC ACTIVITY OF COPPER-DOPED RAHS/TIO_2

The benefit of transition metals doping is improved trapping of electrons to inhabit $e^- - h^+$ recombination during irradiation. Decrease in charge carriers recombination results in enhanced photoactivity. The photo physical mechanism of doped TiO_2 is not yet understood, but the p-type metal ion dopants (with valences lower than that of Ti^{4+}) are believed to act as acceptor centers as opposed to the n-type. Metal ion implantation with ions such as copper was found to cause large shift in the absorption band of titanium dioxide toward the visible region. There is no red shift in the case of TiO_2 implanted with Si which indicates that the bathochromic shift is not due to the implantation itself.

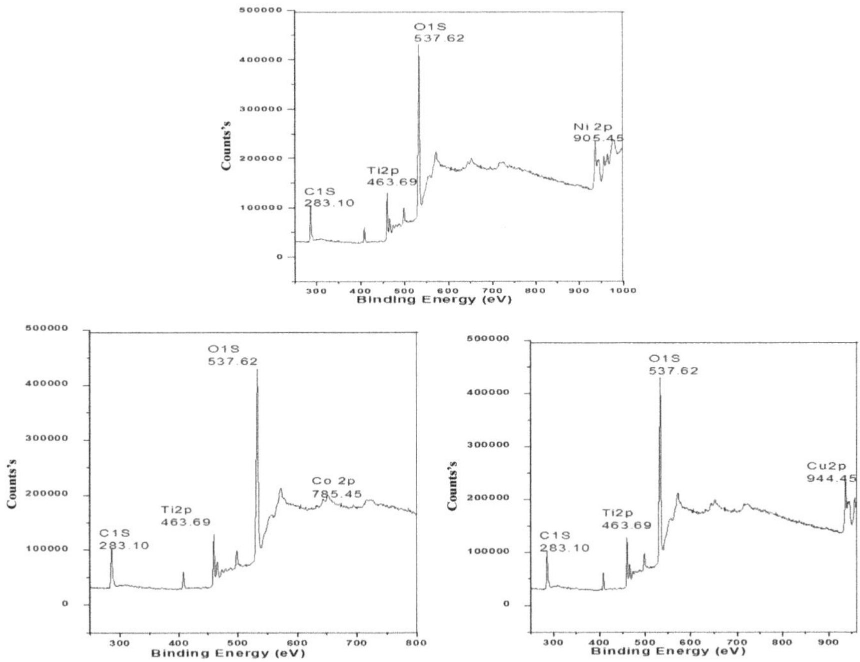

FIGURE 9.3 XPS spectrum of survey spectrum of (a) RaHS/TiO_2/Ni, (b) RaHS/TiO_2/Co, (c) RaHS/TiO_2/Cu.

The relevant reactions at the CuO surface causing the degradation of dyes can be expressed as follows:

$$CuO + h\nu \text{ (Sunlight)} \longrightarrow CuO\ (e^-_{CB} + h^+_{VB})$$

$$CuO\ (h^+_{VB}) + H_2O \longrightarrow CuO + H^+ + OH^\cdot$$

$$CuO\ (h^+_{CB}) + OH^- \longrightarrow CuO + OH^\cdot$$

$$CuO\ (e^-_{CB}) + O_2 \longrightarrow CuO + O_2^{\cdot}$$

$$O_2^{\cdot} + H^+ \longrightarrow HO_2^{\cdot}$$

Dye + OH ⟶ Degradation products

Dye + h^+_{VB} ⟶ Oxidation products

Dye + e^-_{CB} ⟶ Reduction products

Initially dye is not decolored. As time increased to half an hour, the degradation started. Moreover, the maximum decolorization was observed at 180 minutes. Temporal variation of spectral changes as a function irradiation of time is shown in Figure 9.4 for MB $^+$RaHS/TiO_2/Cu nanocomposites.[9] From the change in intensity as observed, it can be seen that the nanocomposites could reach maximum degradation for 180 minutes of sunlight irradiation. The reaction takes places during the degradation of MB.

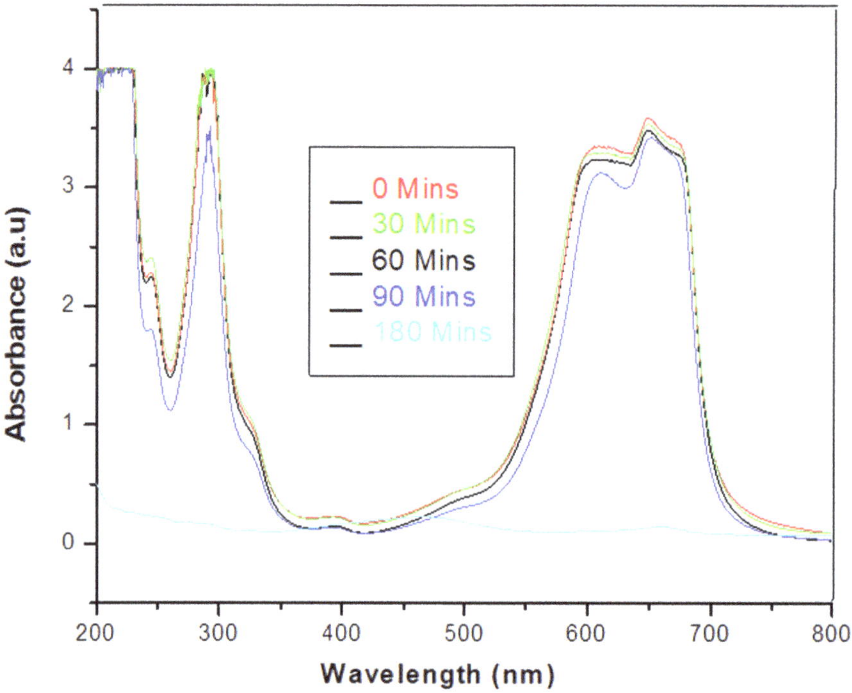

FIGURE 9.4 Photocatalytic degradation of methylene blue by RaHS/TiO_2/Cu under sunlight at different time intervals.

Adsorption mechanism is also involved within the photocatalysis process, but the degradation rate from the adsorption process is not significantly different from the photocatalysis. It means that photon energy is not sufficient to preserve hydroxyl radicals to oxidize MB in the solution. Dye degradation of methylene blue dye was investigated using biosynthesized SiO_2:TiO_2 nanoparticles by dark light at different time intervals (0, 30, 60, and 90 minutes), as shown in Figure 9.4(a). In RaHS/TiO_2 maximum dye degradation was removed by 60 min. The complete removal observed in a very short time can be related to the strong interactions between the MB molecules and the catalysts surface.

In Figure 9.5 methylene blue showed a quick biosorption rate during the initial 30 minutes of contact time because of incredible accessibility of surface region/restricting destinations or enormous number of empty surface locales for color particles to be biosorbed. For the most part color anions will tie to every one of the dynamic locales until they are completely involved or until it achieves harmony.[10] Subsequently, with time, less dynamic locales are accessible and hence decrease how much color being adsorbed or after a pass of time the excess empty surface destinations are hard to be involved because of shocking powers between the solute particles on the strong and mass stages.[11] At the point when the time is expanded, evacuation of productivity is high.

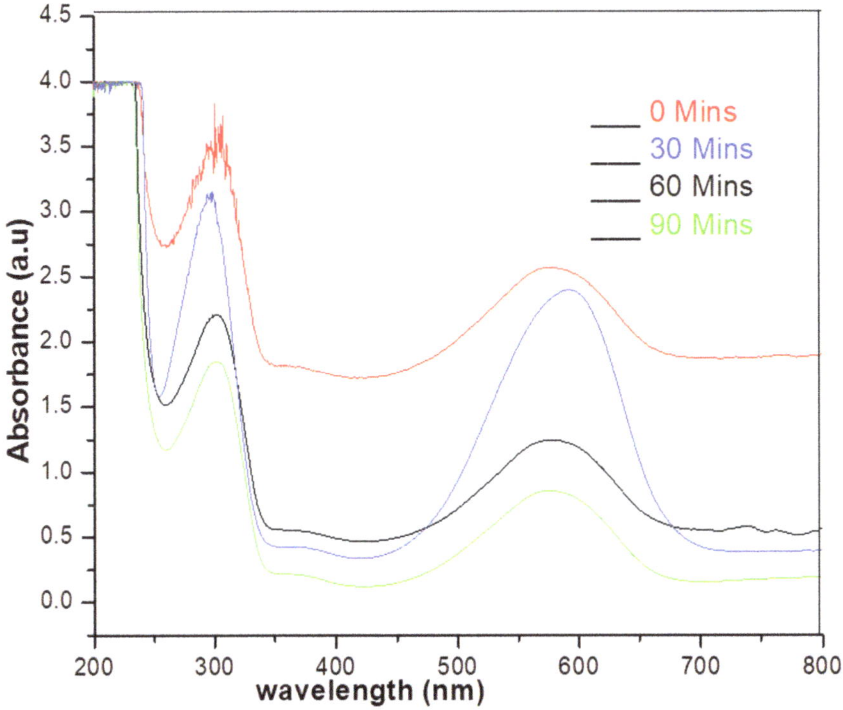

FIGURE 9.5 Adsorbent activity of methylene Blue by RaHS/TiO_2/Cu under dark light at different time intervals.

Mechanism during effect of transition metals (Ni/Co/Cu) doping bio silica incorporated TiO_2 on adsorption of MB:

a) UV radiation can excite electron from the valence band of bio silica: TiO_2 to its conduction band, but due to the recombination of electron–hole pair, the adsorbent activity is low.
b) Doping transition metals into bio silica TiO_2 entraps the excited electron thereby electron–hole recombination.

9.4 CONCLUSIONS

The broad absorption band was found, when bio SiO_2 was synthesized using agro waste (RaH). The highest wavelength UV absorption was obtained for the sample RaHS: TiO_2 (2.5:1) having peak at approximately 310 nm. Many reports concluded that SiO_2 causes the blue shift of TiO_2 by two mechanisms: (i) Reduced particle size of TiO_2 due to silicon framework and (ii) Forming Ti–O–Si bonds leading to the modification of electronic structure.

The absorption spectra of the resulting metal-doped RaHS/TiO_2 nanocomposites showed an increased shift in visible light in comparison with the RaHS/TiO_2. The defects created in the titania network and crystalline structure are responsible for change in the band gap energy. UV–Vis spectra for pure and doped SiO_2/TiO_2 nanocomposites showed absorption peaks slightly shifted to longer wavelength by introducing doping metal ions (Ni^{2+}, Co^{3+}, and Cu^{2+}).

Methylene blue was degraded using transition metal-doped SiO_2/TiO_2 nanocomposites under sunlight irradiation. This study reveals that the photocatalytic activity of the catalysts depends on the metal doping (Ni, Co, and Cu) on SiO_2/TiO_2. The highest degradation efficiency was observed for nickel-doped rice husk silica, cobalt-doped maize husk silica, and copper-doped ragi husk silica. In comparison of all the prepared catalysts, maximum degradation occurred in copper-doped ragi husk silica.

The enhanced photocatalytic activity of transition metal-doped SiO_2/TiO_2 nanocomposites catalyst with increase in pH could be due to reduced hydrogen-related defects, which may have potential applications in the future. It was found that the transition metal-doped SiO_2/TiO_2 nanocomposites prepared by the sol–gel method played a key role in photocatalytic performance through studying the degradation efficiency of transition metal-doped SiO_2/TiO_2 nanocomposites catalyst.

- Adsorption studies showed an effective removal of MB in aqueous medium.
- The removal efficiency was high at smaller particle size of adsorbent material, because as the particle size decreases, the surface area increases and this leads to high removal efficiency.

KEYWORDS

- **ragi husk silica**
- **methylene blue**
- **Ni/Co/Cu-SiO_2**
- **TiO_2 nanocomposite**

REFERENCES

1. Laribi, M.; Bachari, K.; Chebout, R.; Touati, M. Synthesis Characterization and Catalytic Application of the Nickel Containing Mesoporous Silica. *J. Assoc. Arab Univ. Basic Appl. Sci.* **2012,** *12,* 42–47.
2. Hu, G.; Taylor, A. B.; Hart, P. J.; McAlister-Henn, L. Crystallization and Preliminary X-ray Crystallographic Analysis of Yeast NAD+ Specific Isocitrate Dehydrogenese. *Acta Crystallogr. Sect. F. Struct. Biol. Cryst. Commun.* **2005,** *61,* 489–488.
3. Greene, D.; Gareia, R. S.; Govan, J.; Gun'Ko, Y. K. Systhesis Characterization and Photocatalytic Studies of Cobalt Ferrite Silica—Titania Nanocomposites. *Nanomaterials* **2014,** *4,* 331–343.
4. Zhang, J.; fu, H.; Lv, X.; Tang, J.; Xu, X. Removal of Cu (II) from Aqueous Solution Using the Rice Husk Carbons Prepared by the Physical Activation Processes. *Bioenergy* **2011,** *35,* 464–472.
5. Koremsan, H.; Ghasemi, Z.; Sei, H. Y. 'Synthesis of Nanozeolite Sodalite from RHA without Organic Additives. *J. Chem. Eng.* **2011,** *89* (3), 601–608.
6. Khansanlou, R.; Ahmed, M. B.; Shameliand, K.; Kalanalai, K. Synthesis and Characterization of Rice Straw/Fe3O4 Nanocomposite by Aquick Precipitation Method . *Molecules* **2013,** *18,* 6597–6607.
7. Kundavoic, L., Stephanopoulos, M. F. Cu and Ag Modified Cerium Oxide for Methane Oxidation. *J. Catal.* **1998,** *179,* 203–221.
8. Yang. J., Lu, Y.; Han, R.; Wang, Y.; Han, P.; Shi, J. Removal of MB from Aqueous Solution by Chaff in Batchmode. *J. Hazad Mater.* **2004,** *137*, 550–557.
9. Oluwafunmilola, O.; Maroto-Valer, M. M. Copper based Tio_2 Honeycomb Monoliths for Co_2 Photoreduction. *Catal. Sci. Technol.* **2014,** *4,* 1631–1637.

10. Farook, A.; Appaturi, J. N. A Facile and Efficient Synthesis of Styrene Carbonate via Cycloaddition of CO_2 to Styreneoxide over Ordered Mesoporous MCM-41-1mi/Br Catalyst. *Appl. Catal. B: Environ.* **2013,** *136,* 150–159.
11. Yang. J., Lu, Y.; Han, R.; Wang, Y.; Han, P.; Shi, J. Removal of MB from Aqueous, Solution by Chaff in Batch Mode. *J. Hazad Mater.* **2004,** *137,* 550–557.

CHAPTER 10

Effect of Zinc Oxide Nanoparticles on Polymerization Shrinkage and Mechanical Properties of ORMO-48

LIZYMOL P. P., VIBHA C., DEEPU D. R., and SONALILAXMAN WAGHMARE

Division of Dental Products, BMT Wing, Sree Chitra Tirunal Institute of Medical Sciences and Technology (An Institute of National Importance under D.S.T., Govt. of India), Thiruvananthapuram, Kerala, India

ABSTRACT

The present study involves the investigation of effect of zinc oxide nanoparticle on polymerization shrinkage and diametral tensile strength (DTS) of ORMO-48. ORMO-48 is an organically modified ceramic resin-based aesthetically acceptable visible light cure (VLC) dental composite which undergoes rapid curing with minimum shrinkage and better cytocompatibility.

10.1 INTRODUCTION

The focus of conservative dental materials research is on developing materials with better biocompatibility, radiopacity, antimicrobial properties, excellent shade matching, enhanced mechanical properties, good bioactivity, and low polymerization shrinkage compared to Bis–GMA-based materials. Bis–GMA free novel resin materials are attempted to overcome this difficulty. Faster curing, better depth of cure, biocompatibility, and superior mechanical properties compared to contemporary materials made the composite the material of choice today. Composites have been considered the mainstay treatment modality for treating damaged tooth since last few decades; although

Bio-Based Polymers and Composites: Properties, Durability, and Applications. Amadou Belal Gueye, Sabu Thomas, Nandakumar Kalarikkal, & Modou Fall (Eds.)

it is having reduced longevity due to high fracture rate, marginal leakage associated with secondary caries and bacterial proliferation on the surface of the restoration.[1] Secondary caries and *S. mutans* are the major reason for the failure of dental resin composite restoration and its replacement.[2]

The use of multifunctional alkoxy-modified inorganic organic hybrid resin as the monomer in dental composite preparation is a recent strategy for reducing the polymerization shrinkage of dental restorative composite. Incorporation of hybrid fillers including bioactive hydroxy apatite (HAP), pyrogenic silica, silanated quartz, and radiopaque glass alone or in combination was attempted to improve the mechanical properties, bioactivity, and radiopacity. Many attempts were going on worldwide to develop aesthetically acceptable visible light cure (VLC) dental composite which undergoes rapid curing with minimum shrinkage and better cytocompatibility along with antimicrobial properties.[3–19] However, recent studies pay attention to the antimicrobial activity of dental composites to reduce the threat of recurrent decay. To impart the antibacterial activity into the resin-based dental composites and adhesives, various approaches are possible.[20–23] The first approach involves the incorporation of antibacterial agents like fluoride and chlorhexidine into the resin matrix.[20,21] They release the antimicrobial agents over the bacterial growth; however, it has reduced life span. The second approach is the incorporation of quaternary ammonium functionality in the resin matrix/additive which is one of the promising methods for improving the life span of the antimicrobial activity of the dental composites.[22] The final approach is the inclusion of metal oxides/ions into the dental composites.[23] Salts of silver and copper as antimicrobial precursor were attempted in composite.[24] Though the shelf life, toxicity, and mechanical strength of the dental composites are not commendable, the leachable low molecular organic compounds exhibited instant antibacterial effect.[25]

ZnO particles have proven antimicrobial properties which seek greater attention due to their stability in assorted environmental conditions and their green synthesis at low temperature.[26] Several microorganisms in the oral cavity which contribute to the formation of caries have been found to be inactivated by metal oxide powders such as ZnO. ZnO nanoparticles have been proven to be more efficient against both Gram-negative and Gram-positive bacteria than bulk particles.[27,28]

The present study involves the investigation of effect of zinc oxide nanoparticle on polymerization shrinkage and diametral tensile strength of ORMO-48. ORMO-48 is an organically modified ceramic resin-based composite with low polymerization shrinkage and good biocompatibility.

Synthesis, characterization, composite preparation, and evaluation of ORMO-48 was studied extensively and already reported.[4,5,17]

10.2 MATERIALS AND METHODS

10.2.1 PREPARATION OF ZINC OXIDE NANOSTRUCTURE (ZNONPS)

A chemical process was used to create zinc oxide nanostructures.[29] Zinc acetate dehydrate [$Zn\ (O_2CCH_3)_2(H_2O)_2$] and sodium hydroxide (NaOH) were used as precursors in the synthesis of ZnO nanostructure. In two beakers containing 100 mL distilled water, zinc acetate dihydrate (27.43 g) and sodium hydroxide (10 g) were added. To obtain clear solutions, the solutions were magnetically stirred until all of the precursors were completely dissolved. The sodium hydroxide solution was then added to the zinc acetate dihydrate solution, drop by drop, using a burette, while continuously stirring. After the reaction a white precipitate is formed; this is washed with distilled water and dried to obtain ZnO nanostructures. Surface modification of zinc oxide nanostructure was carried out by the reported procedure.[14]

10.2.2 PREPARATION OF RESIN MIXTURE USING ORMO-48

Organically modified ceramic resin (50 parts) is diluted with TEGDMA (50 parts) which acts like a crosslinking agent. To this mixture photoinitiator (CQ) 0.25% concentration of the total weight of the mixture (TW); 0.09% of TW inhibitor (BHT) and photosensitizer (SALOL), 0.04% of TW activator (UV-9), and 0.25% of TW amine activator (DMAPEA) were added. The prepared mixture was stirred to form a homogenetically dispersed consistent resin to ensure good mechanical properties. The prepared resin matrix (ORMO-48) is stored in dark container (prevent reaction to light) to undergo preradical polymerization reaction.

10.2.3 PREPARATION OF DENTAL COMPOSITE PASTE (ORMO-48) USING RESIN MATRIX AND QUARTZ FILLER

The filler quartz (350 phr of the weight of the resin) and fumed silica with the proportion of (12% of the weight of the resin matrix) was mixed with the resin mixture. Quartz filler acts as a reinforcement agent for resin matrix and fumed silica helps in filling the voids. The filler percentage was adjusted to get paste like consistency which is prepared by the method of mastication. Zinc oxide (0.5–3%) was incorporated during the preparation of the paste in the agate mortar.

10.2.4 PREPARATION OF DENTAL COMPOSITE

The prepared paste is divided into five parts and mixed with different concentrations (0, 0.5%, 1%, 2%, and 3%) of ZnO nanoparticles. The DTS specimens are prepared and photocured (halogen light) for one minute on each side. These specimens are then tested for shrinkage and diametral tensile strength.

10.2.5 DETERMINATION OF PARTICLE SIZE OF SYNTHESIZED ZNO NANOPARTICLES

Dynamic light scattering (DLS) (Malvern Instruments, Zetasizer Nano ZS, UK) was employed for determining the size of particles of synthesized ZnO nanoparticles and the particles have been dispersed in distilled water.

10.2.6 EVALUATION OF DENTAL COMPOSITE

The effect of ZnO nanoparticles in ORMO-48 dental composites was investigated by evaluating polymerization shrinkage and diametral tensile strength.

10.2.6.1 POLYMERIZATION SHRINKAGE OF THE DENTAL COMPOSITES

Specimens of dimensions 6-mm diameter × 3-mm thickness were prepared for evaluating polymerization shrinkage of the dental composite. The composite paste was packed into the desired mold and subjected to photo-curing for 60 s on either side. The cured samples were separated from the mold and accurately measured the diameter of the sample and mold using digital Vernier calliper (Mitutoyo, Japan).

Shrinkage is tested by comparing the diameter of the specimen and the mold and calculated by using eq 10.1

$$\frac{\text{inner diameter of mold - diameter of specimen}}{\text{inner diameter of mold}} * 100 \tag{10.1}$$

10.2.6.2 DIAMETRAL TENSILE STRENGTH (DTS)

Diametral Tensile Strength (DTS) testing for the dental composite is carried out as per standard ADA no. 27 (American Dental Association)[30] with a samples dimension of 6-mm diameter × 3-mm thickness. The paste was packed into the specific mold and exposed to photocuring for 60 s on both sides. The photocured samples were removed from the mold and kept at incubator (37 ± 1 °C) for 23 ± 1 h. The samples were conditioned 1 h before testing by keeping it in 22 ± 2 °C after taking from the incubator. The diametral tensile strength was evaluated using Universal Testing Machine (Instron, Model 3365, UK) with a crosshead speed of 10 mm/min and 2 kN load cell was used. The load at which break occurs was noted and the diametral tensile strength (DTS) was calculated using eq 10.2.

$$DTS(MPa) = {2P}/{\pi DL} \tag{10.2}$$

where P = load (N), D = diameter, L = thickness (mm)

10.2.6.3 STATISTICAL ANALYSIS

The statistical evaluation of the present study was done using one-way analysis of variance (ANOVA) to determine the significance difference among the samples having p value less than 0.005.

10.3 RESULTS AND DISCUSSION

The synthesized Zinc Oxide nanostructure (ZnONPs) showed an average size of 129 nm using distilled as a solvent with a PdI value 0.164 was used in the present study (Figure 10.1).

ORMO-48 paste was used as the control material as its physico-chemical, mechanical, cyto-compatibility, and in vivo biocompatibility studies were previously reported.[2–7] Both surface-modified and non-treatedzinc oxide nanostructure were used to incorporate with the ORMO-48 during paste preparation. The effect of the incorporation of surface-modified as well as nontreated ZnONPs in ORMO-48 was investigated by evaluating the polymerization shrinkage and DTS of the prepared composites.

The effect of nontreated zinc oxide nanostructure on polymerization shrinkage is tabulated in Table 10.1, and compared the result with the control ORMO-48 composite (without ZnONPs) (Figure 10.2). The p-values of the samples were tabulated in Table 10.2 using ANOVA.

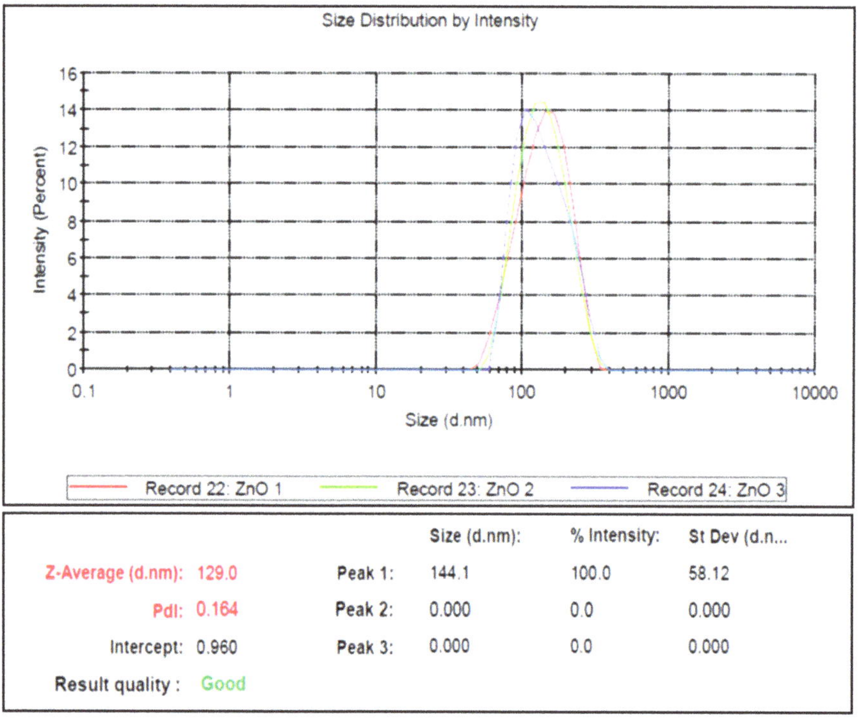

FIGURE 10.1 Particle size distribution of ZnONPs using DLS.

With the incorporation of high concentration of ZnO nanoparticles enhanced opacity of the paste occurred, which adversely affects photocuring and finally leads to the increment in polymerization shrinkage and deterioration in mechanical properties of the dental composites.[31] The same trend was observed for the present study, as the polymerization shrinkage amplified with the increase in the concentration of ZnONPs. The increase in shrinkage of the composites was not consistent as it may be due to nonuniform dispersion ofZnONPs in the composite paste of ORMO-48. However, the increase in polymerization shrinkage of ORMO-48 composite with respect to ZnONPs

TABLE 10.1 Polymerization Shrinkage studies of ORMO-48.

ZnO nanoparticles (ZnONP)	Inner diameter of mold in mm (M)	Specimen diameter in mm (S)	Shrinkage in % $\frac{M-S}{S}*100$	Mean shrinkage (%)	Shrinkage standard deviation
0% (control)	5.98	5.93	0.8361	0.7804	0.0965
	5.98	5.93	0.8361		
	5.98	5.94	0.6689		
0.5%	6.00	5.94	1.0000	0.9459	0.0951
	5.99	5.93	1.0016		
	5.98	5.93	0.8361		
1%	6.00	5.94	1.0000	1.2762	0.3429
	6.00	5.90	1.6600		
	5.99	5.92	1.1686		
2%	5.98	5.91	1.1705	1.2249	0.0958
	5.99	5.92	1.1686		
	5.99	5.91	1.3355		
3%	5.98	5.93	0.8361	0.8361	0
	5.98	5.93	0.8361		
	5.98	5.93	0.8361		

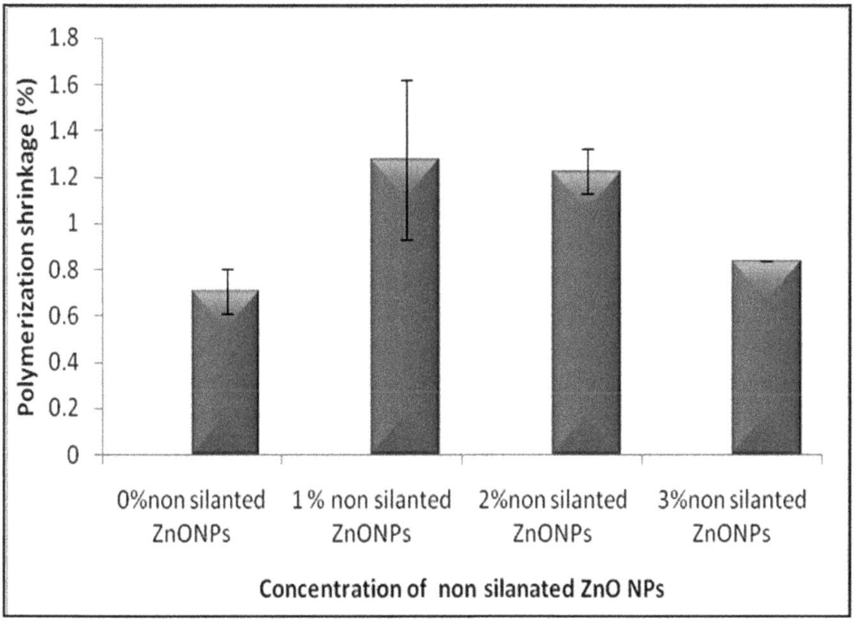

FIGURE 10.2 Effect of non-silanted ZnONPs on polymerization shrinkage of ORMO-48.

incorporated ORMO-48 composites was not significant ($p < 0.005$), except for the concentration of 2% ZnONPs (Table 10.2). The filler content has significant influence in polymerization shrinkage.[32] So the untreated ZnONPs may not be effectively bonded with the resin matrix resulting in the anomalous polymerization shrinkage of these newly prepared dental composites. Recent studies[33–35] used strategies such as ladder structured siloxane network in the backbone and side chains, hybrid inorganic organic hybrid.

TABLE 10.2 Calculation of Polymerization Shrinkage Significant Difference, i.e., p-Value for ZnONPs and ORMO-48.

ORMO-48	0.5% ZnONP	1% ZnONP	2% ZnONP	3% ZnONP
0.8361	1.0016	1.0000	1.1686	0.8361
0.8361	1.0000	1.6600	1.3355	0.8361
0.6689	0.8361	1.1686	1.1705	0.8361

p-value for ORMO-48 and 0.5% ZnONP = 0.101811

p-value for ORMO-48 and 1.0% ZnONP = 0.073491

p-value for ORMO-48 and 2.0% ZnONP = 0.004801

p-value for ORMO-48 and 3.0% ZnONP = 0.373901

The diametral tensile strength of nontreated zinc oxide nanostructure incorporated ORMO-48 composite was evaluated and compared with the control ORMO-48 (Figure 10.3). The mechanical properties of the newly prepared composites also exhibited the same scenario as observed in the case of polymerization shrinkage. DTS of all concentrations of nontreated zinc oxide nanostructures incorporated ORMO-48 composites were within the standard deviation of Control material, except the concentration having 2% ZnO nanoparticle and having a significant reduction compared to ORMO-48 composite (p value < 0.005). These results entailed the zinc oxide nanoparticle required surface modification for effective bonding with the resin matrix. So we incorporated surface-modified ZnO nanoparticle in the second phase of our study.

The polymerization shrinkage of silanted ZnONPs incorporated ORMO-48 was significantly high ($p < 0.005$) compared to the control ORMO-48 (Figure 10.4). But the increment is uniform and the anomalous nature as in the former phase was not observed which confirmed that surface modification of nanoparticles can have significant influence in the properties of the dental composites. The modifications in the filler part especially silanation can improve the properties as well as it can be consistent enough to reproduce the results.[23] So the results undoubtedly proved that the incorporation of

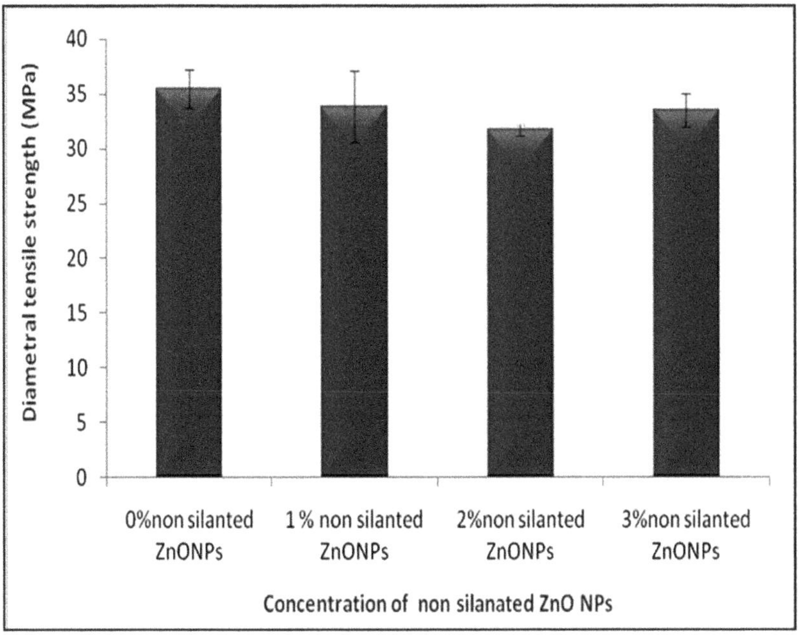

FIGURE 10.3 Effect of non-silanted ZnONPs on DTS of ORMO-48.

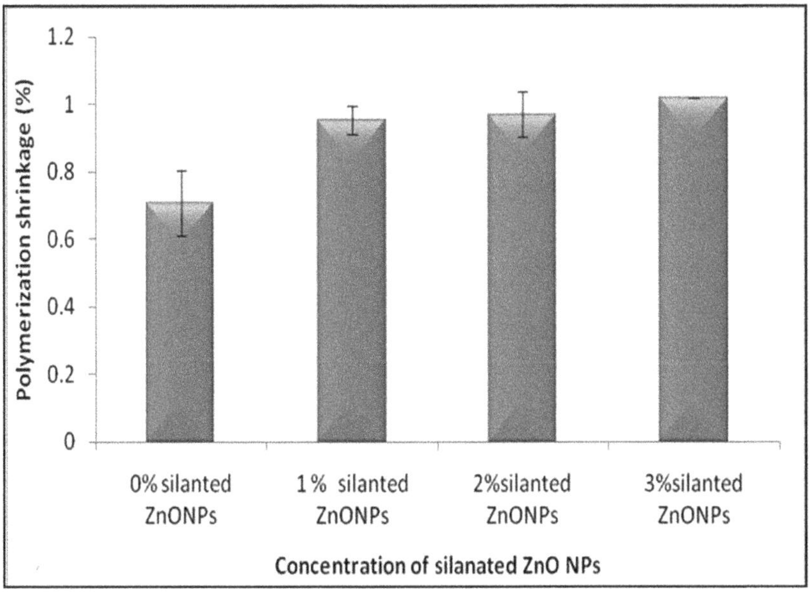

FIGURE 10.4 Effect of silanted ZnONPs on polymerization shrinkage of ORMO-48.

silanted ZnONPs in ORMO-48 has a significant effect in exhibiting a trend for the composites. Moreover, additional investigations using these silanted ZnONPs incorporated ORMO-48 composites are required for generalizing the inclination of properties especially polymerization shrinkage.

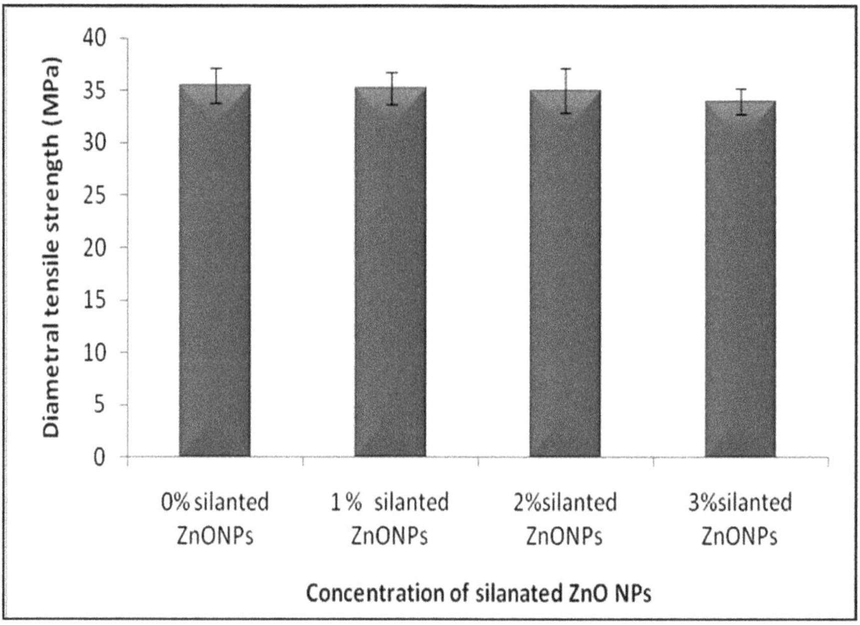

FIGURE 10.5 Effect of silanted ZnONPs on DTS of ORMO-48.

The mechanical properties of silanted ZnONPs incorporated ORMO-48 composites were comparable with the control ORMO-48 composite with a significant p value greater than 0.005 (Figure 10.5). The results showed that the mechanical properties of silanted ZnONPs incorporated ORMO-48 composites were significantly improved than the non-silanated ZnONPs incorporated ORMO-48 composites. The modification of ZnONPs provides exceptional integration as well as it prevents agglomeration leads to a uniform distribution of fillers which is very essential in the applications like the development of dental composite.[36]

The field of dentistry has revolutionized over the past five decades with the development of polymeric dental composites. However, uses of dental composites are still limited by problems such as discoloration in the tooth cavity, polymerization shrinkage, and poor wear resistance. The developments achieved are mainly through synthesis of novel organic monomers, modifications in formulation and filler technologies, light curing equipment with advanced technology, and discovery of photoinitiators, which are more efficient. The physico-chemical, mechanical, in vitro cyto-compatibility, and *in vivo* biocompatibility of ORMO-48 were reported previously.[3–7] Despite these developmental advances, an ideal material with inherent antimicrobial properties is yet to be developed. The use of zinc oxide nanoparticles (ZnO) that have diameter more than 100 nanometers is expected to have low cytotoxicity and silanted ZnONPs incorporated ORMO48 is expected to be a potential material for dental restorative applications.

10.4 CONCLUSION

The exact physical and chemical properties of zinc oxide nanoparticles depend on the method of synthesis. Here, we used ORMO-48, a photo-polymerized composite, for dental restorative application as the control material and incorporated with different concentrations of ZnO nanoparticles, and studied the effects of ZnO nanoparticles on polymerization shrinkage, and diametral tensile strength. ORMO-48 is a visible light cure composite based on organically modified ceramic resin. Present studies showed that with proper surface modification, optimum amount of incorporated zinc oxide nanoparticle has a significant effect on polymerization shrinkage and mechanical properties of ORMO-48.

ACKNOWLEDGMENTS

The authors gratefully acknowledge KSCSTE, Government of Kerala, India for financial support. We also thank Director, and Head, BMT Wing, SCTIMST for extending the facilities in BMT Wing.

KEYWORDS

- **zinc oxide nanoparticles**
- **polymerization shrinkage**
- **mechanical properties**
- **ORMO-48**

REFERENCES

1. Zhou, W.; Liu, S.; Zhou, X.; et al. Modifying Adhesive Materials to Improve the Longevity of Resinous Restorations. *Int. J. Mol. Sci.* **2019,** *20* (3), 723.
2. Bourbia, M.; Ma, D.; Cvitkovitch, D. G.; Santerre, J. P.; Finer Y. Cariogenic Bacteria Degrade Dental Resin Composites and Adhesives. *J. Dent. Res.* **2013,** *92* (11), 989–994.
3. Kalliyanakrishnan, V.; Lizymol, P. P.; Jacob, K. A. Synthesis and Characterization of a Urethane Dimethacrylate Resin with Application in Dentistry. *J. Polym. Mater.* **2004,** *21* (2), 137–144.
4. Mohanan, P. V.; Lizymol, P. P. Cytogenetic Evaluation of the Physiological Saline Extract of a Newly Developed Dental Material ORMO-48. *Toxicol. Int.* **2011,** *18,* 155.
5. Lizymol, P. P.; Mohanan, P. V.; Shabareeswaran, A.; Krishnan V. K. Biological Evaluation of a New Organically Modified Ceramic Based Dental Restorative Resin. *J. Appl. Polym. Sci.* **2012,** 125, 620–629.
6. Lizymol, P. P. Novel Inorganic Organic Hybrid Resins as Biomaterials "Recent Advances in Material Science; Series Journal. In *Composites and Nanocomposites,* Vol. 4; Haghi, A. K., Ed.; Apple Academic Press (Publisher CRC Press): Canada, 2013; chapter 8 2013/3/1, p 89.
7. Lizymol, P. P. Physical and Mechanical Properties of Visible Light Cure Composites Based on Novel Organically Modified Ceramic Resins. *Adv. Mat. Res.* **2013,** *685,* 50–53.
8. Lizymol, P. P.; Kalliyanakrishnan V. A Comparative Study on the Efficiency of Two Photo Initiators for Polymerization of Dental Light Cure Composite Resins. *J. Appl. Polym. Sci.* **2008,** *107* (5), 3337–3342.
9. Vibha, C.; Lizymol, P. P. A Bioinspired Bioactive Multi-Component Polymerizable Material for Dental Restorative Applications: Effects of Calcium Content on Physico Mechanical Properties. *Adv. Mater. Lett.* **2017,** *8* (1), 58–64.
10. Vibha, C.; Lizymol, P. P. Effect of pH During Synthesis on Molecular Weight and Polymerization Shrinkage of Photo Polymerized Hybrid Resins for Biomedical Applications. *J. Bionanosci.* **2016,** *10* (5), 414–417.
11. Vibha, C.; Lizymol, P. P. Development of Hydroxyapatite-Reinforced Biocomposites Based on Polymerizable Multifunctional Strontium Containing Inorganic-Organic Hybrid Resins for Biomedical Applications. *Mater. Lett.* **2017,** *197,* 63–66.
12. Kalliaynakrishnan, V.; Lizymol, P. P.; Sindhu, P. N. Urethane Tetramethacrylates: Novel Substitutes as Resin Matrix in Radiopaque Dental Composites. *J. Appl. Polym. Sci.* **1999,** *74,* 735–746.
13. Hass, K.-H. Hybrid Inorganic-Organic Polymers Based on Organically Modified Si-Alkoxides. *Adv. Eng. Mater.* **2000,** *2* (9), 571–582.
14. Lizymol, P. P. Studies on New Organically Modified Ceramics Based Dental Restorative Resins. *J. Appl. Polym. Sci.* **2010,** *116,* 509–517.
15. 15. Lizymol, P. P. Effects of Diluent Concentration Upon the Properties of Organically Modified Ceramics Based Composites for Application in Dentistry. *J. Appl. Polym. Sci.* **2004,** *94,* 469–473.
16. Lizymol, P. P. Thermal Studies: A Comparison of the Thermal Properties of Different Oligomers by Thermogravimetric Techniques. *J. Appl. Polym. Sci.* **2004,** *93,* 977–985.

17. Lizymol, P. P. Studies on Shrinkage, Depth of Cure and Cytotoxic Behavior of Novel Organically Modified Ceramic Based Dental Restorative Resins. *J. Appl. Polym. Sci.* **2010,** *116,* 2645–2650.
18. Lizymol, P. P.; Kalliyana, K. V. Aging Effects of Dental Restorative Materials upon Surface Hardness. *J. Polym. Mater.* **2009,** *26* (2), 207–214.
19. Vibha, C.; Lizymol, P. P. Synthesis and Characterization of a Novel Radiopaque Dimethacrylatezirconium Containing Pre-Polymer for Biomedical Applications. *Mater. Lett.* 2019, *237,* 294–297.
20. Wiegand, A.; Buchalla, W.; Attin T. Review on Fluoride-Releasing Restorative Materials—Fluoride Release and Uptake Characteristics, Antibacterial Activity and Influence on Caries Formation. *Dent. Mater.* **2007,** *23,* 343–362.
21. Leung, D.; Spratt, D. A.; Pratten, J.; Gulabivala, K.; Mordan, N. J.; Young, A. M. Chlorhexidine-Releasing Methacrylate Dental Composite Materials. *Biomaterials* **2005,** *26,* 7145–7153.
22. Gong, S. Q.; Niu, L. N.; Kemp, L. K.; et al. Quaternary Ammonium Silane-Functionalized, Methacrylate Resin Composition with Antimicrobial Activities and Self-Repair Potential. *ActaBiomater.* **2012,** *8* (9), 3270–3282.
23. TavassoliHojati, S.; Alaghemand, H.; Hamze, F.; AhmadianBabaki, F.; Rajab-Nia, R.; Rezvani, M. B.; Kaviani, M.; Atai, M. Antibacterial, Physical and Mechanical Properties of Flowable Resin Composites Containing Zinc Oxide Nanoparticles. *Dent. Mater.* **2013,** *29* (5), 495–505.
24. Jaynes, B. S.; Gande, M. E.; Fenton, R. J.; Stadler, U.; Mamak, M.; Choi, S. An Antimicrobial Metal Composite. WO2012158702A9, 2012.
25. Jin, X. Method and Antibacterial/Antimicrobial Compositions in Dental Compositions, US 20120328553, 2012.
26. Tiwari, V.; Mishra, N.; Gadani, K.; Solanki, P. S.; Shah, N. A.; Tiwari, M. Mechanism of Anti-bacterial Activity of Zinc Oxide Nanoparticle Against Carbapenem-Resistant Acinetobacterbaumannii. *Front. Microbiol.* **2018,** *9,* 1218.
27. Sirelkhatim, A.; Mahmud, S.; Seeni, A.; et al. Review on Zinc Oxide Nanoparticles: Antibacterial Activity and Toxicity Mechanism. *Nanomicro Lett.* **2015,** *7* (3), 219–242.
28. Angel Villegas, N.; Silvero Compagnucci, M. J.; Sainz Ajá, M.; Rocca, D. M.; Becerra, M. C.; Fabián Molina, G.; Palma, S. D.Novel Antibacterial Resin-Based Filling Material Containing Nanoparticles for the Potential One-Step Treatment of Caries. *J. Healthc. Eng.* **2019,** *6367919,* 1–8.
29. Hasnidawani, J. N.; Azlina, H. N.; Norita, H.; Bonnia, N. N.; Ratim, S.; Ali, E. S. Synthesis of ZnO Nanostructures Using Sol-Gel Method. *Procedia Chem.* 2016, *19,* 211–216.
30. America Dental Association: ADA27: Resin-based Filling Materials, 1993.
31. Miyazaki, M.; Hinoura, K.; Onose, H.; Moore, B. K. Effect of Filler Content of Light-Cured Composites on Bond Strength to Bovine Dentine. *J. Dent.* **1991,** *19* (5), 301–303.
32. Zanchi, C. H.; Ogliari, F. A.; Marques e Silva, R.; et al. Effect of the Silane Concentration on the Selected Properties of an Experimental Microfilled Composite Resin. *Appl. Adhes. Sci.* **2015,** *3,* 27.
33. Jeyatha, W. B.; Mani, S.; Paul, W.; Lizymol, P. P. Synthesis and Characterization of Ladder Structured Ormocer Resin of Siloxane Backbone and Methacrylate Side Chain. *Mater. Lett.* **2022,** *310,* 131192.

34. Vibha, C.; Lizymol, P. P. Synthesis and Characterization of a Novel Radiopaque Dimethacrylate Zirconium Containing Pre-Polymer for Biomedical Applications. *Mater. Lett.* **2019,** *237,* 294–297.
35. Lizymol, P. P.; Paul, W. Bioactive Composites for Hard Tissue Regeneration. *Trends Biomater. Artif. Organs* **2020,** *34* (3), 84–85.
36. Amdjadi, P.; Ghasemi, A.; Najafi, F.; Nojehdehian, H.; Pivotal Role of Filler/Matrix Interface in Dental Composites: Review. *Biomed. Res.* **2017,** *28* (3), 1054–1065.

CHAPTER 11

Structural and Optical Characterization of PVP-Stabilized Ag-Doped CuO Nanoflakes via Hydrothermal Method

P. S. RAMESH[1] and D. GEETHA[2]

[1]*Thiru Kolanjiappar Government Arts college, Virudhachalam, Tamil Nadu, India*

[2]*Department of Physics, Annamalai University, Tamil Nadu, India*

ABSTRACT

Ag-doped CuO ($CuAg_xO$, where x = (8, 10, 12, and 14%)) nanocrystals (NC) have been synthesized by the hydrothermal method. Copper nitrate tri hydrate [Cu $(NO_3)_2 \cdot 3H_2O$], silver nitrate [$AgNO_3$], sodium hydroxide (NaOH), and nonionic surfactant (PVP) $(C_6H_9NO)_n$ were used as precursors, stabilizer, and reducing chemicals. The morphology, optical and microstructure were determined by scanning electron microscopy (SEM) with energy-dispersive spectrum (EDS), UV-visible spectrum, photoluminescence spectra (PL), X-ray diffractometry (XRD), atomic force microscopy (AFM), and Fourier transformer infrared spectroscopy (FT-IR). The presence of active functional groups and chemical bonding is confirmed by FTIR. The PL spectra of the $CuAg_xO$ systems show near band edge (NBE) UV emission at 311 nm and a very small shift in green band (GB) emission from 500 to 501 nm with reduction in PL intensity as increasing dopant (Ag) concentration which confirms the substitution of Ag ion into the CuO lattice. The X-ray line broadening method was used to calculate the average particle size. X-ray analyses reveal nanosized Ag (hexagonal phase) doped CuO crystallites in monoclinic structure with varying lattice parameter values. With the incorporation of

Bio-Based Polymers and Composites: Properties, Durability, and Applications. Amadou Belal Gueye, Sabu Thomas, Nandakumar Kalarikkal, & Modou Fall (Eds.)

Ag^{2+} in the place of As the concentration of dopants rose, Cu^{2}+ caused an increase in nanocrystal size. The EDS result shows that the amount of Ag in the product is about 17%; the Ag-doped hexagonal NC exhibits a blue shift and UV emission peak, which may be induced by Ag-doping of different concentrations (8, 10, 12, and 14%).

11.1 INTRODUCTION

CuO is a narrow band gap p-type semiconductor. Its interesting properties make it an important material for a variety of practical applications, such as catalysis, batteries, solar energy conversion, gas sensing, and field emitter. A number of different fabrication techniques were reported for CuO. Synthesis of spherical structures consisting of self-assembled CuO nanoribbons was reported by Liu.[1] CuO nanowires were synthesized using thermal oxidation of Cu substrates.[2] The growth of CuO nanofibers from electrodeposited Cu nuclei was reported.[3,4] CuO nanotubes and nanorods were synthesized by the hydrothermal treatment of Cu $(OH)_4^{2-}$ precursor.[5] CuO nanorods were also prepared by thermal decomposition[6] or dehydration[7] of Cu $(OH)_2$. CuO nanoribbons obtained from sacrificial templates of $Cu_2(OH)_2CO_3$ and $Cu_2Cl(OH)_3$ were also reported. CuO nanorods and nanoribbons were also synthesized at moderate temperature (77–82 °C) in water–ethanol solution. Among various synthesis methods, hydrothermal or soft chemical reaction methods are of interest, because they are safe and performed at moderate (<200 °C) temperatures. In particular, low synthesis temperature is a significant advantage for organic optoelectronic device fabrication.

This inspired the development of an intriguing new field of study for creating electrical and photocatalytic applications: the alteration of CuO with noble metals.[8] Charge transfer from the CuO to the Ag nanoparticles happened when CuO was coated with Ag nanoparticles, preventing the combination of charge carriers and producing a noticeably improved photocatalytic performance. Herein, a simple and novel hydrothermal process was proposed for the synthesis of Ag-doped CuO particles with high crystallinity and uniform morphology, in which double deionized water (DW) was used as a hydrothermal solvent and polyvinylpyrrolidone (PVP) as a protective agent. So far, there were only limited studies discussing the influences of hydrothermal parameters on the properties of noble metal-doped crystalline CuO. In this work, the synthesis conditions including dopant (Ag) concentration were altered to control the particle size. The obtained samples were well characterized by means of XRD, SEM/EDS, TG/DTA, and optical spectra.

11.2 EXPERIMENTAL SECTION

11.2.1 EXPERIMENTAL MATERIALS

Copper nitrate tri hydrate [Cu $(NO_3)_2 \cdot 3H_2O$], silver nitrate [$AgNO_3$], sodium hydroxide (NaOH), and nonionic surfactant [$(C_6H_9NO)_n$ – PVP] were used as chemicals in this work. All were AR grade with 99% purity and purchased from Merck and SISCO Pvt., Ltd., Chemicals Company of India. They were not even further purified before usage. All dilution and sample preparation procedures employed double distilled water.

11.2.2 PREPARATION OF CUAG$_X$O NANOCRYSTALS

$CuAg_xO$ nanocrystals were fabricated via the facile hydrothermal method. For the synthesis of Ag-doped CuO NCs, analytical grade copper nitrate tri hydrate [Cu $(NO_3)_2 \cdot 3H_2O$], silver nitrate [$AgNO_3$], sodium hydroxide (NaOH), and Poly (vinyl pyrrolidone) (PVP) were used. All the reagents were used as received (Merck) without further purification. In a typical reaction process for the growth Ag-doped CuO, 1.245 g Copper nitrate tri hydrate, 0.212 g of silver nitrate, and 0.1 M of PVP were dissolved in 100 mL ultra pure water and stirred for 30 min. Simultaneously, a 5 mL NaOH (5 M) was added dropwise to this aqueous copper nitrate tri hydrate, silver nitrate, and PVP solution at room temperature to produce a mixer of milky black, solution. After stirring, the solution was transferred into a 200 ml glass beaker and sealed and maintained at 75 °C for 2 h, and then cooled down to room temperature. The resulting black colored precipitate was filtered out by Whatmann filter paper, repeatedly rinsed with deionized water and absolute ethanol to remove organic impurities, and finally dried in a hot-air oven at 100 °C for 15 h. The black precipitates were collected and ground in an agate mortar. Finally, nonionic surfactant-assisted Ag-doped CuO nanocrystallite was obtained in powder form. The same procedure is repeated for the remaining samples synthesized with nominal composition of $CuAg_xO$ with x = 8, 10, 12, and 14%.

11.2.3 CHARACTERIZATION DETAILS

The crystalline structure and crystallite size of the samples were characterized by X-ray Diffractometer (X′ PERT PRO-PANalytical, PHILIPS) in the scanning range of 10–80° (2θ) using Cu Kα radiation having a wavelength of 1.5406 Å at the scanning rate of 15.50 s with the measurement temperature 25 °C. The FTIR spectra were recorded using "Nicolet iS5"—FTIR spectrometer in the range of 4000–400 cm^{-1} with spectral resolution of 1 cm^{-1} in the transmittance mode. The UV-Visible absorption spectra were recorded using a (SHIMADZU, UV-1650 PC) spectrophotometer in the region 200–700 nm, ethanol as the reference liquid. Photoluminescence (PL) studies were done by VARIAN-DFC2500 spectrophotometer. The morphology of the samples was investigated by JEOL – JSM – 5610 MLV with INCA EDS with the resolution of 3.0 nm (HV), 4.5 nm (LV).

11.3 RESULT AND DISCUSSION

11.3.1 *STRUCTURAL INVESTIGATION*

11.3.1.1 XRD ANALYSIS

The successful preparation of $CuAg_xO$ NCs with various AgNO3 concentrations was confirmed by the measurement of XRD. Figure 11.1(a-d) shows the XRD pattern of Ag-doped CuO ($CuAg_xO$), where x = 8, 10, 12, and 14%. With the use of lattice constants, the diffraction peaks may be indexed to the monoclinic CuO phase of a = 4.685 Å, b = 3.426 Å, c = 5.130 Å, and β = 99° 24° (JCPDS 05-0661); a dominant diffraction peak at 34.6° (2θ) was characterized as the tenorite CuO (002) plane diffraction. It should be noted that the spectra show four additional weak peaks at 17.09°, 44.21°, 55.84°, and 77.33° (2θ), which can be referred to the (002), (103), (006), and (201) crystal planes of hexagonal close packed phase of Ag (JCPDS File 87-0598 with lattice constants of a = 2.886 Å and c = 10.000 Å).[9] The spectra showed no peaks for further contaminants, demonstrating the excellent purity of all products.

According to these results, pure silver was attached to the CuO surfaces. Additionally, when concentration increases Ag_2NO_3, the intensities of those peaks that are indexed to Ag progressively grow, and the fine peaks suggest that the nanostructures are highly crystallized. Unit cell parameters and cell volumes of the products were calculated using the unit cell program method. By comparing the reference data of pure CuO with Ag doped CuO samples, owing to hexagonal phase (a = 2.886, c = 10.00 Å) of Ag^{2+} ions, we can see decreasing and increasing trend in "a," "b," and "c" values, respectively.

Figure 11.1(a-d) clearly shows that there is shift and intensity variation in the peak locations. It is concluded from the cell parameter values given in Table 11.1 that Ag^{2+} ions (ionic radius 0.94 Å) enter into the structure replacing Cu^{2+} ions (ionic radius 0.73 Å). Silver (Ag) is a strong catalyst and possesses a strong affinity for alkoxide; it adsorbs the organics and acts as a capping agent.[10,11] Increasing the concentration of silver to 14%, a number of lower intensity peaks in the powder Xrd results are observed in addition to the main CuO peaks. In the case of lower concentrations of silver doping, the XRD findings demonstrate a highly crystallized monoclinic structure and homogenous phase development.

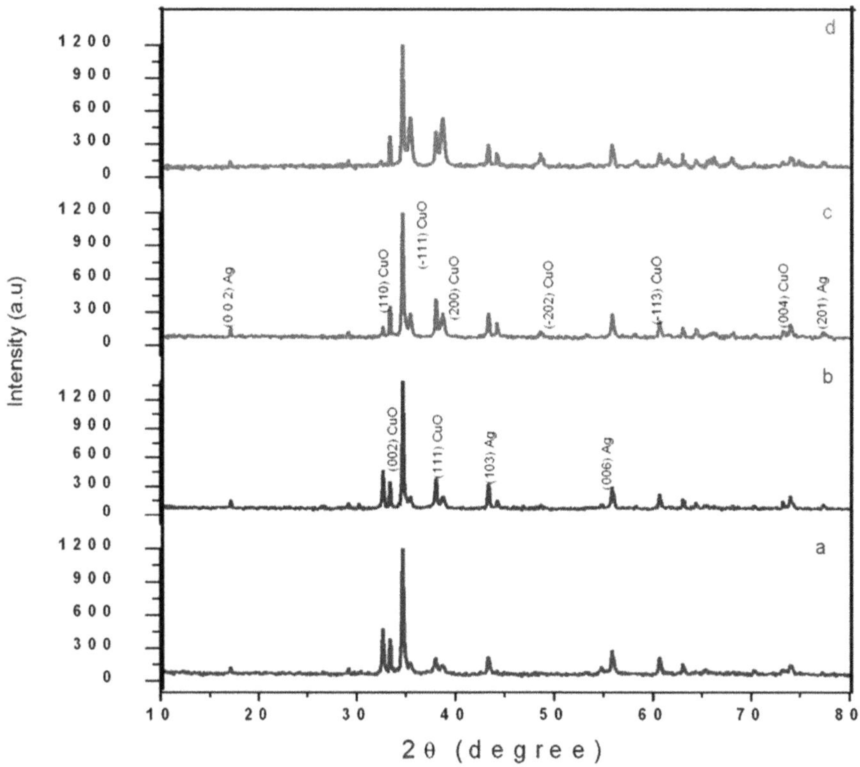

FIGURE 11.1 XRD pattern of (0.1 M) PVP-assisted Ag (a) 8, (b) 10, (c) 12, and (d) 14% doped CuO nanoflakes.

The increasing trend of lattice parameters clearly indicates that Ag^{2+} ions are substituted for Cu^{2+} in the CuO structure. Such a significant lattice can be understood qualitatively considering the size of the ions and their local coordination. Substitution of 0.94 Å sized Ag^{2+} ions instead of 0.73 Å sized Cu^{2+} is expected to enlarge the inter atomic spacing significantly. The difference in ionic radii between Ag^{2+} and Cu^{2+} in tetrahedral coordination is large. As a result, changes in lattice parameter and cell volume with Ag^{2+} substitution are also high. Increase in lattice parameter due to transition metal (TM) doping has been observed by the authors.[12] Average crystallite size of the synthesized samples has been calculated for the intense reflection along (002) by Debye–Scherer's equation.

$$D = k\,\lambda/\beta\cos\theta \tag{11.1}$$

It was discovered that the size grows as the doping concentration rises. The average crystallite sizes calculated from X-ray line broadening were 61, 65, 69, and 56 nm for Ag doping of (x = 8, 10, 12, and 14%), respectively. This result suggests that doping inhibits crystallite growth as observed in other systems. Internal lattice micro-strain of the synthesized samples has been evaluated by using the formula, $\epsilon = (\beta \cos\theta)/4$ as reported by Kathirvel et *al*. Dislocation density (δ) is calculated using the relation.

$$\delta = 1 / D^2.$$

Table 11.1 shows the average crystallite size increases with Ag doping concentration up to 12% and tends to decrease afterward. Probably, there is the substitution of oxygen by silver anions in the CuO sites, leading to a partial healing of the crystalline structure.[13] This may be probably due to the effect of crystallization process with doping.[14] There is a decrease in crystallite size with increase in doping concentration observed after 12%, which may be due to super saturation of atoms.

It is seen that as Ag content increases, the dislocation density and strain generated in the NCs decreases and NCs synthesized at 12% Ag dopant are found to have minimum value of dislocation density and strain; thereafter, the dislocation density and strain increase. There is a decrease in crystallite size with increase in doping concentration observed after 12%, which may be due to super saturation of atoms. It is seen that as Ag content increases, the dislocation density and strain generated in the NCs decrease with Ag content and NCs synthesized at 12% are found to have minimum value of dislocation density and strain; thereafter, the dislocation density and strain increase.

11.3.2 SPECTRAL ANALYSIS

Figure 11.2 reveals the FT-IR spectra of Ag-doped CuO NPs in the presence of nonionic surfactant PVP. The recorded FT-IR spectra for the PVP capped CuO NPs show characteristic functional groups of metal–oxygen (Cu-Ag-O)

TABLE 11.1 Comparison of 2θ, FWHM, d-spacing, Crystallite Size, Lattice Parameters "a," "b," and "c," c/a Ratio, Cell Volume, β, ϵ and δ of Ag (8, 10,12, and 14%) Doped CuO Nanoflakes.

Parameters	CuO with (8%) Ag	CuO with (10%) Ag	CuO with (12%) Ag	CuO with (14%) Ag	Reference Data (JCPDS Card No:05-0661)
2θ (degree)	34.669	34.666	34.689	34.670	35.450
FWHM (degree)	0.136	0.127	0.12	0.15	_
d [Å]	2.585	2.585	2.583	2.585	2.530
Crystallite size (nm)	61.22	65.56	69.39	55.51	_
(h k l)	(0 0 2)	(0 0 2)	(0 0 2)	(0 0 2)	_
a[Å]	3.564	3.567	3.567	4.566	4.685
b[Å]	3.265	3.264	3.265	2.984	3.426
c[Å]	7.717	7.720	7.718	7.497	5.130
c/a ratio	2.164	2.163	2.163	1.641	1.094
volume[$Å^3$]	87.57	87.66	87.65	101.49	81.16
β[beta]	102.85	102.87	102.86	83.36	99.24
Strain (ϵ) 10^{-4}	5.661	5.287	4.995	6.244	_
Dislocation density (δ) 10^{14} $line^2/m^2$	2.668	2.326	2.076	3.24	_

bonding in the low-frequency region as shown in Table 11.2. In the present investigation, the low-frequency region 400–700 cm^{-1} assigned the characteristic stretching vibration properties of Cu-O nanostructures corresponding to 416, 417, 418, 419, 489, 493, 496, and 497 cm^{-1}.[15] Furthermore, transmittance peaks at 597 cm^{1} 608, 611, and 613 cm^{-1} have been referred to Cu-O monoclinic phase along the [−202] direction. Transmittance peaks at 843, 839 cm^{-1} correspond to stretching mode of N-H carboxylate group. In addition, the absorption peaks correspond to OH stretching and bending vibrations at 3500–1600 cm^{-1} which are stronger due to the presence of OH group in the PVP molecules.[16] Around 3387, 3405, 3409, and 3413 cm^{-1}, a broad absorption band occurs which may be attributed to the stretching vibrations and the bending vibrations of absorption water molecule observed around 1632, 1640, 1643, and 1654 cm^{-1}. The absorption regions around 1382 and 1383 cm^{-1} are connected to longitudinal phonon vibrations, which share characteristics with nanoparticles in general.

TABLE 11.2 FT-IR Peaks and Vibrational Assignments of Ag-Doped CuO Nanoflakes.

Wave number in cm^{-1}	Functional group Assignments
3387, 3413, 3405, 3409	OH stretching of H_2O
2915, 2916, 2848, 2926	C-H alkyl stretching
1731	CO_2 in air
640, 1632, 1643, 1654	OH bending mode of H_2O (or) C=O stretching of PVP molecules
1383, 1382	COO^- stretching vibration (or) vibrations of longitudinal phonon related to nanoparticles
1019, 1018, 1008	Stretching vibration of C-O
843, 839	Stretching of N-H carboxylate group
611, 608, 613, 597	Stretching mode of Cu-O along [−202] direction
497,419, 496,418, 493, 417, 489,416	Characteristics stretching vibration of Cu-O band

In nanoscale samples, there may be a lot of suspend bonds that are upright to the particle surface, which makes the stretching vibrations of suspend bonds more active. Figure 11.2 shows the band referred to the stretching frequency at 497 cm^{-1} for CuO with Ag = 8% which is shifted to lower frequency as 496 cm^{-1} for Ag = 10%, 493 cm^{-1} for Ag = 12%, and 489 cm^{-1} for Ag = 14%. Ag atoms are slightly heavier than Cu atoms. So, the incorporation of Ag atoms into CuO lattice may cause the shifts in wave numbers and changes in transmittance intensities.

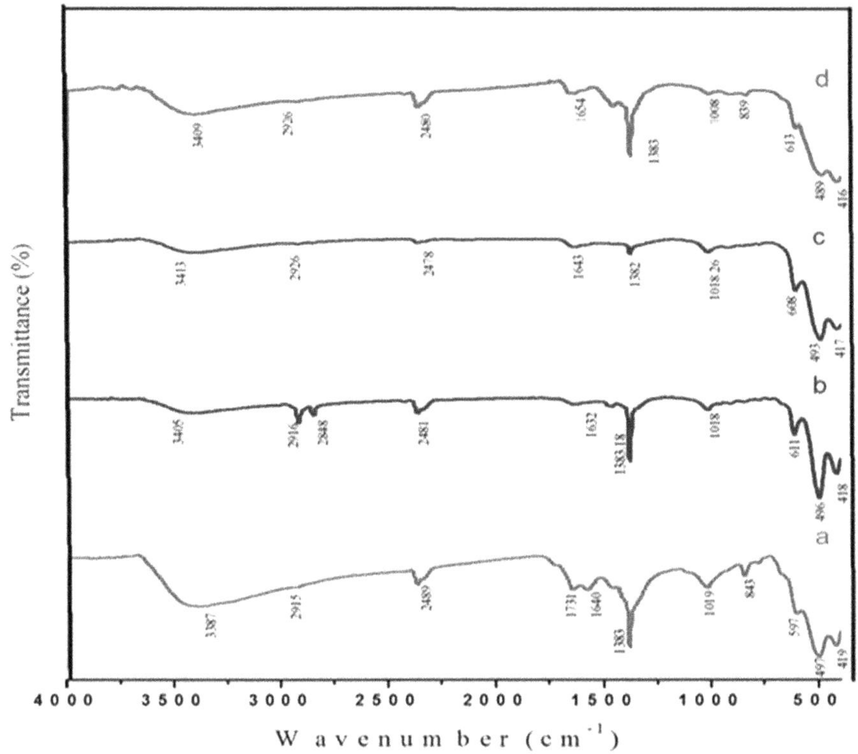

FIGURE 11.2 FTIR transmission spectra of (0.1M) PVP-assisted Ag (a) 8, (b) 10, (c) 12, and (d) 14% doped CuO nanoflakes.

11.3.3 *OPTICAL ANALYSIS*

11.3.3.1 UV–VISIBLE ABSORPTION SPECTRUM

Figure 11.3 shows the UV–visible spectra of the as-formed $CuAg_xO$ Nps dispersed in ethanol exhibiting the maximum absorption peaks at 305 nm. In the spectrum, the peaks at 294–305 nm are due to surface plasmon absorption of metal oxide. The collective oscillation of the free conduction band electrons, which is energized by the input electromagnetic radiation, is the cause of the surface plasmon absorption in the metal oxide nanoparticles. When the wavelength of the incoming light is significantly greater than the particle diameter, this form of resonance is observed.

Surface plasmon absorption bands with a maximum at 294–305 nm indicate the formation of CuO nanoparticles.[17] UV–visible spectroscopic studies were carried out at room temperature to estimate the band gap energies and also to study the effect of Ag doping on CuO in the range 200–800 nm. The UV–Vis absorption spectrum of Ag-doped CuO sample is shown in Figure 11.3. A red shift in the absorption edge is observed which may be related to the formation of shallow levels inside the band gap due to doping up to 12% of silver concentration.

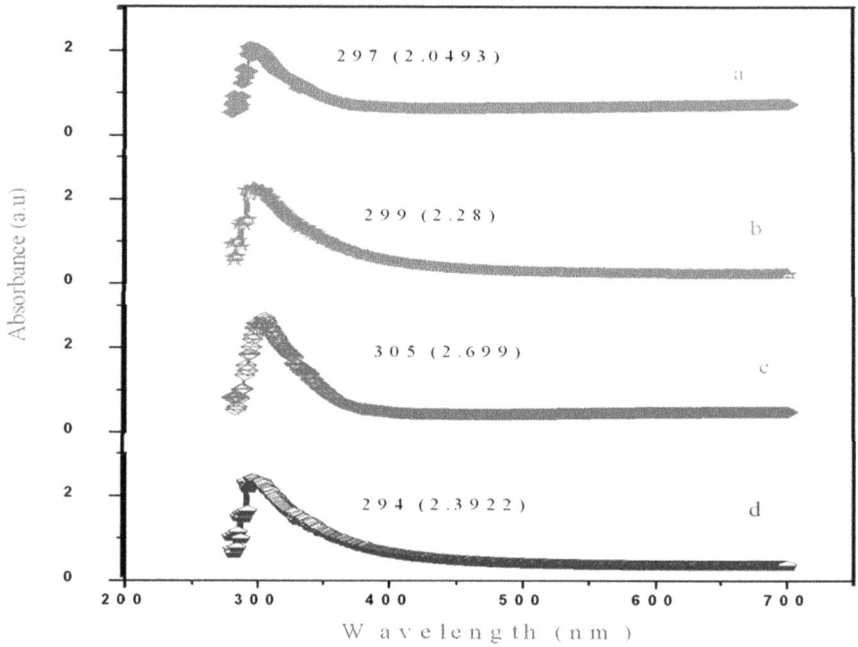

FIGURE 11.3 UV–visible absorption spectra of PVP-assisted Ag-doped (a) 8, (b) 10, (c) 12, and (c) 14% CuO nanoflakes.

Optical energy band gap E_g of the samples was calculated using the Tauc's relation

$$\alpha h\upsilon = A\,(h\upsilon - E_g)^{1/n} \qquad (11.2)$$

where A is a constant, Eg is an optical band gap, h is a value of photon energy, and is a value of the absorption coefficient close to the fundamental absorption edge, and exponent term "n" depends on the character of the transition, n = 1/2, 2, 3/2, or 3 y for permissible direct, permissible non-direct, prohibitive direct, and prohibitive non-direct transitions, respectively. Figure 11.4 shows the plot of $(\alpha h\nu)^2$ versus (hv) according to eq 2; the optical energy gaps of the Ag (8, 10, 12, and 14 wt %) doped CuO nanocrystallite were determined as 2.505, 2.473, 2.413, and 2.452 eV respectively, from the intersection of the straight line at $(\alpha h\upsilon)^2 = 0$. This shows that in the synthesized samples, the direct allowed transition is dominant.

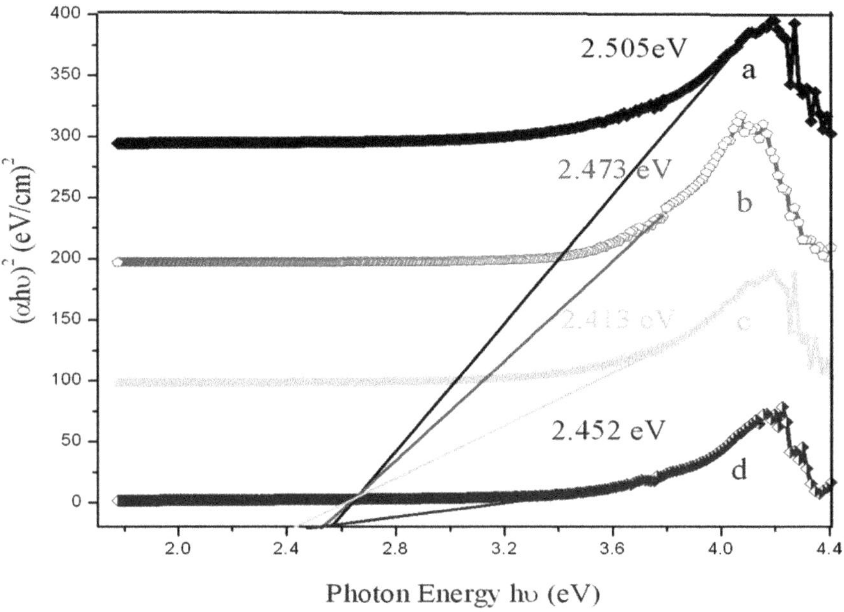

FIGURE 11.4 Plots of the $(\alpha h\nu)^2$ Vs $(h\nu)$ of PVP-assisted Ag-doped (a) 8, (b) 10, (c) 12, and (c) 14% of CuO nanoflakes.

TABLE 11.3 Comparison of E_g and D Values of 8, 10, 12, and 14% Ag-Doped CuO Nanoflakes.

S. No	Samples	Band gap Energy E_g (eV)	Crystallite Size (D) (nm)
1	CuO with (8% Ag)	2.505	61
2	CuO with (10% Ag)	2.473	65
3	CuO with (12% Ag)	2.413	69
4	CuO with (14% Ag)	2.452	56

The obtained UV–Vis spectroscopic measurements are in good agreement with the XRD analysis. The absorption ($\alpha \geq 10^4$ cm^{-1}) is related to direct band transitions. Narrow absorption peaks at 297, 299, 305, and 294 nm correspond to the band gap energy values 2.50, 2.47, 2.41, and 2.45 eV respectively, which are larger than the reported value for CuO (E_g = 1.67 eV), but less than the reported value for CuO nanocrystals in earlier study by[18] from the UV–visible spectra which confirms CuO crystals having semiconductor character. It is shown to be an industrially important material that can be widely used in applications such as solar energy transformation. The optical band gap energy decreases with increase in particle size, which can be attributed to the improvement in morphological and crystallinity of the samples. This shows that the crystallite size increases slightly with the increase of Ag content. By increasing the molecular concentration at the crystal surface, the grain growth is enhanced. The outcomes are much in line with previously published material. It has been observed from Table 11.3 that there is decreasing style in band gap energy with increasing silver concentration up to 12%, the tendency of an increase in band gap energy with increase in doping concentration at 14% which may be due to super saturation of silver atoms. Copper oxide's P-type semiconductor is a crucial functional component of photovoltaic cells. CuO has a high solar absorbance and a low thermal emittance, making it an appealing choice for selective solar absorption.

11.3.4 LUMINESCENCE STUDY

Since PL emission results from the recombination of free carriers, it was possible to use photoluminescence emission spectra to study the effects of the Ag doping on the properties of charge carrier trapping, immigration, and transfer as well as comprehend the fate of electron–hole pairs in CuO/Ag nanoflakes. Figure 11.5 illustrates the room temperature PL spectra of the as-prepared CuO/Ag samples. The excitation energy wavelength was 250 nm, which originated from a Xe-arc lamp.

PL signals result from the recombination of photoinduced charge carriers; so PL measurements were carried out to confirm the charge separation behavior and efficiency in the CuO/Ag system. It is found that all individual curves similarly show two emission bands, including a near broad UV emission at around 285–335 having the center at 312 nm, corresponding to the near band edge (NBE) emission of CuO, and a weak green emission at around 490–520 nm whose center is at 502 nm. It can be seen that the PL intensities of these samples vary in the following order: CuO/Ag (8%) < CuO/Ag (10%) < CuO/Ag (12%) > CuO/Ag (14%). The order of PL intensities of the samples was indicating the existence of charge separation effect.[19] In these samples, CuO/Ag nanoflakes were synthesized under the same experimental conditions (reaction time, concentration of surfactant, and copper source). As a result,

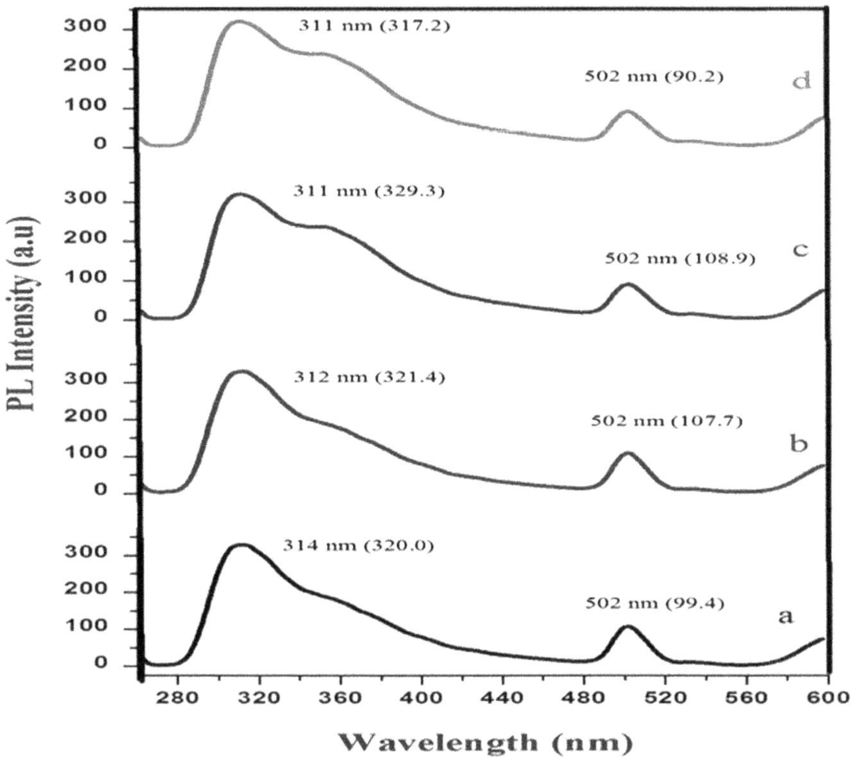

FIGURE 11.5 Room temperature photoluminescence spectra of Ag-doped (a) 8, (b) 10, (c) 12, and (c) 14% of CuO nanoflakes.

the differences in PL intensities of the samples were mainly caused by the change of Ag content in different samples. It is generally believed that a lower excitonic PL intensity means an enhanced separation and transfer of photoinduced electrons. Therefore, it is reasonable to say that the PL intensities of Ag-loaded (14%) sample can have good photocatalytic activity. From Figure 11.5(c), it can be seen that the CuO/Ag (14%) sample exhibits a much lower emission intensity than other products, which results in that CuO/Ag (14%) nanoparticles promote the separation of holes and photoinduced electrons. And it is expected that the photocatalytic activity will be enhanced. Moreover, with the increase of the concentration of Ag nanoparticles, the PL intensity of CuO/Ag (14%) evidently decreases. It is possible to attribute the PL quenching to the following causes. First, the quenching takes place on smaller metal nanoparticles according to the radiating plasmon concept. That is to say, it is caused by the tiny Ag nanoparticles that are affixed to the CuO nanoflakes' surface. Second, electron transfer from the semiconductor to the metal ions, which is consistent with the outcome of photocatalytic, may be used to quench PL emission. Specifically, during the PL measurement, the excitation light excites electrons in the CuO to move from the valence band to the conduction band. The formation and growth of Ag nanoparticles may be affected by surface defects that can trap photoinduced electrons and holes. This improves the efficiency of charge separation in the CuO near the surface and, in turn, lessens the likelihood that photoinduced electrons and holes will recombine, lowering the CuO's PL intensity. Furthermore, it is thought that the high Ag concentration may lead to the creation of new electron–hole pairs recombination centers, which would negate its beneficial effects on preventing recombination.

11.3.5 SURFACE ANALYSIS

11.3.5.1 MORPHOLOGICAL STUDY

Investigating the morphological characteristics and particle dispersion properties, the CuAgO nanocomposites were subjected to microscopic analysis. Figure 11.6(a, b) shows typical SEM micrographs of CuO/Ag (12%) and CuO/Ag (14%) samples and exhibits flakes like structures with large-scale quantity. Figure 11.6(a) seems to have larger particle size and less surface roughness, since it is prepared by CuO/Ag (12%). For sample CuO/Ag (14%), smaller particle size, greater surface smoothness, and a considerable distribution in crystallite size Ag nanoparticles are formed on the surface of CuO nanoflakes. By tuning the reaction concentration, the quantity of Ag nanoparticles on CuO's surface may be carefully managed.

On the edges of the CuO nanoflakes, there were a few distributed Ag nanoparticles, when the concentration is increased from 12% to 14%, which in turn decreases the particle size from 69 nm to 56 nm as indicated in XRD results. In order to determine the exact elemental composition of the synthesized products, the energy-dispersive spectroscopy (EDS) detector equipped with SEM analysis was carried out. The EDS analysis shows the strong signal in the silver region and confirms the presence of Ag NPs. Metallic silver nano crystals generally show typical optical absorption peak approximately at 2.983 keV due to surface plasmon resonance.

In addition, the nanoparticles are observed to aggregate in the form of clusters as shown in Figure 11.6(a) at a micrometric scale with many air gaps among the clusters, signifying a poor adhesion between the CuO nanoparticle and the Ag dopant. However, in composites obtained by increasing the silver concentration to 14% shown in Figure 11.6(b), a smooth surface and well-dispersed nanocomposites were observed, indicating an improved interaction between the CuO NPs and the Ag dopant.

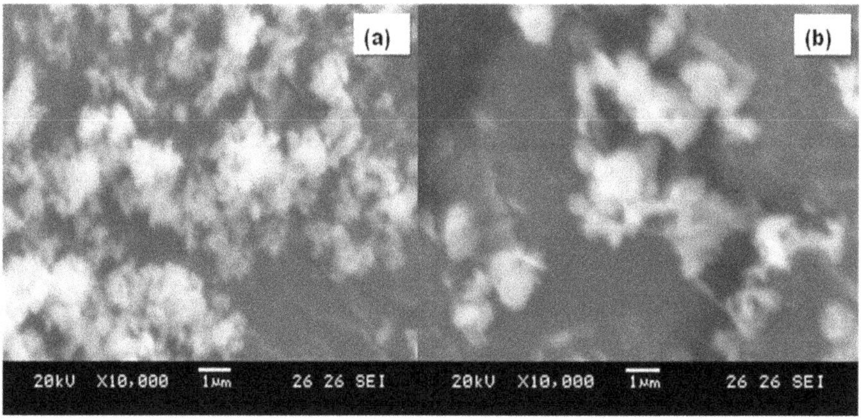

FIGURE 11.6 (i) SEM micrographs of Ag (a) 12 and (b) 14% doped CuO nanoflakes.

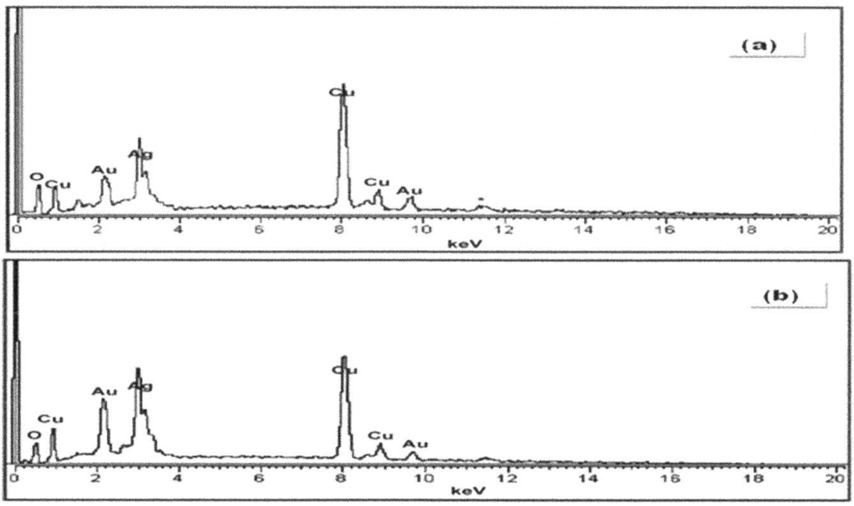

FIGURE 11.7 (ii) EDS images of Ag (a) 12 and (b) 14% doped CuO nanoflakes.

Energy-dispersive X-ray analysis (EDS) of CuO/Ag NPs reveals strong signal in the silver region and thus confirms the deposition of Ag nanoparticles on the surface of CuO nanoflakes. The peaks in Figure 11.7(a, b) correspond to Cu, O, and Ag, respectively. The Au signal arises from the substrate. Table 11.4 shows the atomic weight of the different elements (Cu, Ag, and O) present in the synthesized product.

TABLE 11.4 Elemental Compositions Present in the Ag (12 and 14%) Doped CuO Nanoflakes from EDS Spectra.

S. No	Elements	Atomic weight ratio for CuO/Ag (12%) sample	Atomic weight ratio for CuO/Ag (14%) sample
1	O	44.80	40.53
2	Cu	41.54	43.85
3	Ag	13.66	15.62

11.4 CONCLUSION

High quality of Ag-doped CuO semiconducting metal oxide nanocomposites was successfully synthesized by the hydrothermal method and investigated by different spectroscopic techniques which support the nanomaterial structure, morphology, and optical properties of the prepared samples. The XRD and SEM measurements revealed the monoclinic and hexagonal phases of CuO and Ag nanoparticles, respectively. The enhancement of lattice constants "a," "b," "β," cell volume and reduction of lattice constant "c," micro-strain, dislocation density, the slight shift of XRD peaks indicate that Ag has really doped into the Cu–O lattice. According to PL results, it is proposed that CuO with Ag (12%) doped sample may give a better result for photocatalytic activities. The presence of functional groups and the chemical bonding with Ag are confirmed by FTIR spectra. The PL spectra of the $CuAg_xO$ system described quenching in PL intensities and small shifts in near band edge (NBE) UV emission from 285 to 335 nm and a shift in green band (GB) emission from 490 to 520 nm which confirms the substitution of Ag into the CuO lattice. The strong UV and weak green bands emphasized that the Ag-doped CuO has good crystal surface. The AFM study indicates that the formation of Ag^{2+} ion-doped CuO nanocrystals has smooth surface and mean particle size of 57.8 nm. Hydrogenated samples have good crystalline structure and better optical properties are appreciable for the fabrication of nano-optoelectronic devices like tunable light-emitting diode.

KEYWORDS

- **Ag-doped CuO**
- **nonionic surfactant (PVP)**
- **hydrothermal synthesis**

REFERENCES

1. Lin, H. H.; Wang, C. Y.; Shih, H. C.; Chen, J. M.; Hsieh, C. T. Characterizing Well-Ordered CuO Nanofibrils Synthesized Through Gas-Solid Reactions. *J. Appl. Phys.* **2004,** *95* (10), 5889–5895.
2. Jiang, X.; Herricks, T.; Xia, Y. CuO Nanowires Can Be Synthesized by Heating Copper Substrates in Air. *Nano Lett.* **2002,** *2* (12), 1333–1338.
3. Gao, X. P.; Bao, J. L.; Pan, G. L.; Zhu, H. Y.; Huang, P. X.; Wu, F.; Song, D. Y. Preparation and Electrochemical Performance of Polycrystalline and Single Crystalline CuO Nanorods as Anode Materials for Li Ion Battery. *J. Phys.* 2004.
4. Zhu, C. L.; Chen, C. N.; Hao, L. Y.; Hu, Y.; Chen, Z. Y. In-situ Preparation of 1D CuO Nanostructures Using $Cu_2(OH)_2CO_3$ Nanoribbons as Precursor for Sacrifice-Template Via Heat-Treatment. *Chem. Solid State Commun.* **2004,** *130,* 681–686.
5. Zhang, Y. C.; Xing, R.; Hu, X. Y. A Green Hydrothermal Route to Copper Nanocrystallites. *J. Cryst. Growth* **2004,** *273* (1–2), 280–284.
6. Yu, Y.; Zhao, Y.; Sun, H.; Ahmad, M. Au Nanoparticles Decorated CuO Nanowire Arrays with Enhanced Photocatalytic Properties. *Mater Lett.* **2013,** *108,* 41–45.
7. Liu, X.; Li, Z.; Zhao, C. X.; Zhao, W.; Yang, J.; Wang, Y.; Li, F. Facile Synthesis of Core–Shell CuO/Ag Nanowires with Enhanced Photocatalytic and Enhancement in Photocurrent. *J. Colloids Interface Sci.* **2014,** *419,* 9–16.
8. Foresti. M. L.; Innocenti, M.; Hamelin, A. Adsorption Behavior of n-Hexanol on Ag (111) from Aqueous 0.05 M KClO4. *Langmuir* **1995,** *11* (2), 498–505.
9. Huang, G. J.; Wang, J. B.; Zhong, X. L.; Zhou, G. C.; Yan, H. L. Synthesis, Structure, and Room-Temperature Ferromagnetism of Ni-doped ZnO Nanoparticles. *J. Mater. Sci.* **2007,** *42,* 6464–6468.
10. Shinde, S. S.; Korade, A. P.; Bhosale, C. H.; Rajpure, K. Y. Influence of Tin Doping onto Structural, Morphological, Optoelectronic and Impedance Properties of Sprayed ZnO Thin Films. *J. Alloys Compd.* 2013, *551,* 688–693.
11. Brinker, C. J.; Scherer, G. W. *Sol-Gel Science: The Physics and Chemistry of Sol-Gel Proessing.* Academic press: San Diego, 1990.
12. Kathirvel, P.; Manoharan, D.; Mohan, S. M.; Kumar, S. Spectral Investigations of Chemical Bath Deposited Zinc Oxide Thin Films—Ammonia Gas Sensor. *J. Optoelectron. Biomed. Mater.* **2009,** *1* (1), 25–33.
13. Mushtaq, A.; Nam, D. S. H.; Kim, Y. S.; Kim, W. B. Synthesis, Characterization, and Electrochemical Properties of Self-assembled Leaf-Like CuO Nanostructures. *J. Solid State Electrochem.* **2010,** *14* (9), 1719–1726.

14. Makkar, M.; Bhatti, H. S. Inquisition of Reaction Parameters on the Growth and Optical Properties of ZnO Nanoparticles Synthesized via Low Temperature Reaction Route. *Chem. Phys. Lett.* **2011,** *507* (1–3), 122–127.
15. Raja Naika, H.; Lingaraju, K.; Manjunath, K.; Kumar, D.; Nagaraju, G.; Suresh, D.; Nagabhushana, H. Green Synthesis of CuO Nanoparticles Using *Gloriosa superba L.* Extract and their Antibacterial Activity. *J. Taibah Univ. Sci.* **2015,** *9* (1), 7–12.
16. Al-Gaashani, R.; Radiman, S.; Tabet, N.; Daud, A. R. Synthesis and Optical Properties of CuO Nanostructures Obtained via a Novel Thermal Decomposition Method. *J. Alloys Comp.* **2011,** *509,* 8761–8769.
17. Sonia, S.; Poongodi, S.; Kumar, P. S.; Mangalaraj, D.; Ponpandian, N.; Viswanathan, C. Hydrothermal Synthesis of Highly Stable CuO Nanostructures for Efficient Photo-catalytic Degradation of Organic Dyes. *Mater. Sci. Semicond. Proc.* **2015,** *30*, 585–591.
18. Ren, C.; Yang, B.; Wu, M.; Xu, J.; Fu, Z.; Lv, Y.; Uo, T.; Zhao, Y.; Zhu, C. Synthesis of Ag/ZnO Nanorods Array with Enhanced Photocatalytic Performance. *J. Hazard. Mater.* **2010,** *182,* 123–129.
19. Yang, J.; Li, Z.; Zhao, W.; Zhao, C.; Wang, Y.; Liu, X. Controllable Synthesis of Ag–CuO Composite Nanosheets with Enhanced Photocatalytic Property. *Mater. Lett.* **2014,** *120,* 16–19.

CHAPTER 12

PVA-Stabilized Zn-Doped CoS/Graphene Nanocomposites as Active Electrode Materials for Supercapacitors and Excellent Nano-Photocatalists

P. PRIYA[1], D. GEETHA[1] and P. S. RAMESH[2]

[1]*Department of Physics, Annamalai University, Chidambaram, Tamil Nadu, India*

[2]*Thiru Kolanjiappar Govt. Arts and Science College, Virudhachalam, Tamil Nadu, India*

ABSTRACT

Zn-CoS/Graphene nanocomposite was synthesized using a simple hydrothermal method. Ternary transition metal sulfides are promising electrode materials due to their better application of electrochemical energy storage property; environmental remediation is promising for future energy and environment. The morphology and structure analysis of Zn-CoS/Graphene were investigated by scanning electron microscope (SEM) and energy-dispersive spectromerty (EDS) and X-ray diffraction (XRD). The application of the Zn-CoS/Graphene NCs as nano photocatalyst was examined via the degradation of organic dye methylene blue (MB) under natural sunlight. The maximum degradation of MB reached about 80% after 120 min. Furthermore, electrochemical performance was investigated by means of cyclic voltammetry, and impedance spectra. The obtained results confirm the product providing excellent electron transfer path and high conduction, which enable the fast surface redox reaction. Supercapacitor based on

Bio-Based Polymers and Composites: Properties, Durability, and Applications. Amadou Belal Gueye, Sabu Thomas, Nandakumar Kalarikkal, & Modou Fall (Eds.)

Zn-CoS /Graphene nanocomposite shows a specific capacitance of 356.45 F/g at a current density of 5 mV s^{-1}. The result indicates that as-prepared materials could potentially be applied in high-performance super capacitor.

12.1 INTRODUCTION

For the last few years, the major problem is the environmental pollution through the water contamination and energy deficiency. Organic dyes are widely used in food, coloring, pharmaceuticals, printing, textile, cosmetics, paints, rubber, and leather industries. Dyes which are not fixed to fabric get washed out as textile effluents will cause environmental pollution if not well treated. The colored effluent discharged from these industries not only affects the esthetic value of the environment but also causes severe imbalance in the water ecosystem. The materials under study are PVA-stabilized Zn-doped CoS/G which are important nanomaterials that have been widely studied due to their interesting optical properties and photocatalytic activity.[1–3]

Developing low-cost sustainable energy sources is the need of the day to meet out energy crisis. Supercapacitors are more promising because of their unique characteristic viz., specific energy, rapid charging and discharging rates, low cost, and environment-friendly nature.[4–6]

Apart from the individual transition metal sulfide, the grapheme-doped binary metal sulfide has good specific capacitance. PVP is used as a template material and Zn-doped CoS nanocomposite was selected to enhance the electrochemical performance. The Zn-doped CoS have high capacitance, but cycling stability was found unsatisfactory. To overcome this Zn-CoS was combined with grapheme selected to stabilize the new material.

12.2 EXPERIMENTAL

12.2.1 SYNTHESIS OF ZN-DOPED CoS/GRAPHENE NANOCOMPOSITE

All the reagents used in the experiment were analytically pure and used without further purification. Typically, different molarity of Zn.4H_2O (1mM, 2 mM and 3 mM); 5 mM of Co·4H_2O and 10 mM thiourea in the ratio of (1:2) were dissolved in 40 mL water. A prepared 3 mg mL^{-1} solution of GO was dispersed in 15 mL of distilled water by ultrasonication for 3 h to obtain a homogeneous dispersion. The total reaction mixture was transferred into a Teflon-lined autoclave and heated to 120 °C for 12 h. The final product was washed and dried at 80 °C for 6 hrs to harvest Zn-CoS/G nanocomposite further calcined at 600 °C for 3 h.

12.2.2 CHARACTERIZATION DETAILS

The crystalline structure and crystallite size of the samples were characterized by X-ray Diffractometer (X′ PERT PRO-PANalytical, PHILIPS) in the scanning range of 10–80° (2θ) using Cu Kα radiation having a wavelength of 1.5406 Å at the scanning rate of 15.50 sec with the measurement temperature 25 °C. UV-Visible absorption spectra were recorded using a (SHIMADZU, UV-1650 PC) spectrophotometer in the region 200–700 nm, ethanol as the reference liquid. The morphology of the samples was investigated by JEOL–JSM–5610 MLV with INCA EDS with the resolution of 3.0 nm (HV), 4.5 nm (LV).

12.2.3 ELECTRODE PREPARATION

The PVA-stabilized Zn-doped CoS/graphene were dispersed in ethanol to obtain slurry and then resultant slurry was sonicated for 10 min and it was placed onto the surface of the glassy carbon electrode using a micropipette, and then dried at 60 °C for 1 h under vacuum to evaporate the solvent. The Zn-CoS/G-modified GCE was the working electrode, a saturated calomel electrode (SCE) as the reference electrode, and a Platinum wire (Pt) acted as the counter electrode. The electrochemical characterization of the prepared samples was carried out in 2 M KOH electrolyte solution an electrochemical workstation (CHI 660C, USA) by cyclic voltammetry (CV), and electrochemical impedance spectroscopy (EIS) using a three-electrode design.

12.2.4 SAMPLE PREPARATION FOR PHOTOCATALYTIC MEASUREMENT

The evaluation of photocatalytic activity of the prepared samples for photocatalytic decolorization of MB solution was performed at 30 °C temperature. The MB aqueous solution of 200 mL with the required concentration of 3 ppm was put into a glass beaker. The prepared sample of 5 mg was added into solution and stirred under very bright sunlight. The solution of 10 mL was taken from glass beaker every 30 minutes to examine the decolorization of MB.

The efficiency of the dye degradation percentage [degradation efficiency] was calculated according to:

$$\text{Degradation efficiency (\%)} = [1\text{-}(C_t/C_o] \times 100 \qquad (12.1)$$

where

C_t is the concentration after time "t"
C_o is the initial concentration

12.2.5 SAMPLE PREPARATION FOR ADSORPTION STUDIES

In general, the adsorption studies of MB were performed over 5 mg catalyst in 20 ML of dye solution at room temperature. The suspension was stirred in the dark. Small portions of the solution were filtered at intervals every half an hour's (0, 30, 60, 90, and 120) after discarding the first part of the filtrate to avoid the effects of dye adsorption. The dark reaction was carried out without catalyst which was taken as the control.

12.3 RESULT AND DISCUSSION

12.3.1 XRD ANALYSIS

The crystal purity and structure of Zn-CoS/GO was investigated by XRD analysis, and the corresponding X-ray diffraction pattern is given in Figure 12.1. The characteristic peaks at $2\theta = 28.7, 47.6, 56.0$, and 77.0 correspond to the (111), (220), (311), and (331) plans for Zn, CoS (JCPDS card no. 47-1656) [Ref]. An additional peak at 26.43 corresponds to the (002) plane of graphene oxide.[7]

12.3.2 OPTICAL STUDIES (UV-VIS SPECTROSCOPY)

The UV-Vis absorption spectra of the Ni-CoS/Graphene are shown in Figure 12.2(a); there are two major absorbance peaks observed. The peaks at 223 nm and 500 nm exhibit high optical transparency up to the UV region with a small absorption band peaking at 223 nm due to π–π * interactions of the aromatic ring.[8] The band gap of 4.4 eV was obtained for all samples. In the case of composite the band gap value decreases; this could be because the graphene sheet provides better electrical conductivity in the composite.

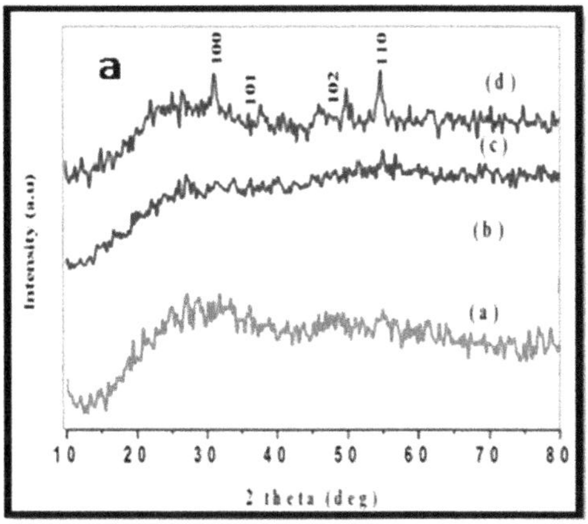

FIGURE 12.1 XRD pattern of PVA-stabilized Zn-doped nanocomposites, CoS on graphene nanocomposites: (a) Zn(0.0), (b) 1 mM, (c) 2 mM, (d) 3 mM.

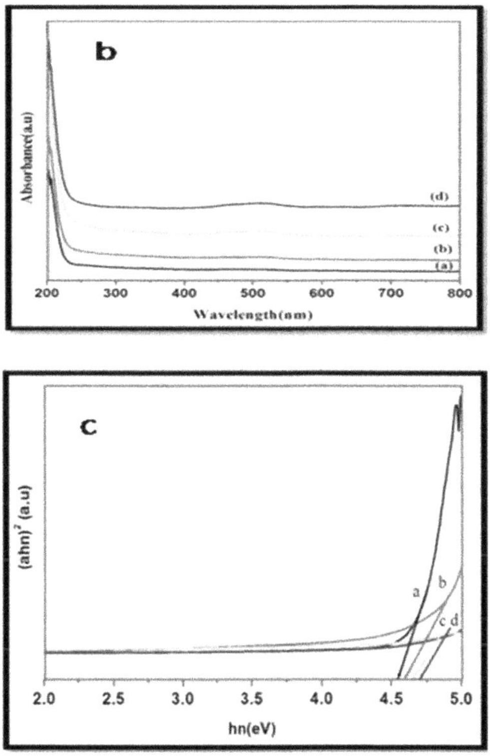

FIGURE 12.2 (a), (b) UV-vis absorption spectrum of PVA-stabilized Zn-doped CoS on graphene nanocomposites and the band gap energy.

12.3.3 SEM ANALYSIS

The surface morphology study reveals the formation of PVA-stabilized Zn-CoS/Graphene nanostructures to a size of 500 and 1 μm. The scanning electron nanostructures (SEM) images shown in Figure 12.3(a, b) nanostructures clearly show that evenly dispersed with minimal agglomerated cluster formation of PVA-stabilized Zn-CoS/Graphene nanostructures. The energy-dispersive spectrometer (EDS) characterization was performed to examine the elemental composition of CoS/Graphene and Zn-CoS/Graphene nanostructures.[9] The EDX spectra of Zn-CoS/Graphene microstructures are shown in Figure 12.3(c), confirming the presence of Zn, Co, S, and C-O peaks only. As expected, the actual doping concentration of Zn ion into the CoS host is slightly lower than the initial concentration during the reaction. No peaks related to any other impurity were detected in the spectrum.

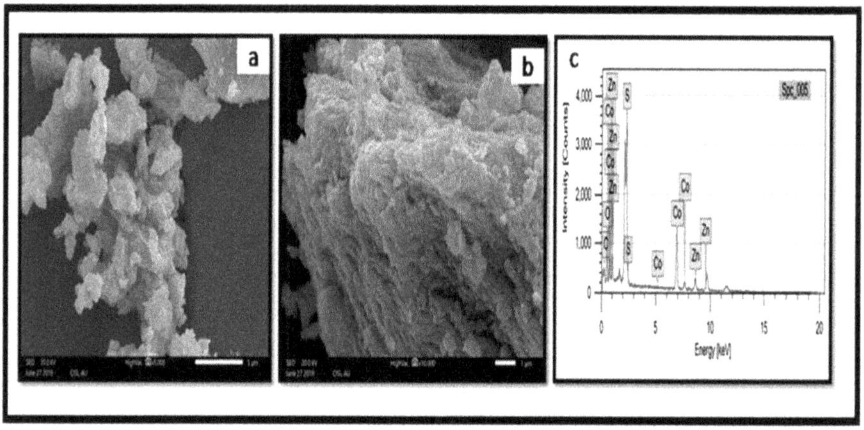

FIGURE 12.3 SEM images and EDS spectrum of PVA-stabilized (3 mM) Zn-doped CoS/Graphene nanocomposites.

12.3.4 PHOTOCATALYTIC ACTIVITY

The photocatalytic performance of the PVA-stabilized Zn-CoS/Graphene nanocomposites in terms of photodegradation of MB under natural sunlight irradiation was investigated. Figure 12.4(a-b) shows that after irradiation of 120 min, the PVA-stabilized Zn-CoS/Graphene nanocomposite was degraded by 80%.

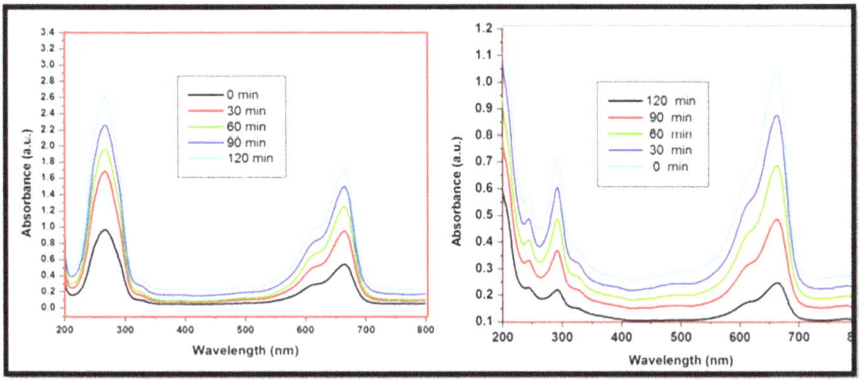

FIGURE 12.4 (a) Photocatalytic response and (b) adsorbent activity of Zn-CoS/G nanocomposites on methelen blue with time.

As expected, the decrease in the concentration and the excellent catalytic effect can be attributed to the homogenous distribution of PVA-stabilized Zn-CoS/Graphene nanocomposite surface. The metallic components of the graphene sheet generate a Schottky barrier at the interface with PVA-stabilized Zn-CoS/Graphene nanocomposite; hence, trapped electrons will transfer and separate the excited molecules absorbed at the interface of the PVA-stabilized Zn-CoS/Graphene nanocomposite and the graphene sheets by retaining the recombination process.[10] This may lead to the possibility of enhancing the photocatalytic degradation efficiency.

12.3.5 ELECTROCHEMICAL STUDY

To investigate capacitive behavior of the composite material were assessed by CV and EIS measurement. Figure 12.5(a-b) shows the CV curves of PVA-stabilized Zn-CoS/Graphene with voltage range of −2.0 to 2.0 V and various scan rates ranging from 5 to 50 mVs^{-1}. The CV curves of PVA-stabilized Zn-CoS/Graphene have demonstrated the CV curves of modified electrode roughly rectangular shape superimposed with broad faradaic curve. This indicates the coexistence of electrical double layer and Faradic pseudocapacitances for modified (2 mM) Ni-CoS/Graphene electrode.[11] When increasing the scan rate, the redox peaks increase, which indicates the good kinetic electrochemical reversibility, which was quite different from the electric double-layer capacitance with a rectangular CV shape.

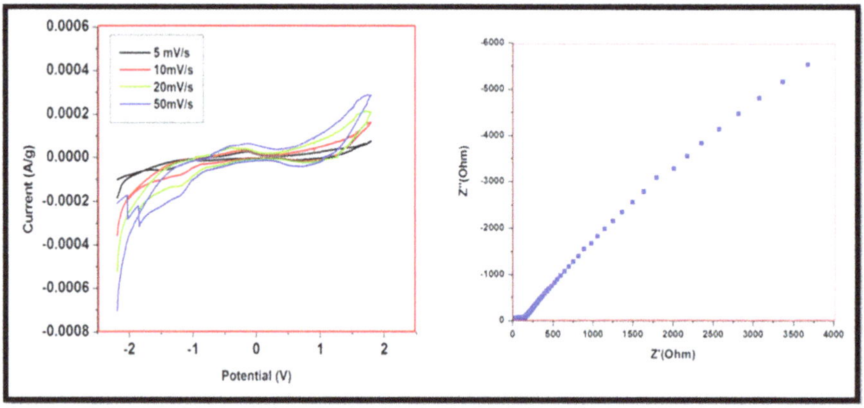

FIGURE 12.5 (a) CV curves for Zn-CoS/G nanocomposite. (b) Nyquist plot of modified Zn-CoS/G electrode.

The specific capacitance of the three PVA-stabilized Zn-CoS/Graphene electrodes can be calculated by CV curves using a standard equation as stated in the experimental section. Based on the above equation, the specific capacitance of the PVA-stabilized Zn-CoS/Graphene was found to be 356.45 Fg^{-1} at 5 mV s^{-1} scan rate respectively; this highest capacitance could be attributed to the well porous nanostructure of PVA-stabilized (3 mM) Zn-CoS/Graphene sample. Porous structure of nanomaterials can significantly develop the utilization and the specific capacitance of active electrode materials.

To understand the electrochemical behavior of the PVA stabilized Zn-CoS/Graphene electrodes were further investigated by EIS in the frequency range from 100 mHz to 100 kHz. The Nyquist plot of the modified PVA-stabilized Zn-CoS/Graphene electrodes is shown in Figure 12.5(b); the semicircle obtained in the high-frequency region belongs to the charge transfer resistance (Rct) of the electrode. This Rct is mainly related to the electroactive surface area of the electrode, owing to the Faradic redox process of the PVA-stabilized Zn-CoS/Graphene involving in the exchange of OH ions.[12] The straight slopping line obtained in the low-frequency region corresponds to Warburg impedance (Zw) of the PVA-stabilized Zn-CoS/Graphene nanocomposite during the redox process. From this electrochemical impedance analysis, the modified PVA-stabilized Zn-CoS/Graphene electrode reveals suitable electrode material for supercapacitor application.[13]

12.4 CONCLUSIONS

In summary, PVA-stabilized Zn-CoS/Graphene have been synthesized successfully using the mild hydrothermal route for the production of large scale with homogeneous uniform sphere like morphology. The crystalline structure and particle size were characterized by XRD. The functional groups were characterized by FTIR. When used as electrode materials for supercapacitor exhibit the highest specific capacitance mainly due to porous nature of the sample. At 5 mV/s scan rate, the specific capacitance was 356.45 Fg^{-1} for (1 mM) Zn-CoS/Graphene. UV-vis reflectance spectra suggest that PVA-stabilized Zn-CoS/Graphene is capable of absorbing visible light and subsequently photocatalytic degradation of organic dyes MB was explored under visible light. The photodegradation efficiency was found to be more than 80%. These nanocomposites could achieve both large specific area and good photocatalytic activity due to the positive synergistic effect of PVA-stabilized Zn-CoS/Graphene NCs. The current study has provided a new route for grapheme-based materials to be used for environmental remediation.

KEYWORDS

- **PVA**
- **Zn-CoS/Graphene nanocomposite**
- **specific capacitance**
- **nanophotocatalyst**

REFERENCES

1. Simon, P.; Gogotsi, Y. Materials for Electrochemical Capacitors. Nanoscience and Technology: A Collection of Reviews from Nature Journals. *World Sci.* **2010,** 320–329.
2. Wang, G.; Zhang, L.; Zhang, J. A Review of Electrode Materials for Electrochemical Super Capacitors. *Chem. Soc. Rev.* **2012,** *41,* 797–828.
3. Zhong, J.-H.; Wang, A.-L.; Li, G.-R.; Wang, J.-W.; Ou, Y.-N.; Tong, Y.-X. Co_3O_4/ Ni $(OH)_2$ Composite Mesoporous Nanosheet Networks as a Promising Electrode for Supercapacitor Applications. *J. Mater. Chem.* **2012,** *22,* 5656–5665.
4. Lijuan, J.; Yajun, W.; Changgen, F. Application of Photocatalytic Technology in Environmental Safety. *Procedia Eng.* **2012,** *45,* 993–997.

5. Khan, A. A.; Kumari, S.; Choudhury, A.; Hussain, S. Phase Tuned Originated Dual Properties of Cobalt Sulphide Nanostructures as Photocatalyst and Adsorbent for Removal of Dye Pollutants. *ACS Appl.Nano Mater.* **2018,** 1223.
6. Byungwoo, K.; Haegeun, C.; Woong, K. High-Performance Supercapacitors Based on Vertically Aligned Carbon Nanotubes and Nonaqueous Electrolytes. *Nanotechnology* **2012,** *23,* 155401.
7. Ali, A.; Oh, W.-C. Photocatalytic Performance of CoS2-Graphene-TiO2 Ternary Composites for Reactive Black B(RBB) Degradation. *J. Korean Ceram. Soc.* **2017,** *54* (4), 308–313.
8. Chang, K.; Chen, W. L-Cysteine-Assisted Synthesis of Layered MoS_2/Graphene Composites with Excellent Electrochemical Performances for Lithium Ion Batteries. *ACS Nano* **2011,** *5,* 4720–4728.
9. Baby Suneetha, R., Selvi, P.; Vedhi, C. Synthesis, Structural and Electrochemical Characterization of Zn Doped Iron Oxide/Grapheneoxide/Chitosan Nanocomposite for Supercapacitor Application, *Vacuum* **2019,** *164,* 396–404.
10. Ji, Z.; Li, N.; Xie, M.; Shen, X.; Dai, W.; Liu, K.; Xu, K.; Zhu, G. High-Performance Hybrid Supercapacitor Realized by Nitrogen-Doped Carbon Dots Modified Cobalt Sulfide and Reducedgraphene Oxide. *Electrochim. Acta* **2020,** 1–16.
11. Maity, C. K.; Goswami, N.; Verma, K.; Sahoo, S.; Nayak, G. C. A Facile Synthesis of Boron Nitride Supported Zinc Cobalt Sulfide Nano Hybridas High-Performance Pseudocapacitive Electrode Material Forasymmetric Supercapacitors. *J. Energy Storage* **2020,** *32,* 101993.
12. Pramanik, A.; Maiti, S.; Dhawa, T.; Sreemany, M.; Mahanty, S. High Faradaic Charge Storage in ZnCo2S4 Film on Ni-Foam with a Hetero-Dimensional Microstructure for Hybrid Supercapacitor. *Mater. Today Energy* **2018,** *9,* 416–427.
13. Xin, N.; Liu, Y.; Niu, H.; Bai, H.; Shi, W. In-situ Construction of Metal Organic Frameworks Derived Co/Zn–S Sandwiched Graphene Film as Free-Standing Electrodes for Ultra-High Energy Density Supercapacitors, *J. Power Sour.* **2020,** *451,* 227772.

CHAPTER 13

Antimicrobial Activity of Poly Vinyl Pyrrolidine-Assisted Pure and Transition Metals (Fe/Zn/Ag)-Doped CuO Nanocomposites

D. GEETHA[1], P. S. RAMESH[2], and K. KANNAKI[3]

[1]*Department of Physics, Annamalai University, Chidambaram, Tamil Nadu, India*

[2]*Thiru Kolanjiappar Government Arts College, Virudhachalam, Tamil Nadu, India*

[3]*Government for Women College, Bargur, Tamil Nadu, India*

ABSTRACT

CuO is an important transition metal oxide due to its unique, unusual, interesting physicochemical and biological properties. Recently, CuO has been used as an antimicrobial agent. The antimicrobial activity was evaluated using PVP stabilized undoped and transition metals (Fe, Zn, and Ag)-doped CuO nanoparticles against human pathogens like *S. aureus* (Gram-positive), *P. aeruginosa* (Gram-negative) bacterial, *P. notatum*, and *A. Niger* (fungal) strains. The antimicrobial response of these nanocomposite materials was explained by interactions of released metal ions with the human pathogens. The antimicrobial efficiency of the prepared undoped CuO is less when compared to transition metal (Zn and Ag) ions-doped CuO NPs, but higher than Fe-doped CuO significantly. The free metal ions play a vital role and the growth inhibition was identified in the order of Ag/CuO>Zn/CuO>CuO>Fe/CuO.

Bio-Based Polymers and Composites: Properties, Durability, and Applications. Amadou Belal Gueye, Sabu Thomas, Nandakumar Kalarikkal, & Modou Fall (Eds.)

13.1 INTRODUCTION

In modern days, nanostructures of metal oxides have gained vast attention from material scientists and engineers owing to their diverse properties compared with their bulk counterparts, which are essential to utilize these materials in various fields of technology. Due to the outbreak of infectious diseases caused by different pathogenic bacteria and fungi, there is a growing concern in the healthcare and food industry to develop effective antimicrobial agents to inhibit microbial growth. So far, organic compounds, essential oils, bacteria-originated antibacterial proteins (bacteriocins), enzymes, and fruit extracts, etc., have been used as biocidal agents. However, the growth of new resistant strains of bacteria and fungi has led to the searching of new bactericides and fungicides that can effectively reduce the harmful effects of microorganisms. In this context, inorganic nanocrystallite has elicited considerable interest as effective antimicrobial agents in controlling the microbes. The key advantages of inorganic nanomaterials are of safety and stability, less toxicity, and heat resistance.[1] Besides, the inorganic nanomaterials contain mineral elements essential to human beings and exhibit strong antibacterial and antifungal activities to pathogenic microbes when used in small concentrations. Many reports are available on the antimicrobial activity of copper oxide (CuO) nanoparticles. Nevertheless, the antimicrobial activity of transition metals-doped CuO was scarcely reported.

Polymers are the macromolecules that have played an essential role in the development of drug delivery technology by providing controlled discharge of therapeutic agents in cyclic dosage, constant doses over long periods, and tunable release of both hydrophobic and hydrophilic drugs.[2]

PVP a biocompatible polymer has important features like (1) stabilizer, (2) growth controller, (3) adhere, and (4) structure directing agent; in the preparation of metal oxide nanoparticles. In this report, PVP acts as both a size regulator and a capping agent because it hinders the nucleus of NPs from agglomeration via the polar groups. When used, this kind of PVP-assisted undoped and transition metals-doped CuO NPs in antimicrobial test, herein PVP acts as a binder which makes adhesion between the microorganism and surface of the NPs.[3,4] In this review, it is highlighted the preparation and characterization of PVP-assisted undoped and transition metals-doped CuO. The latest developments in polymers-assisted metal oxide NPs are capable of molecular identification to illustrate areas of research advancing the frontiers of drug delivery.

The main objective of this study was to synthesize non-toxic inorganic luminescent nanoparticles of CuO doped with Fe, Zn, and Ag and to create an interface between PVP-assisted inorganic materials for possible use as nanoscale antimicrobial agents for potential biological, medical, and pharmaceutical applications. For example, targeted drug delivery also addresses the mechanism of interaction between nano-sized metal oxide particles and microbes. It is focused on the factors of microbial species, metal compositions, and NP's physico-chemical properties. The anti-microbial activities of NPs can be explored for potential air purification applications, air quality control, and public health.

To realize the antimicrobial activity of pure CuO and transition metals (Fe, Zn, and Ag)-doped CuO NPs, we synthesized different-sized nanocomposites by changing the dopant concentration during the synthesis process. The anti-microbial response of the synthesized nanostructures against *S. aureus* (G^{+ve}), *P. aeruginosa* (G^{-ve}) bacterial strains, and two fungi (*P. notatum* and *A. niger*) was investigated.

13.2 MODEL ORGANISMS

The model bacterial and fungal microorganisms used in this study were *S. aureus* (G^{+ve}), *P. aeruginosa* (G^{-ve}) bacteria, and *P. notatum*, *A. niger* (fungi), respectively. They were acquired from the Department of Microbiology, Annamalai University, Chidambaram, Tamil Nadu, India. Cultures were prepared at 35°C ± 2°C in darkness overnight using MHA and RBA broth.

13.2.1 SAMPLE SELECTION CRITERIA

Photoluminescence (PL) intensity strongly depends on the crystallite size and the surface chemistry of the NPs, lower PL intensity leads to smaller particle size. Small crystals having a large surface-to-volume ratio may increase the antimicrobial activity. Among all the synthesized samples PVP-assisted undoped and transition metals (Fe, Zn, and Ag (14%))-doped CuO samples were selected for antimicrobial test because they possess the smallest size with less PL intensity among the various concentrations. Many reports showed that the antimicrobial activity of NPs is size, shape, luminescence intensity, and photocatalytic activity-dependent, etc.

The inclusion of metals into semiconducting oxides enhances the PL intensity, which increases the crystallite size up to 12% doping, when the concentration is tuned to 14% PL intensity is quenched which in turn decreases the size, which leads to induce more charge separation or reduces the recombination of charge carriers which lead to increase the antimicrobial activity of the nanosystem.

13.2.2 MATERIALS

PVP-assisted undoped and transition metals (Fe, Zn, and Ag)-doped CuO NPs synthesized via a facile hydrothermal route, de-ionized water, human pathogens viz., *S. aureus* (G^{+ve}), *P. aeruginosa* (G^{-ve}) bacterial, and *P. notatum and A. niger* fungal strains, Muller Hinton Agar (MHA), Rose Bengal Agar (RBA).

13.2.3 PREPARATION FOR ANTIMICROBIAL STUDY-KIRBY BAUER DISK DIFFUSION (KBDD) TECHNIQUE

Antimicrobial activities of the synthesized PVP-assisted CuO and transition metals (Fe, Zn, and Ag)-doped CuO nanostructures were done against the selected micro-organisms using KBDD technique. The clean cultures of bacterial organisms were subcultured in Muller-Hinton (MH) broth and fungal organisms were in Rose Bengal (RB) broth at 35°C ± 2°C on a rotary shaker at 160 rpm. For microbial growth, a sod of culture was prepared by spreading the 100 μL pure culture having 10 CFU/mL of every experimental organism on nutrient agar dishes using a sterilized glass-rod spreader. Agar dishes were left over for 10 min to allow the culture to get captivated. Disposal dishes inoculated with the tested microbes at a concentration of 1 mL solution were used for the test. Zone of inhibition was obtained and measured after 24 h for bacterial strain and 48 h for fungal strains at 35°C.

13.3 RESULT AND DISCUSSION

The eco-toxicity of four samples, namely PVP-assisted pure CuO and transition metals (Fe, Zn, and Ag)-doped CuO was studied in the aqueous medium. Four types of microorganisms as stated in the materials part have been chosen to investigate the antimicrobial activities of NPs against them.[5] Nanoparticles toxicity was mainly associated with their nano-size effect. Further, various literature data showed that antimicrobial properties of NPs depend on the basis of synthesized conditions (concentration of the precursors, temperature, PH level, size and shape of the NPs, optical properties like band gap, luminescence intensity, and photocatalytic, etc.).[6]

13.3.1 ANTIBACTERIAL PROPERTIES

The antibacterial activities were studied using the modified KBDD method by observing an inhibition zone around the NPs. After 24-h incubation at 37°C, bacterial growth in the MHA agar was visualized in Figure 13.1(a and b). The inhibition zone diameter was measured and reported in Table 13.1.

The differences in responsiveness appear to be related to the nature of their cell wall structures. *S. aureus* is composed of several layers of peptidoglycan with various pores that were more susceptible to the intracellular transduction caused by the transition metal oxide CuO NPs, leading to cell wall interruption.[7]

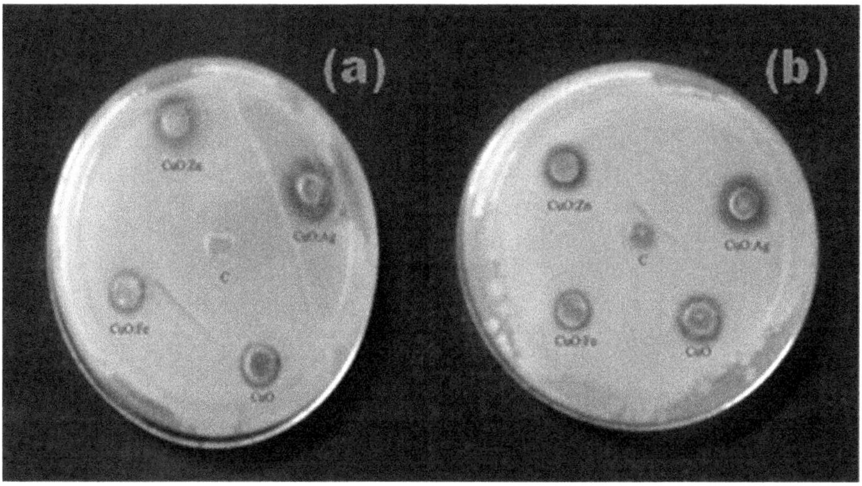

FIGURE 13.1 Zone of Inhibition produced by PVP-assisted undoped and transition metals (Fe, Zn, and Ag)-doped copper oxide NPs against (a) *S. aureus* (b) *P. aeruginosa* bacterial strains.

In contrast, the cell walls of *P. aeruginosa* (G^{-ve}) bacteria are relatively thin, mainly consisting of peptidoglycan and outer layers of lipopolysaccharide, lipoprotein, and phospholipids, which would be less prone to be attacked by the CuO NPs.

In summary, CuO and transition metal-doped CuO NPs have higher antibacterial activity against *S. aureus* than *P. aeruginosa*. Figure 13.2(a and b) shows bar graphs representation of the inhibition zone produced by prepared nano samples against *S. aureus* and *P. aeruginosa* bacterial strains, respectively. Among the various metal oxides used in this study, iron-doped

CuO shows poor antibacterial activity against both G^{+ve} and G^{-ve} strains. Silver-doped CuO nanoparticles show an excellent antibacterial activity.

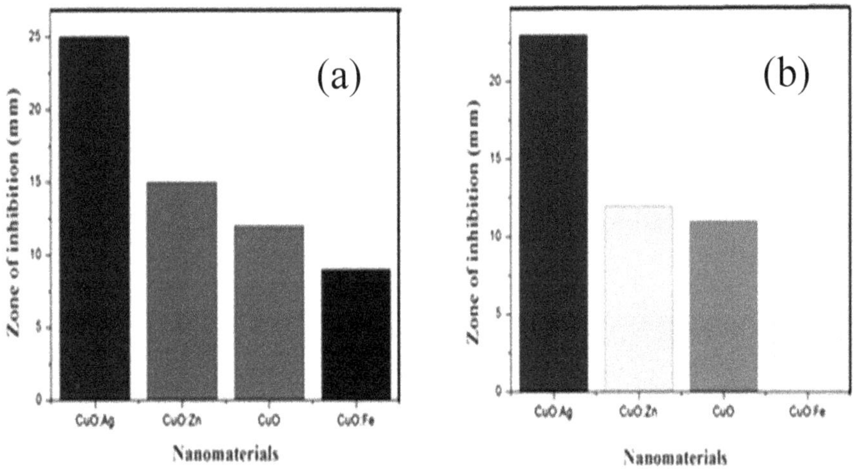

FIGURE 13.2 Graphs showing zone of inhibition formed by PVP-assisted undoped and transition metals (Fe, Zn, and Ag)-doped copper oxide NPs against (a) *S. aureus* (b) *P. aeruginosa* bacterial strains.

13.3.2 ANTIFUNGAL ACTIVITY

The antimicrobial activities were also studied using the KBDD technique by observing an inhibition zone around the NPs. *P. notatum* and *A. niger* were chosen as antifungal strains. Figure 13.3(a and b) shows the antifungal activities of PVP-assisted pure and transition metals-doped CuO against the selected fungal strains.[8,9]

After 48-h incubation at 37°C, fungal growth in the RBA agar was visualized as shown in Figure 13.3(a and b); diameters of the inhibition zones around the NPs in the order of $10 \leq 10 < 18 < 30$ mm for (*P. notatum*) and $10 < 12 < 17 < 24$ mm (*A. niger*) for $\text{CuO:Fe} < \text{CuO} < \text{CuO:Zn} < \text{CuO:Ag}$, respectively. Their different sensitivities appear to be related to the nature of cell wall structures of fungi.

The antifungal effects of the transition metals-doped CuO NPs were explained by interactions of NPs with the surface of fungi. Among the various metal oxides used in this study, iron-doped CuO shows the least antifungal activity against both *P. notatum* and *A. niger*. Silver-doped CuO nanoparticles show excellent antifungal activity.

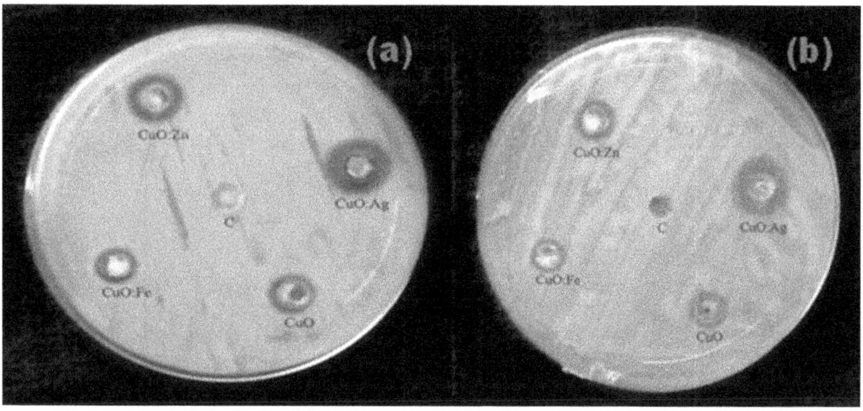

FIGURE 13.3 Zone of inhibition produced by PVP-assisted undoped and transition metals (Fe, Zn, and Ag)-doped copper oxide NPs against (a) *P. notatum* (b) *A. niger* fungal strain.

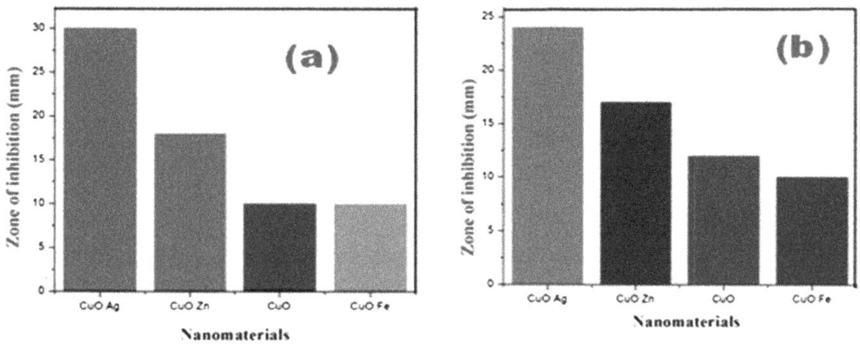

FIGURE 13.4 Graphs showing zone of inhibition formed by PVP-assisted undoped and transition metals (Fe, Zn, and Ag)-doped copper oxide NPs against (a) *P. notatum* (b) *A. niger*.

13.3.3 COMPARATIVE STUDY OF ANTIBACTERIAL AND ANTIFUNGAL ACTIVITY

From the overall view of the results, as shown in Figure 13.5, it is concluded that the antimicrobial activity of copper oxide (CuO) nanoparticles is associated with the toxicity of doped metal ions (Fe^{2+}, Zn^{2+}, and Ag^{2+}). Toxicity induced by these dissolved ions was considerable to be used in controlling antimicrobial activity.

The antibacterial and antifungal activity of transition metals-doped CuO nanostructures were studied by measuring the diameter of the inhibition zone and the values are given in Table 13.1.

The sizes of inhibition zone are different according to the type of bacteria. CuO nanostructures exhibit improved antibacterial activity for G^{+ve} bacteria, when compared to G^{-ve} bacteria.

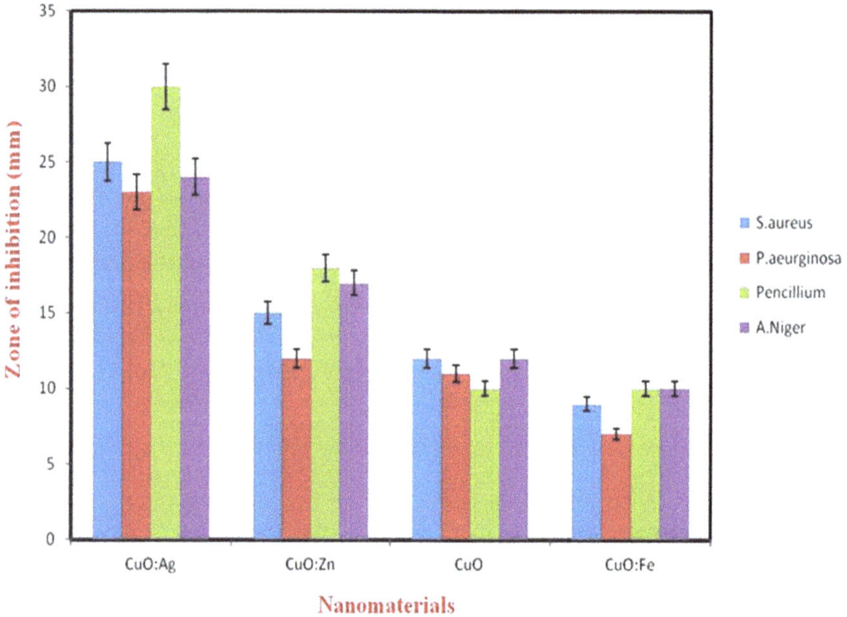

FIGURE 13.5 Graphs showing zone of inhibition formed by PVP-assisted undoped and transition metals (Fe, Zn, and Ag)-doped copper oxide NPs against *S. aureus* (G^{+ve}) and *P. aeruginosa* (G^{-ve}) bacterial strains and *P. notatum* and *A. niger* fungal strains.

TABLE 13.1 Inhibition Zone Data of PVP-Assisted Undoped and Transition Metals-Doped CuO Nanostructures for G^{+ve} and G^{-ve} Bacterial and Fungal Strains.

Nanomaterials	Bacterial strain		Fungal strain	
CuO	12	11	10	12
CuO:Fe	09	07	10	10
CuO:Zn	15	12	18	17
CuO:Ag	25	23	30	24

The cell wall structure, cell physiology, metabolism, or degrees of contact are the factors determining the difference in activity against these two types of bacteria. The cell wall of the G^{-ve} bacteria is composed of a thin layer of peptidoglycan, whereas G^{+ve} bacteria are composed of a thick layer of peptidoglycan, consisting of linear polysaccharide chains cross-linked by short peptides, thus forming more rigid structures.[10] It can be observed that the viability of the fungi has been suppressed considerably around the discs, which obviously demonstrates the antifungal activity of CuO nanostructures.

TABLE 13.2 Undoped and Various Transition Metals-Doped CuO NPs and Their Crystallite Size, and Optical Band Gap Energy Used in Antimicrobial Test (I-Intensity).

S. No	Doping metal (14%) with PVP assisted CuO	Band gap energy (eV)	Crystallite size from XRD (nm)	PL wavelength with intensity λ (I)
1	Undoped CuO	2.74	17	460 (370)
2	Fe	2.33	17	535 (109)
3	Zn	1.96	8	500 (447)
4	Ag	2.45	55	502 (90.2)

The results observed from this study, suggest that the doping of metals on the surface of metal oxide CuO nanoparticle size increases the value of the electron-hole charge separation via reducing the optical band-gap energy as shown in Table 13.2, which results in a delay in the recombination rate and causes an excellent antimicrobial activity. Among all the synthesized metal oxide nanomaterial's, Ag-doped CuO NPs exhibit the best antimicrobial activity against G^{+ve} and negative microbial strains used in this study, whereas Fe-doped CuO nanoparticles exhibit the least antimicrobial activity. It was observed from the results that Ag-doped CuO nanoparticles have an excellent antimicrobial potential, because they can inhibit the enzymatic system of the respiratory chain, thus changing the DNA of microbes. It is also believed that Ag NPs penetrate through the bacterial cell membrane, which triggers huge cell damage.

The inclusion of transition metals into copper oxide semiconductors leads to increase in crystallite size which in turn decreases the optical band gap of the various metal oxide nanopowders up to 12% of doping. Among them small-size crystallites have been chosen for antimicrobial activity, due to their direct and narrow band gap energies; also they show appreciable antimicrobial activity. The PVP-assisted pure CuO nanoparticles showed moderate antimicrobial activity, while doping of silver and zinc ions improves the efficiency, whereas doping of iron reduces the efficiency of antimicrobial activity of the parent nanosystem. Table 13.2 shows various transition metals-doped CuO NPs used in this study to investigate their effect on different microbes with their size and optical band gap energies.[11] It has been confirmed that the antimicrobial activity of metal oxide compounds in aquatic system was mainly caused by soluble ions. This is because the NPs often aggregate in the aqueous medium; their nano-associated particle effect on cellular function is reduced.

13.4 CONCLUSION

In this study, a cost-effective hydrothermally synthesized PVP-assisted CuO and transition metals-doped CuO nanostructures have been tested against various microorganisms. Among all the tested metal oxide nanomaterial's, Ag-doped CuO exhibited the best antimicrobial activity against all the microbes used in this investigation. It was concluded that Ag-doped CuO nanoparticles have an excellent antimicrobial potential; while Fe-doped CuO nanoparticles exhibited the least antimicrobial activity. Results demonstrate that flakes like Ag-doped CuO nanostructures exhibit good antimicrobial activities against the tested bacteria and fungi. It can be foreseen that CuO nanoparticles can be explored as a good biomaterial in the near future. G^{+ve} bacterial and fungal strains are more sensitive than G^{-ve} strains against the NPs tested. In the present study, the synthesized nanostructures showed remarkable antibacterial activity against both G^{+ve} and G^{-ve} bacteria. The extent of inhibition of bacterial growth observed in this study is found to be variable and size-dependent.

The biopolymer-templated CuO nanoparticles synthesized in ambient conditions show considerable potential for applications in a number of niches. The nanoparticles so formed are stable for about 1 month as observed. These nanoparticles with biocompatibility and antimicrobial potency may be exploited for various biomedical applications including the development of catheters, topical antimicrobial gel formulations, food packaging materials, food processing equipment, and so on.

KEYWORDS

- **antimicrobial activity**
- **transitional metal oxide**
- ***S. aureus***
- ***P. aeruginosa***
- ***P. notatum***
- ***A. Niger***

REFERENCES

1. Tankhiwale. R.; Bajpai, S. K. Preparation Characterization and Antibacterial Applications of ZnO-Nanocrystallite Coated Polyethylene Films for Food Packaging. *Colloids Surf. B* **2012,** *90*, 16–20.
2. R.; Karthik, S.; Shivashangari, K. S.; Ravikumar, V. Anticancer Activity of Ficus Religiosa Engineered Copper Oxide Nanoparticles. *Mater. Sci. Eng. C* **2014,** *44*, 234–239.
3. Nehra, P.; Chauhan, R. P.; Garg, N.; Verma, K. Antibacterial and Antifungal Activity of Chitosan Coated Iron Oxide Nanoparticles. *Br. J. Biomed. Sci.* **2018,** *75* (1), 13–18.
4. Prasanna, A. P. S.; Niranjan, R.; Kaushik, M.; Devasena, T.; Kumar, J.; Chelliah, R.; Swaminathan, S. Metal Oxide Curcumin Incorporated Polymer Patches for Wound Healing. *Appl. Surf. Sci.* **2018,** *449*, 603–609.
5. Maddinedi, S. B.; Mandal, B. K.; Pappu, G.; Anna, K. K.; Ghosh, A. R.; Reddy, P. S. Synthesis of CuO Nanosheets and its Applications Towards Catalysis and Antimicrobial Activity. *J. Indian Chem. Soc.* **2015,** *92*, 1–6.
6. Nouri, A.; Yaraki, M. T.; Ghorbanpour, M.; Agarwal, S.; Gupta, V. K. Enhanced Antibacterial Effect of Chitosan Film Using Montmorillonite/CuO Nanocomposite. *Int. J. Biol. Macromol.* **2018,** *109*, 1219–1231.
7. Ananth, A.; Dharaneedharan, S.; Heo, M. S.; Mok Y. S. Copper Oxide Nanomaterials: Synthesis, Characterization and Structure-Specific Antibacterial Performance. *Chem. Eng. J.* **2015,** *262*, 179–188.
8. Alavi, M.; Kennedy, J. F. Recent Advances of Fabricated and Modified Ag, Cu, CuO and ZnO Nanoparticles by Herbal Secondary Metabolites, Cellulose and Pectin Polymers for Antimicrobial Applications. *Cellulose* **2021,** *28* (6), 3297–3310.
9. Safaei, M.; Taran, M.; Imani, M. M. Preparation, Structural Characterization, Thermal Properties and Antifungal Activity of Alginate-CuO Bionanocomposite. *Mater. Sci. Eng. C* **2019,** *101*, 323–329.
10. Cho, K.; Park, J.; Osaka, T.; Park, S. The Study of Antimicrobial Activity and Preservative Effects of Nano Silver Ingredient. *Electrochem. Acta* **2005,** *51*, 956–960.
11. Renuga, D.; Jeyasundari, J.; Athithan, A. S.; Jacob, Y. B. A. Synthesis and Characterization of Copper Oxide Nanoparticles Using Brassica oleracea var. Italic Extract for its Antifungal Application. *Mater. Res. Express* **2020,** *7* (4), 045007.

CHAPTER 14

Effect of Sulfur Sources on the Properties of PVA-Stabilized Indium Sulfide for Photo Degradation Application

LILLY MARY[1], D. GEETHA[1], and P. S. RAMESH[2]

[1]*Department of Physics, Annamalai University, Chidambaram, Tamil Nadu, India*

[2]*Thiru Kolanjiappar Government Arts College, Virudhachalam, Tamil Nadu, India*

ABSTRACT

The preparation of Indium sulfide nanostructures is from the hydrothermal route at constant temperature (120°C) by using four different sulfur sources. $InCl_3$ and SC $(NH_2)_2$, TAA, Na_2So_3, L-cystine as precursors, Poly vinyl alcohol, and Oxalic acid are used as stabilizing and reducing agents, respectively. The investigation took place on the effect of change on the structural, morphological, and optical properties of Indium sulfide. It has been identified as a tetragonal structure by the X-ray diffraction patterns. The result shows that among the prepared samples $InCl_3$ with SC $(NH_2)_2$ sulfur sources have been improved structural and morphological properties. The result of the examined work on the effect of crystallinity, morphology, and optical studies with different sulfur source stoichiometry were analyzed. The best final product $InCl_3$ with SC $(NH_2)_2$ sulfur source was used for photodegradation of MO. The degradation response of (MO) organic dye under natural sunlight was investigated. The maximum degradation of MO reached about 90% after 150 min.

Bio-Based Polymers and Composites: Properties, Durability, and Applications. Amadou Belal Gueye, Sabu Thomas, Nandakumar Kalarikkal, & Modou Fall (Eds.)

14.1 INTRODUCTION

Photocatalytic is the same as to artificial photosynthesis process; water is separated into hydrogen and oxygen. Photocatalytic activity of nano-sized Indium sulfide is used for hydrogen production and also Indium sulfide Nanostructures have belonged to III–VI materials for various applications in optoelectronic devices. Technology has been a wide interest among these materials in recent years.[1] Specifically, metal sulfide nanomaterial exhibits electrical and optical properties with potential applications in photovoltaic, display devices, and batteries.[2] Among different metal chalcogenide nanomaterials, metal sulfide nanomaterials are given more attention because of their high carrier mobility, large band gap, and good photovoltaic properties.[3,4] Indium sulfide Nanostructures are shown to be potential materials due to their stability, photo conductive behavior, transparence to visible light, and also they are used as the photocatalytic material for the degradation of dyes.[5] Indium sulfide has tremendous advantages, due to its excellent illumination stability, superior visible light activity, stable chemical and physical characteristics, and low toxicity.[6] It is well-known that the size and morphology of nanomaterials have an important influence on their properties, such as the movement of electrons and holes and the transportation related to phonons.[7]

In general, the photocatalytic activity of the Indium sulfide depends on several factors such as crystalline phase, crystal size, and specific surface area as well as on other factors, such as production capacity of electron-hole pair, the separation efficiency of the photogene-rated charge pair, the transfer efficiency examined due to compounds adsorbed on the Indium sulfide surface.[8–10] In addition, the photoactivity of Indium sulfide also depends on the preparation method, thermal treatment, concentration of the pollutant, and concentration of the photocatalyst.[11] Numerous synthesis methods have already been reported that fabricate Indium sulfide with a variety of morphologies, such as hydrothermal, solvent thermal, chemical bath deposition, laser-induced synthesis, microwave synthesis route, and sonochemistry of indium organic or inorganic salts.[12,13] The preparation method describes the structure, grain size, morphology, and composition of the photocatalyst.[14] The different structures have been prepared using various methods but the hydrothermal method is used for low-temperature synthesis, cost-effectiveness, and large scale of production.[15]

Different precursors have been employed for the synthesis of Indium sulfide nanoparticles such as indium nitrate [$In(NO_3)_3$], indium acetylacetonate, and indium chloride ($InCl_3 \cdot 4H_2O$).[16] For Environmental clean, the need of alternative energy sources is to deliver by nano-sized materials to replace fossil fuels. The next step of newer photocatalytic nanomaterials will enable to produce the energy by solar in the form of Hydrogen.[17]

In the synthesis of Indium sulfide, the most common sulfur sources which are sulfoxide, thioglygolic acid, and thiols. Here, different sulfur sources are used to synthesize Indium sulfide, viz., thiourea, TAA, Na_2So_3, and L-cystine in the aqueous solution.[18] The photocatalytic properties are studied of as synthesized Indium sulfides are examined by the degradation of methyl orange (MO).[19] On the other hand, removal of organic pollutants, especially, different dye indicators like MO, rhodamine B, crystal violet (CV), etc., which are extensively used in dyeing, foodstuff, pulp, paper, and textile industries through the use of semiconductor-based photocatalysts has been a very active research topic in recent years.[20] In addition, the photocatalytic properties of synthesized Indium sulfide were thoroughly characterized and examined. The synthesized Indium Sulfide is used in rapid dye adsorption and excellent photodegradation capacity in aqueous solution is rapidly scrutinized.

14.2 EXPERIMENTAL METHODS

14.2.1 MATERIALS

In the purification process, most of the chemical reagents were used without doubt. $InCl_3$ (purity 97% Sigma Aldrich) was used as a starting to prepare crystalline Indium sulfide nanoparticles. $SC((NH_2)_2)$, Na_2So_3, TAA, L-cystein, and oxalic acid were precursors purchased from sigma–Aldrich. De-ionized water was used to prepare Indium sulfide nanoparticles.

14.2.2 PREPARATION OF INDIUM SULFIDE PHTOCATALYST WITH DIFFERENT SULFUR SOURCES

Indium sulfide nanoparticles were synthesized through a hydrothermal procedure. The 25 mmol $InCl_3$ and 35 mmol thiourea were, respectively, dissolved in 25 mL deionized water at room temperature for 10 min. After fully dissolved, 0.126 g of oxalic acid and poly vinyl alcohol were used as reducing and stabilizing agents. Ammonia as a complex agent was added to the solution and stirred vigorously for 30 min. The same procedure was followed for different sulfur sources like Na_2So_3, TAA, and L-cystine in the same concentration and transferred into a Teflon-lined autoclave, followed by a hydrothermal treatment at 240°C for 5 h. When the autoclave cooled at room temperature, the as-synthesized Indium sulfide precipitate from different concentrations was collected and washed sequentially with deionized water and ethanol several times by centrifugation and filtration. Then obtained as-synthesized precipitates were dried in a hot air oven at 80°C for 3 h.

14.2.3 CHARACTERIZATION

The crystal structure of the product was measured by using a powder X-ray diffractometer (XRD) (pw 3040/60) which operated at 40 kv and 30 mA using Cu Kα X-ray (1.54060 A°) in the 2θ range of 10–80° at a scan rate of 1° min^{-1}. The morphology of the synthesized product was examined with SEM. The optical property was measured by dispersing the as-synthesized Indium sulfide nanoparticles in ethanol and the transmittance as a function of wavelength was collected by using Perkin Elmer Lambda 750 UV-vis spectrophotometer.

14.3 RESULT AND DISCUSSIONS

14.3.1 X-RAY DIFFRACTION METHOD

Figure 14.1(A, B, C, and D) shown in the experimental XRD pattern are obtained for products of Indium sulfide from $InCl_3$ & $SC((NH_2)_2$, $InCl_3$ & Na_2So_3, $InCl_3$ & TAA, and $InCl_3$ with L-cystine four different sulfur sources are as precursors. The peaks for thiourea, Na_2So_3, TAA, and L-cystine are plotted.

The above figure indicates that Indium sulfide has the main peak after an hour of annealing at 300°C. Among the prepared four samples, $InCl_3$ and Thoiurea have a tetragonal crystal structure of Indium sulfide and it is well matched with the (JCPDS No: 65-1472) of the Indium sulfide nanoparticles (110) (101) (121) (221) (250). The cubic $In(OH)_3$ phase disappeared after annealing. Instead, the XRD pattern of cubic In_2O_3 phase (JCPDS no. 65-3170), as revealed by two intense XRD peaks of (222) and (440) planes appearing at 30.52 and 50.90, could be obviously observed. The weight fractions of In_2O_3 and Indium sulfide had the same change trends with those of $In(OH)_3$ and Indium sulfide. The diameter of the Indium sulfide was calculated from the full-width half maximum (FWHM), which was obtained from the XRD, using Scherer's equation

$$D = K\lambda/\beta\cos\theta - \text{Scherer equation}$$

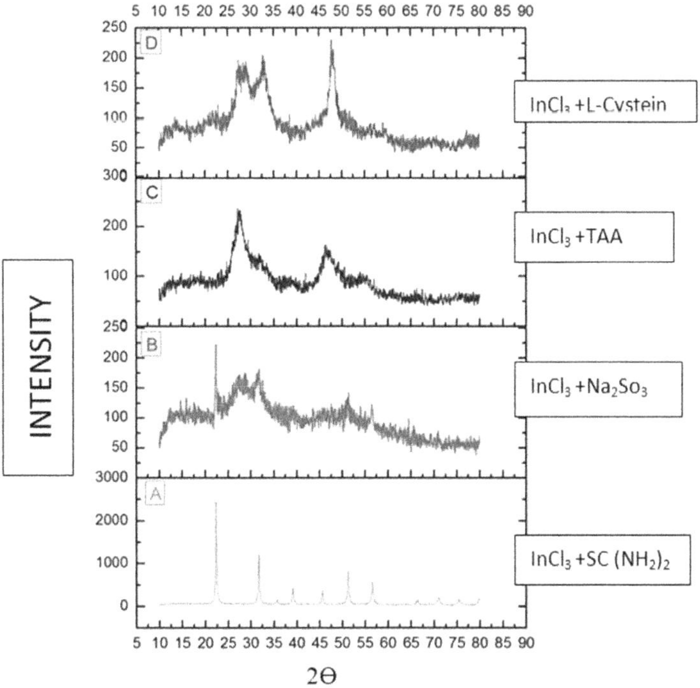

FIGURE 14.1 XRD patterns of indium sulfide nanoparticles (A) thiourea, (B) Na_2So_3, (C) TAA, (D) L-cystine.

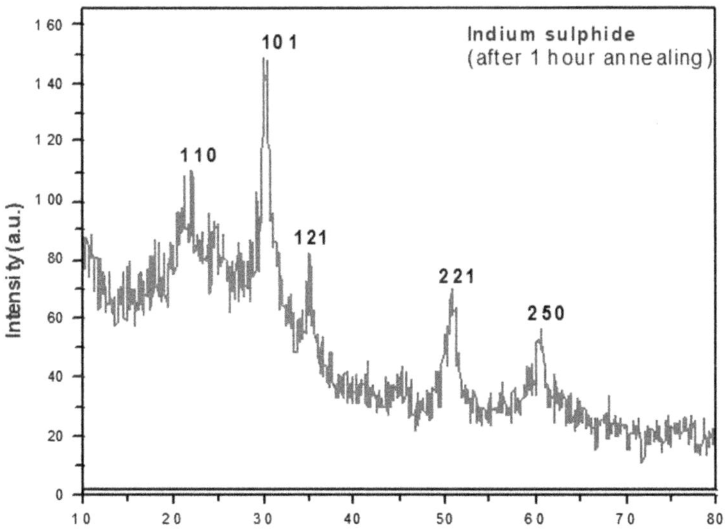

Where "β" is the breadth of the observed diffraction line at the level of half its maximum intensity, "λ" is the wavelength of the X-ray source which is used in XRD and "K" is the so-called shape factor, which usually takes a value of approximately 0.9.

14.3.2 SEM ANALYSIS

Figure 14.2 shows the Indium sulfide photocatalyst was investigated by SEM analysis. The morphological study of Indium sulfide nanoparticles is shown in Figure 14.2. The nature of Indium sulfide was observed in uniformed-sized particles arranged homogeneously. The preparation of Indium sulfide nanostructures thiourea was used as the major sulfur source, oxalic acid, and PVA as shape controlling agent. Morphologies of Indium sulfide nanoparticles were shown in pure form, which observed the catalyst in crystalline nature. The sizes of particles were shown as higher due to agglomeration in SEM analysis.

FIGURE 14.2 SEM image of indium sulfide ($InCl_3$ + Thoiurea).

14.3.3 PHOTO CATALYTIC EXPERIMENT

The experiment analyzed Indium sulfide nanoparticles with the photodegradation of an organic dye such as MO under the wavelength of λ > 400 nm in photocatalytic experiment. The photo catalytic study of the mixture 0.0005 g of MO and 0.05 g of the Indium sulfide from thiourea is observed which acts as a target pollutant in the solution. The solution mixture is aerated for 150 min with UV-vis, and under the photocatalytic study at adsorption equilibrium, there is no change of MO after 150 min. photo catalytic activity performed at room temperature when the mixture by placing 15 cm vessel inside the photo reactor under the visible light source. The degradation rate is calculated after centrifugation and analyzed using UV-vis spectroscopy.

$$\text{Degradation rate (\%)} = A_0 - A\ A_0 \times 100$$

Where A_0 and A are the absorbance of solution at A_0 and A min, respectively.

14.3.4 OPTICAL PROPERTIES

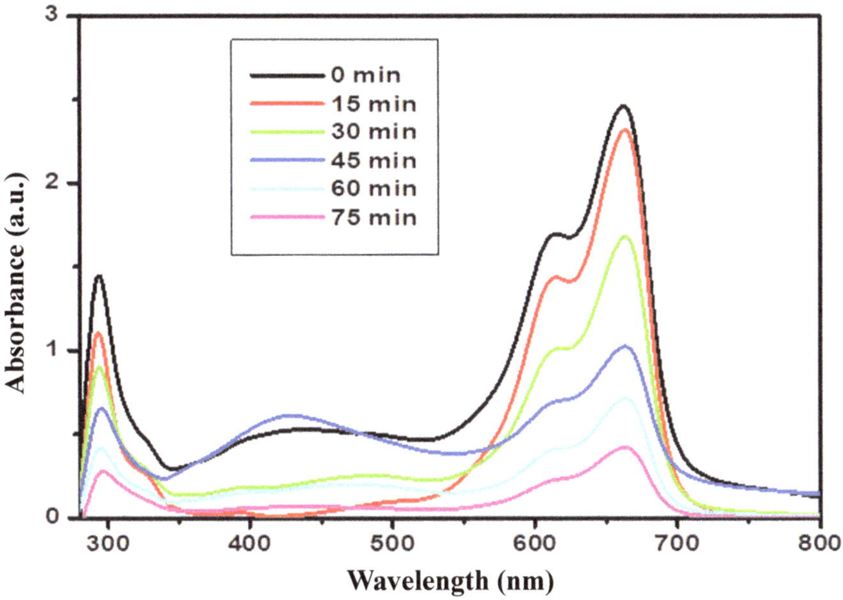

FIGURE 14.3 UV–visible spectra of MO degradation by indium sulfide.

The optical properties of nanoparticles Indium sulfide were studied by UV-vis absorption spectroscopy which was obtained from $InCl_3$ and Thoiurea using different sulfur sources as precursors. Indium sulfide nanostructures are synthesized using Indium chloride and different precursors show the UV-vis region without a clear band edge in strong absorption. Indium sulfide acts as a visible light photocatalyst at the visible region of electromagnetic radiation.

14.4 CONCLUSION

In summary, the hydrothermal method was employed successfully to synthesize Indium sulfide nanostructures with $InCl_3$ and with different sulfur sources and their XRD revealed pure orthorhombic crystalline in Indium sulfide from $InCl_3$ and Thoiurea in their sharpen peaks. This study examined the morphology and optical properties that were influenced by the different sulfur sources in Indium sulfide. From the observation of the I–V curves, Indium sulfide nanoparticles have the property of photocatalytic activity. Thiourea was used as a major source for the synthesis of Indium sulfide In the visible light after 150-min irradiation, methyl orange was degraded 90% approximately. From the results, the effects of sulfide sources were better applicable photocatalyst.

KEYWORDS

- **sulfur**
- **PVA**
- **indium sulfide**
- **photo-catalytic**
- **photo-degradation**

REFERENCES

1. Nayak, A. K.; Lee, S.; Sohn, Y.; Pradhan, D. Synthesis of In 2 S 3 Microspheres Using a Template-Free and Surfactant-Less Hydrothermal Process and Their Visible Light Photocatalysis. *CrystEngComm*. **2014,** *16* (34), 8064–8072.
2. Botha, N. L.; Ajibade, P. A. Optical and Structural Characterization of Copper Sulphide Nanoparticles from Copper (II) Piperidine Dithiocarbamate. *Opt. Quantum Electron.* **2020,** *52* (7), 1–11.
3. Geetha, D.; Ramesh, P. S. Highly Stable Low Cost Anode Interfacial Material Preparation and Characterization via Hydrothermal Route for Organic Solar Cell. *Mater. Today Proc.* **2017,** *4* (2), 4319-–4328.
4. Wei, C.; Guo, W.; Yang, J.; Fan, H.; Zhang, J.; Zheng, W. Facile Solvothermal Synthesis of 3D Flowerlike β-In 2 S 3 Microspheres and their Photocatalytic Activity Performance. *RSC Adv.* **2014,** *4* (92), 50456–50463.
5. Ai, C.; Wang, Q.; Lei, Y.; Shao, X. Solar Photocatalytic Activity of Indium Sulfide Nanocrystal and Degradation of the Tetracycline by In2S3. *Desalin. Water Treat.* **2016,** *57* (45), 21428–21436.

6. Ameta, R.; Soni, D.; Benjamin, S.; Chouhan, N.; Ameta, S. C. Nano-Sized Photocatalytic Materials for Solar Energy Conversion and Storage. In *Materials Science Forum*; Trans Tech Publications Ltd., 2016; vol 855, pp 58–77.
7. Sharma, R. K.; Chouryal, Y. N.; Chaudhari, S.; Saravanakumar, J.; Dey, S. R.; Ghosh, P. Adsorption-Driven Catalytic and Photocatalytic Activity of Phase Tuned In 2S3 Nanocrystals Synthesized via Ionic Liquids. *ACS Appl. Mater. Interfaces* **2017,** *9* (13), 11651–1166
8. Feng, X.; Chen, Y.; Wang, M.; Guo, L. Hydrothermal Synthesis of Pyramid-like In2S3 Film for Efficient Photoelectrochemical Hydrogen Generation. *Int. J. Hydrog. Energy* **2017,** *42* (22), 15085–15095.
9. Wei, X.; Wang, C.; Ding, S.; Yang, K.; Tian, F.; Li, F. One-Step Synthesis of Ag Nanoparticles/Carbon dots/TiO2 Nanotube Arrays Composite Photocatalyst with Enhanced Photocatalytic Activity. *J. Environ. Chem. Eng.* **2021,** *9* (1), 104729.
10. Xie, L.; Ren, Z.; Zhu, P.; Xu, J.; Luo, D.; Lin, J. A Novel CeO2–TiO2/PANI/NiFe2O4 Magnetic Photocatalyst: Preparation, Characterization and Photodegradation of Tetracycline Hydrochloride Under Visible Light. *J. Solid State Chem.* **2021,** *300*, 122208.
11. Chen, Y.; Jiang, Z.; Xu, L.; Liu, C.; Cheng, Y.; Zou, Y.; Zhang, Q. Composite Magnetic Photocatalyst Bi24O31Br10/NiFe2O4: Hydrothermal Preparation, Characterization and Photocatalytic Mechanism. *Mater. Sci. Semicond. Process.***2021,** *126*, 105669.
12. Yan, T.; Feng, Y.; Ren, X.; Li, J.; Lu, Y.; Sun, M.; Yan, L.; Wei, Q.; Ju, H. Fabrication of CDs Hybrid MIL-68 (In) Derived In2O3In2S3 Hollow Tubular Heterojunction and Their Exceptional Self-Powered PEC Aptasensing Properties for Ampicillin Detecting. *J. Materiomics* **2021,** *7* (4), 721–727.
13. Wang, L.; Zhou, Y.; Xu, Q.; Wu, X. K.; Xu, G. D.; Guo, J. X.; Fang, D. Fabrication of In2S3/MIL-68 (In) Heterojunction Composite Photocatalysts for Degradation of Rhodamine B and Hydrogen Evolution. *J. Porous Mater.* **2022,** *29* (1), 181–192.
14. Li, S.; Liao, J.; Dong, Y.; Fu, Y.; Zhu, Y. Enhanced Photocatalytic Activity of Ternary g-C3N4/NaTaO3/Biomass Carbon Composite Photocatalysts Under Visible-Light Radiation. *J. Mater. Sci. Mater. Electron.* **2020,** *31* (22), 19613–19622.
15. Mora-Hernandez, J. M.; Huerta-Flores, A. M.; Torres-Martínez, L. M. Tailoring Charge Transport in BaBiO3/NaTaO3 Heterojunction Interface for Enhanced Photocatalytic and Photoelectrochemical H2 Generation. *J. Photochem. Photobiol. A Chem.* **2020,** *391*, 112363.
16. da Silva, M. P.; de Souza, A. C. A.; de Lima Ferreira, L. E.; Neto, L. M. P.; Nascimento, B. F.; de Araújo, C. M. B.; Fraga, T. J. M.; da Motta Sobrinho, M. A.; Ghislandi, M. G. Photodegradation of Reactive Black 5 and Raw Textile Wastewater by Heterogeneous Photo-Fenton Reaction Using Amino-Fe3O4-Functionalized Graphene Oxide as Nanocatalyst. *Environ. Adv.* **2021,** *4*, 100064.
17. Lai, X.; Zhu, F.; Wu, Y.; Huang, R.; Wu, X.; Zhang, Q.; Yang, K.; Qin, S. New High-Pressure Polymorph of In2S3 With Defect Th3P4-Type Structure. *J. Solid State Chem.* **2014,** *210* (1), 155–159.
18. Horani, F.; Lifshitz, E. Unraveling the Growth Mechanism Forming Stable γ-In2S3 and β-In2S3 Colloidal Nanoplatelets. *Chem. Mater.* **2019,** *31* (5), 1784–1793.
19. Xue, B.; Xu, F.; Wang, B.; Dong, A. Shape-Controlled Synthesis of β-In 2 S 3 Nanocrystals and their Lithium Storage Properties. *CrystEngComm* **2016,** *18* (2), 250–256.
20. Gao, Y.; Zhang, S.; Bu, X.; Tian, Y. Surface Defect Engineering Via Acid Treatment Improving Photoelectrocatalysis of β-In2S3 Nanoplates for Water Splitting. *Catal. Today* **2019,** *327*, 271–278.

CHAPTER 15

PLA Stabilized/Graphene-Based Copper Sulfide-Hole Transport Material for Organic Photovoltaics

V. MENAKA[1], D. GEETHA[1], and P. S. RAMESH[2]

[1]*Department of Physics, Annamalai University, Chidambaram, Tamil Nadu, India*

[2]*Thiru Kolanjiappar Government Arts and Science College, Vriddhachalam, Tamil Nadu, India*

ABSTRACT

The syntheses of poly lactic acid (PLA)-stabilized and graphene-based Cu-S nanoparticles were successfully synthesized by a simple hydrothermal route. The obtained nanostructure was characterized by XRD, UV-Vis, and SEM/EDS studies. The XRD pattern reveals that the nanostructure showed a crystalline nature and which increases with an increase in concentration in CuS on graphene. The effect of polymer (PLA) on the morphology of the copper sulfide was studied via SEM/EDS. Morphological analysis showed that the nanoparticles were uniform and EDS analysis confirmed the components present in the final product. An increase in PLA concentration resulted in a smaller copper sulfide size, decreasing from 32 to 24 nm. The prepared nanomaterial is transferred to glass/Indium tin oxide treated as a hole transport layer in solar cells. The cells attained a maximum power that graphene-based CuS could represent an effective alternative to hole transport/electron layer in organic photovoltaics.

Bio-Based Polymers and Composites: Properties, Durability, and Applications. Amadou Belal Gueye, Sabu Thomas, Nandakumar Kalarikkal, & Modou Fall (Eds.)

15.1 INTRODUCTION

High carrier mobility, mechanical flexibility, and transparency of graphene have received more attention in the research community which is recognized as transparent hole transport material in photovoltaic devices. Copper sulfide nanostructures are of great interest due to their unique properties. A best method adapted to control over partial size, shape is of critical importance. Hence to stabilize the CuS nanostructures, very small concentration of polymer (PLA) is used. This is novel simple and low-cost preparative technique to enhance CuS NPs properties is needed the day to scale up the production of CuS on graphene. Among the semiconducting materials, the combination of graphene-polymer-p-type semiconductors has not been studied. Besides the band gap, other properties of poly lactic acid (PLA)-stabilized CuS-graphene make this material for solar cell applications due to its extraordinary absorption co-effecient, hole mobility, electrical conductivity, and hole concentration. There is a need to minimize electron recombination and modified band gap for photovoltaic applications.

Graphene with semiconductor materials can be prepared by various methods.[1–4] All these materials focus on electrochemical applications.[5] Among all the other methods, the simple hydrothermal method has been adopted for this investigation, because of its versatile advantages viz., reaction kinetics, single crystals growth, controlled temperature, and high purity. Using the hydrothermal process, we have synthesized copper sulfide on graphene with the addition of PLA in the reaction mixture. In these nanostructures, the tuning of optical and electrical properties can be made by inducing changes in the material composition, size, structure, and morphology. Polymeric materials play a vital role due to their ease of processing, solubility, less toxicity, and controlled growth. PLA has been used as a sterilization and reducing agent of the sample.[6,7] The synthesized G-CuS (PLA) exhibit enhanced light absorption with a modified band gap and the obtained results prove this material having potential candidates for solar cell. To the best of our knowledge, this is the first report demonstrating the synthesis of PLA-stabilized CuS-G structures at this temperature (130°C) using polymer (PLA) as a mild template on graphene.

15.2 EXPERIMENTAL

15.2.1 A MODIFIED HUMMERS' METHOD WAS PREVIOUSLY USED TO SUCCESSFULLY PREPARE GO

To begin, 1 g graphite powder and 1 g $NaNO_3$ were combined and added to 50 mL concentrated sulfuric acid, with the temperature of the reaction maintained at 0°C using an ice bath. After that, 6 g KMn04 was added. Slowly reduce the temperature of the reaction to below 20°C. The mixture was then kept at 35°C for 1 h while being constantly stirred. In the aforementioned solution, 80 mL pure water was added dropwise and the temperature was raised to 95°C for 30 min. Finally, 200 mL of distilled water was added. To obtain yellow GO solution, 6 mL 30% H_2O_2 solution was added to water. The concoction was to make the ethanol, it was centrifuged and rinsed multiple times with distilled water, hydrochloric acid, and ethanol the resulting mixture was rinsed repeatedly.

15.2.2 RGO/CuS COMPOSITE SYNTHESIS

In 50 mL of highly dispersed aqueous GO solution (2 mg/mL), 20 mL of $CuCl_2$ solution (1 mmol) was added under continual stirring for 30 min. Then, under magnetic stirring for 30 min, 20 mL of thiourea solution (2 mmol) was dropped into the aforesaid mixture dropwise; the black color precipitate was formed.

15.3 RESULT AND DISCUSSION

15.3.1 STRUCTURAL ANALYSIS (XRD)

The structural properties of PLA-stabilized CuS on graphene were investigated using powder X-ray diffraction patterns are shown in Figure 15.1(a–d) at four different reaction times (6, 8, 10, and 12 h) at a constant temperature of 130°C. When comparing the result of XRD with the JCPDS card details 06-0464 corresponding to synthetic hexagonal, the main peaks coincide with

27.6°, 29.2°, 32.1°, 47.9°, 52.7°, and 59.3° of primitive hexagonal structure of covellite (CuS). There is no significant difference in the main diffraction peaks, but the intensity of some peaks varies as a function of the reaction time. Based on the XRD results, the main peak and intensities match those of covellite, observing that the crystallinity of the particles obtained at 10 h (reaction time) is higher than those obtained at low temperature.

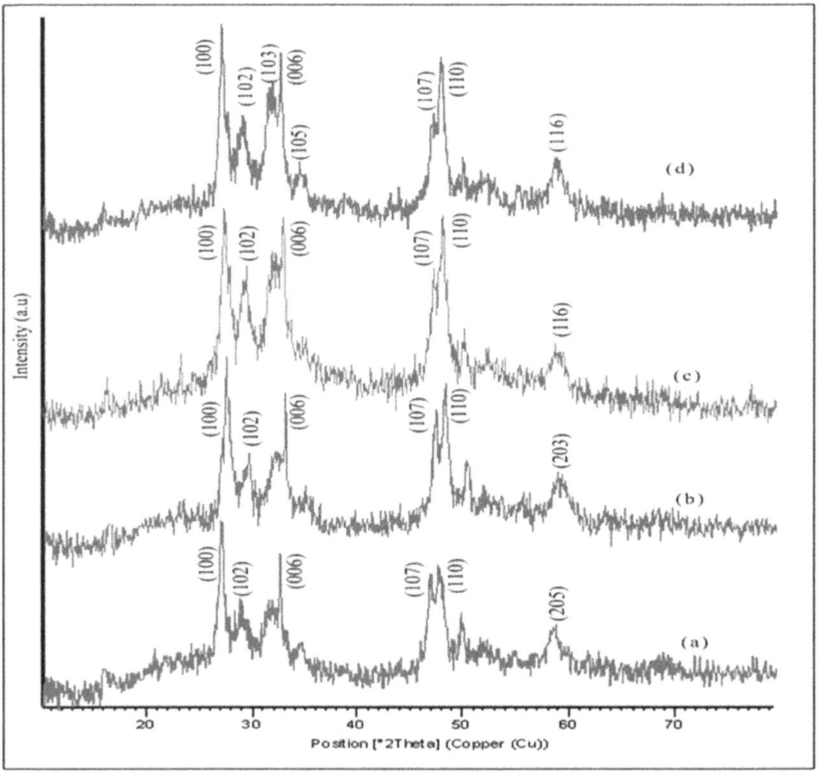

FIGURE 15.1 XRD patterns of the PLA (0.005%) stabilized CuS-G nanostructures CuS1-G, (b) CuS2-G, (c) CuS3 –G, and (d) CuS4-G.

An initially large domain contributing coherently to that respective XRD peak will produce a line only, but the finite resolution of the instrument itself and other factors with broadening in β was considered. Broadening of XRD lines is also associated with small particle size of the coherently diffracting crystallites or strains present within the sample. All the diffraction peaks can be indexed as CuS with hexagonal structure that the sample as (100), (102), (103), (006), (107), (110), and (116) reflections are significantly strong and narrow. Therefore, the synthesized products can easily be assigned to a pure CuS with hexagonal phase. The stronger (100), (103), and (110) reflections are most likely to the sphere-like morphology.

15.3.2 OPTICAL ABSORPTION ANALYSIS (UV-VIS)

The spectroscopic study on copper sulfide with different reaction times is presented in Figures 15.2(a–d). It has been described that the overall structure of the spectrum was not affected by the size of the particle. The UV-vis absorption is much broader in the visible range. It indicates that the structure of CuS-G leads to the broadening of optical absorption. Therefore, the CuS-G structure has great potential in the field of optoelectronics. The λ_{max} spectra of CuS exhibited some difference at the maximum absorption wavelength. The λ_{max} of CuS nanospheres were 346, 314, 321, and 491 nm for 6, 8, 10, and 12 h (CuS samples), respectively. It can be observed from the spectra that the reaction time has a significant influence on the peak orientation of the obtained product. The curve presents a shift of minima absorption to an infrared domain, with increasing reaction time.

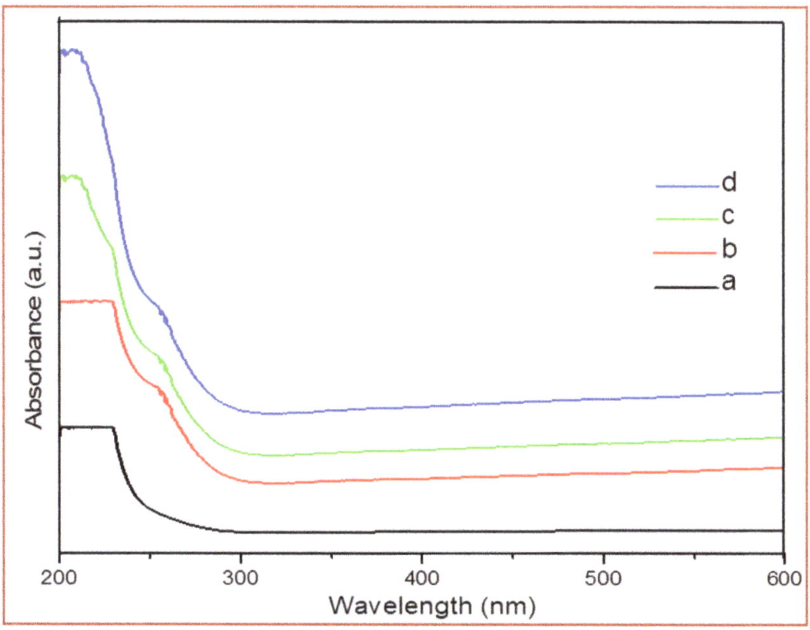

FIGURE 15.2 UV-visible absorption spectra of PLA (0.005%)-stabilized CuS-G nanostructures (a) CuS1-G, (b) CuS2-G, (c) CuS3-G, and (d) CuS4-G.

Optical band gap has been calculated from the UV-vis absorbance data for CuS-G nanoparticles 6, 8, 10, and 12 h. The energy band gap was calculated using the relation given below

$$(\alpha h\nu)^2 = A(h\nu - E_g)$$

The energy absorption gap is of direct type and the band gap energy is found to be 1.6–3.2eV. The absence of absorption bands in the visible region and the wide band gap of the nonmaterials attest to the suitability of the nonmaterials are for photonic and optical applications. The wide energy band gap shows that the defect concentration in the nonmaterials is very low and has large transmittance in the visible region. It is interesting that the UV-visible absorption spectrum has an up slight slope with the wavelength from 314 to 491 nm, which is different from the literature.

15.3.3 SURFACE MORPHOLOGICAL ANALYSIS (SEM/EDS)

To investigate the growth process of the nanosphere-like morphology obtained by hydrothermal route with the temperature value 130C, time-dependent experiments were carried out. The morphology and the size of the synthesized products were characterized by SEM. The SEM image in Figure 15.3(a–d) provides a representative overview of the sample. Figure shows the morphologies of product synthesized under different reaction times. It can be seen that the product consists of agglomerate at this stage (Figure 15.3a). Initially, formed nanospheres were not very clear. When the time was prolonged from 6 to 10 h, more uniform nanospheres were found. It can be seen from Figures 15.3(b–d) that the product obtained after 10 h of reaction time is composed of a large quantity of hierarchical spheres having a mean diameter of 32 nm indicating that a high yield of hierarchical structures can be realized under the adopted experimental conditions. From Figures 15.3c and 15.3d for 10 and 12 h reaction time, the particle size was estimated at 24 and 27 nm but an agglomeration of the samples was observed. Agglomeration of nanocrystals could be removed by adding suitable stabilizer, suitable concentration of templating materials or micron-size particles with displaying irregular shapes.

The factor (reaction time) affecting the morphology of CuS-G can be ascribed to the reaction time. As shown in Figure 15.4a, the average diameter of a nanosphere is 20 nm whereas at 10–12 h, it exhibits clear nanospheres and the average diameter is 35, 27, and 24 nm, respectively. It reveals that different reaction time results in the formation of uniform nanosphere morphology. A typical EDS pattern of prepared CuS nanosphere demonstrates that the sample consists of Cu, carbon, and S. The product obtained from this route can be observed to consist of agglomerated nanocrystals, as well as micron-size particles, displaying irregular shapes.

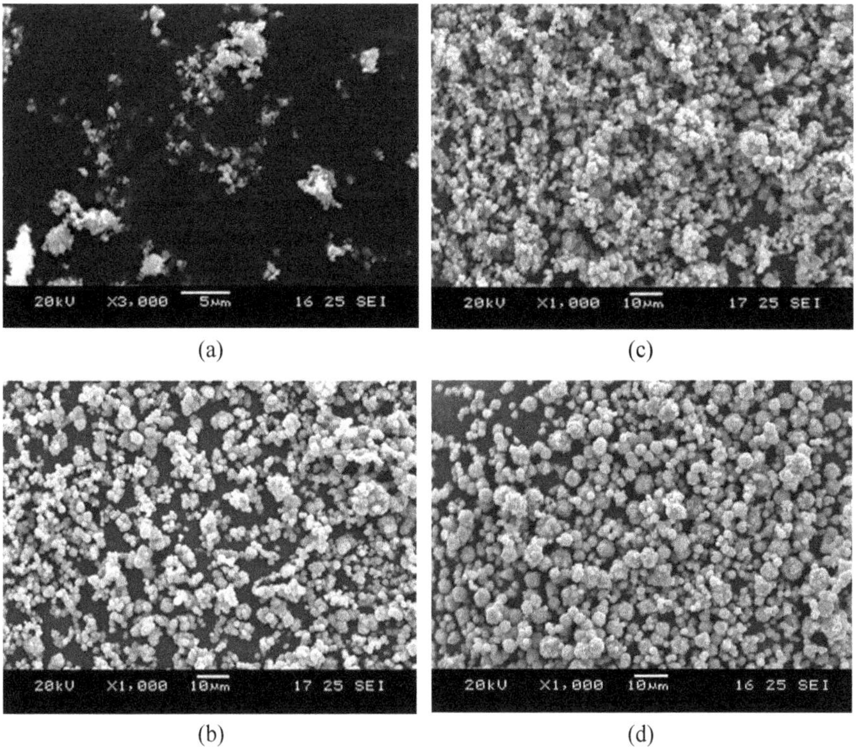

FIGURE 15.3 Surface morphology of PLA (0.005%)-stabilized CuS-G nanostructures (a) CuS1-G, (b) CuS2-G, (c) CuS3-G, and (d) CuS4-G.

15.3.4 EFFECT OF REACTION TIME

A series of experiments was carried out to systematically investigate the influence of experimental parameters such as sulfur (thiourea) and reaction time on the evolution of the product morphology. It can be seen that the product is exclusively nanoparticles with a diameter of about 20–35 nm and the hierarchical structures can occur at 10 h compared to 6 h. In contrast, excess reaction time leads to the fast hydroxylation of thiourea, so that a large amount of CuS-G precipitates was produced and the nucleation and growth of the CuS crystals are too rapidly to be favorable for the formation of the well-defined spheres.

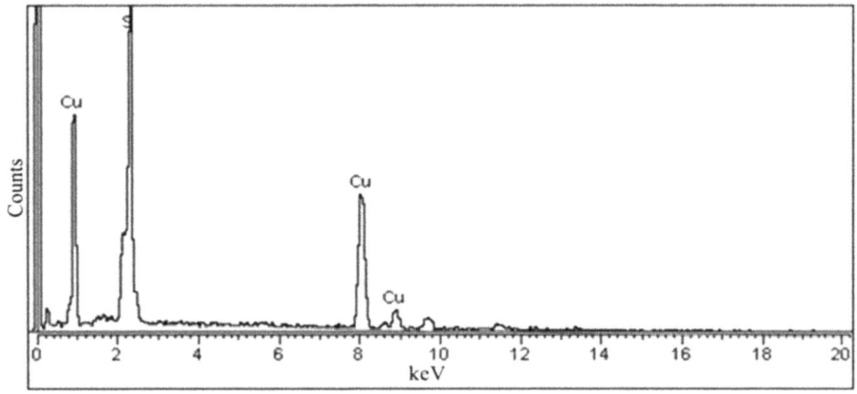

FIGURE 15.4 (a) EDS spectrum of CuS-G nanostructures at 130°C.

15.3.5 PHOTO CONDUCTIVITY STUDIES

The best sample (PLA-stabilized CuS-G (10 h)) was chosen for this study, the prepared film acts as a conductor under 100 W cm^{-2} halogen lamp with bias voltage at room temperature for I–V measurements. CuS-G nanostructures used in between ITO and Ag contacts. The photocurrent generated by the sample, due to the surface-to-volume ratio of the product increases the photon-induced excitation of electrons and its dissociation through the electrostatic interaction in the sample between graphene and CuS. The absorbed G on the surface of CuS helps in trapping the electrons to the electrodes which increase the photocurrent. Hence from this investigation, PLA-stabilized CuS composite photo voltaic devices can be fabricated for solar cell.

15.4 CONCLUSIONS

Synthesis of PLA-stabilized CuS-G was achieved using a simple hydrothermal route. The average particle size was calculated from XRD data which are approximately 20–35 nm. The obtained CuS-G (10 h) with polymer (0.005%) promotes the CuS-G nanostructures with high yield. In general, the final products at other reaction conditions have a lesser significant effect on the particle size. The reaction times have a remarkable influence on the nucleation which leads to the formation of CuS-G nanospheres. For increasing the reaction time, the particle size has decreased. The average value of the forbidden energy gap evaluated by graphical determinations is 3.2eV, corresponding to a semiconductor material. The rise in the photocurrent is due to the increased light absorbance and enhanced electron transport in the material. The method for the synthesis of CuS-G nanoparticles reported in this work is simple and feasible for production at large scale of CuS-G, nanoparticles with potential applications in solar cells.

KEYWORDS

- **PLA-CuS-G**
- **graphene**
- **copper sulfide**
- **organic photo-voltaic**
- **nano-materials**
- **solar cell**
- **hole transport layer**

REFERENCES

1. Su, Y.; Lu, X.; Xie, M.; Geng, H.; Wei, H.; Yang, Z.; Zhang, Y. A One-Pot Synthesis of Reduced Graphene Oxide-Cu2S Quantum Dot Hybrids for Optoelectronic Devices. *Nanoscale* **2013,** *5*, 8889–8893.
2. Myung, Y.; Jang, D. M.; Cho, Y. J.; Kim, H. S.; Park, J.; Kim, J. U.; Lee, C. J. Nonenzymaticamperometric Glucose Sensing of Platinum, Copper Sulfide, and Tin Oxide Nanoparticle–Carbon Nanotube Hybrid Nanostructures. *J. Phys. Chem. C* **2009,** *113*, 1251–1259.
3. Radich, J. G.; Dwyer, R.; Kamat, P. V. Cu2S Reduced Grapheme Oxide Composite for High-Efficiency Quantum Dot Solar Cells. Overcoming the Redox Limitations of S2-/Sn2- at the Counterelectrode. *J. Phys. Chem. Lett.* **2011,** *2*, 2453–2460.
4. Pham, T. A.; Choi, B. C.; Jeong, Y. T. Facile Covalent Immobilization of Cadmium Sulfide Quantum Dots on Graphene Oxide Nanosheets: Preparation, Characterization, and Optical Properties. *Nanotechnology* **2010,** *21* (46), 465603.
5. Zhang, Z.; Zhou, C.; Jia, M.; Fu, Y.; Li, J.; Lai, Y. Synthesis of Copper Tin Sulfide/Reduced Graphene Oxide Composites and Their Electrochemical Properties for Lithium Ion Batteries. *Electrochim. Acta* **2014,** *143*, 305–311.

6. DiLeo, L.; Romano, D.; Schaeffer, L.; Gersten, B.; Foster, C.; Gelabert, M. C. Effect of Complexing Agent on Hydrothermal Growth of ZnO Crystals. *J. Cryst. Growth* **2004,** *271*, 65–73.
7. Xue, F.; Bi, N.; Liang, J.; Han, H. A Simple and Efficient Method for Synthesizing Te Nanowires from CdTe Nanoparticles with EDTA as Shape Controller Under Hydrothermal Condition. *J. Nanotechnol.* **2012,** *2012*, 751519.
8. Usman, M. S.; Ibrahim, N. A.; Shameli, K.; Zainuddin, N.; Yunus, W. M. Z. W. Copper Nanoparticals Mediated by Chitosan: Synthesis and Characterization via Chemical Methods. *Molecules* **2012,** *17* (12), 14928–14936.
9. Bakra, Z. H.; Walia, Q.; Fakharuddinc, A.; Schmidt-Mendec, L.; Browne, T. M.; Josea, R. Advances in Hole Transport Materials Engineering for Stable and Efficient Perovskite Solar Cell. *Nano Energy* **2017,** *34*, 271–305.
10. Matondo, J. T.; Maurice, D. M.; Chen, Q.; Bai, L.; Guli, M. Inorganic Copper-Based Hole Transport Materials for Perovskite Photovoltaics: Challenges in Normally Structured Cells, Advances in Photovoltaic Performance and Device Stability. *Solar Energy Mater. Solar Cells* **2021,** *224*, 111011.
11. Matebese, F.; Taziwa, R.;Mutukwa, D. Progress on the Synthesis and Application of CuSCN Inorganic Hole Transport Material in Perovskite Solar Cells. *Materials* **2018,** *11* (12), 2592.
12. Muthukumarasamy, N.; Santhanama, A.; Asokan, V.; Yuvapragasama, A.; Venkatraman, M. R.; Palanisamyc, S. E.; Sundaramd, S. Velauthapillaie, D. A Review on the Classification of Organic/Inorganic/Carbonaceous Hole Transporting Materials for Perovskite Solar Cell Application. *Arabian J. Chem.***2020,** *13* (1), 2526–2557.
13. Fakharuddin, A.; Vasilopoulou, M.; Soultati, A.; Haider, M. I.; Briscoe, J.; Fotopoulos, V.; Girolamo, D. D.; Davazoglou, D.; Chroneos, A. Robust Inorganic Hole Transport Materials for Organic and Perovskite Solar Cells: Insights into Materials Electronic Properties and Device Performance. *RRL Solar* **2021,** *5* (1), 2000555.
14. Heidariramsheh, M.; Mirhosseini, M.; Taghavinia, **N.**; Abdizadeh, K.;Mahdavi, S. M. Evaluating Cu2SnS3 Nanoparticle Layers as Hole-Transporting Materials in Perovskite Solar Cells. *ACS Appl. Energy Mater.***2021,** *4* (6), 5560–5573.
15. Kartikay, P.; Mokurala, K.; Sharma, B.; Kali, R.; Mukurala, N.; Mishra, D.; Kumar, A.; Mallick, S.; Song, J.; Jin, S. H. Recent Advances and Callenges in Solar Photovoltaic and Energy Storage Materials: Future Directions in Indian Perspective. *J. Phys. Energy* **2021,** *3*, 034018.
16. Nejand, B. A.; Ahmadi, V.; Gharibzadeh, S.; Shahverdi, H. R. Cuprous Oxide as a Potential Low-Cost Hole-Transport Material for Stable Perovskite Solar Cells. *ChemSusChem* **2016,** *9* (3), 302–313.
17. Gharibzadeh, S.; Abdollahi, B.; Moshaii, A.; Mohammadian, N.; Alizadeh, A. H.; Mohammadpour, R.; Ahmadi, V.; Alizadeh, A. Two-Step Physical Deposition of a Compact CuI Hole-Transport Layer and the Formation of an Interfacial Species in Perovskite Solar Cells. *ChemSusChem* **2016,** *9* (15), 1929–1937.
18. Krishna, A.; Sabba, D.; Yin, J.; Bruno, A.; Boix, P. P.; Gao, Y.; Dewi, H. A.; , Gurzadyan, G. G.; Soci, C.; Mhaisalkar, S. G.; Grimsdale, A. C. Facile Synthesis of a Furan-Arylamine Hole-Transporting Material for High-Efficiency, Mesoscopic Perovskite Solar Cells. *Chemistry* **2015,** *21* (43), 15113–15117.
19. Grisorio, R.; Iacobellis, R.; Listorti, A.; Marco, L. D.; Cipolla, M. P.; Manca, M.; Rizzo, A.; Abate, A.; Gigli, G.; Suranna, G. P. Rational Design of Molecular Hole-Transporting

Materials for Perovskite Solar Cells: Direct Versus Inverted Device Configurations. *ACS Appl. Mater. Interfaces* **2017,** *9* (29), 24778–24787.
20. Rajeswari, R.; Mrinalini, M.; Prasanthkumar, S.; Giribabu, L. Emerging of Inorganic Hole Transporting Materials For Perovskite Solar Cells. *Chem. Record* **2017,** *17* (7), 681–699.
21. Teimouri, R.; Mohammadpour, R. Potential Application of $CuSbS_2$ as the Hole Transport Material in Perovskite Solar Cell: A Simulation Study. *Superlattices Microstruct.* **2018,** *118*, 116–122.
22. Cho, H. Y.; Park, E. J.; Kim, J. H.; Park, L. S. Effect of Hole Injection Layer/Hole Transport Layer Polymer and Device Structure on the Properties of White OLED. *J. Nanosci. Nanotechnol.* **2008,** *8* (10), 4916-4922.
23. Li, Y.; Xu, Z.; Zhao, S.; Qiao, B.; Huang, D.; Zhao, L.; Zhao, J.; Wang, P.; Zhu, Y.; Li, X.; Liu, X.; Xu, X. Highly Efficient p-i-n Perovskite Solar Cells Utilizing Novel Low-Temperature Solution-Processed Hole Transport Materials with Linear π-Conjugated Structure. *Small* **2016,** *12* (35), 4902–4908.
24. Luo, H.; Lin, X.; Hou, X.; Pan, L.; Huang, S.; Chen, X. Efficient and Air-Stable Planar Perovskite Solar Cells Formed on Graphene-Oxide-Modified PEDOT:PSS Hole Transport Layer. *Nanomicro Lett.* **2017,** *9* (4), 39.
25. Jung, J. W.; Chueh, C. C.; Jen, A. K. Y. High-Performance Semitransparent Perovskite Solar Cells with 10% Power Conversion Efficiency and 25% Average Visible Transmittance Based on Transparent CuSCN as the Hole-Transporting Material. *Adv. Energy Mater.* **2015,** *5* (17).

CHAPTER 16

Structural Properties of Ag-Doped PVP-Stabilized SnO_2 Engineered with TiO_2 for Gas Sensing Applications

V. VIDHYA[1], P. ANBARASU[1], and D. GEETHA[2]

[1]*Department of Electronics, GAC, Trichy-22, Bharathidasan University, Tiruchirappalli, Tamil Nadu, India*

[2]*Department of Physics, Annamalai University, Annamalai Nagar, Chidambaram, Tamil Nadu, India*

ABSTRACT

In recent years, owing to increasing concern about global environment pollution issue because a lot of waste from industries and human were dumped directly into the atmosphere. Gas sensors play vital role in the part of pollution prevention efforts. In this paper, we have reported that synthesis of Ag-SnO_2:TiO_2 nanocomposites by hydrothermal method and co precipitation method using Stannic chloride, silver nitrate, and titanium isopropoxide as precursors. The resulting nanocomposites crystallization, morphology, element composition, and identification of functional groups and structure were characterized using X-ray diffracrogram (XRD), scanning electron microscopy (SEM), FTIR, and UV-vis spectroscopy analyses. Gas sensor designed to detect gas molecules present in the atmosphere. The prepared sensor device using the films exhibits higher sensitivity to reducing group of gases which accepts electron under the reaction with surface oxygen species. The presence of Ag ions in NCs induces the gas sensing response by the process of spill-over effect and chemical sensitization.

Bio-Based Polymers and Composites: Properties, Durability, and Applications. Amadou Belal Gueye, Sabu Thomas, Nandakumar Kalarikkal, & Modou Fall (Eds.)

16.1 INTRODUCTION

In our Global environment, gas sensors play a vital role in many aspects of technology, in fields such as environmental monitoring, or air quality assurance, industrial processes, automotive technologies, to name just a few, and are increasingly. In the devices of gas sensor, gas sensing layer is the most important element which can be used to sense the gases of various concentrations and generate the change in electrical resistance.[1] The gas sensing layer is basically a chemiresistor which changes its resistance value according to the concentration of particular gas in the environment.[2] So whenever toxic gases are being detected, the current flow through it varies because the resistance of the element changes which represent the change in concentration of the gases.

Several advantages like fabrication method, cost-effective, stability, durability, miniaturization, and low detection limit are satisfied by semiconducting-based metal oxide-based gas sensors.[3,4] There are two types of SMOS. (1) P-type CuO, NiO, and Cr_2O_3, (2) N-type SnO_2, TiO_2, and ZnO. Among these, SnO_2 is the most suitable material for gas sensing because of its good sensing behavior toward various gases. SnO_2 shows the high sensitivity at high temperature[5] but it has the disadvantage of sensitivity and selectivity in low temperature. The surface phenomenon of SnO_2 can be modified by adding dopants which influence their various characteristics like band gap, charge carriers, fermi level, and size of the crystallites.[6] Doping is one of the ion-exchange and adsorption process.

In this report, Ag/TiO_2 doped with SnO_2 which modify the parameter to improve the selectivity and sensitivity at low temperature. Ag/TiO_2:SnO_2 nanocomposites are synthesized by means of a simple hydrothermal with co-precipitation route with polyvinylpyrrolidone (PVP) which controls the size of the material.[7] Here, Ag was impregnated into porous SnO_2:TiO_2 nanostructures. Three various concentration of Ag was doped on SnO_2:TiO_2 for optimizing the sensing material for best sensing performance. The aim of this work is to investigate the effect of Ag content and the supporting material (TiO_2) on reducing gas sensing performance of SnO_2. As expected, Ag-doped SnO_2:TiO_2 nanostructure shows an excellent selectivity to Ammonia gas at room temperature.

16.2 EXPERIMENTAL PROCEDURE

16.2.1 *SYNTHESIS PROCESS OF SnO_2, SnO_2:TiO_2, AG/SnO_2:TiO_2 NANOSTRUCTURES*

Stannic chloride ($SnCl_4.5H_2O$), silver nitrate (Ag_2No_3), titanium isopropoxide ($C_{12}H_{28}O_4Ti$), ethanol, and de-ionized water were purchased from

Sigma Aldrich. Three samples are taken (1) stannic chloride only (2) stannic chloride and titanium isopropoxide (4 mmol:1 mmol) (3) stannic chloride, silver nitrate, and titanium isopropoxide (4 mmol:1 mmol :1 mmol), were dissolved in de-ionized water separately and kept for stirring of 45 min and transferred into Teflon-lined autoclave, sealed, and maintained at 120°C for 6 h and the final sample was transferred into Tubular–Furnace maintained at 500°C for 1 h. The resulting SnO_2, SnO_2:TiO_2, and Ag/SnO_2:TiO_2 samples were filtered washed and dried at 80°C for 6 h.

16.3 RESULT AND DISCUSSION

16.3.1 STRUCTURAL ANALYSIS (XRD)

The diffraction patterns were recorded on X'Pert-PRO using Cu Kα radiation. X-ray diffract meter at a scan rate of 1° over the range 20°–80° (2θ) at room temperature. The XRD patterns of the SnO_2, SnO_2:TiO_2, Ag/SnO_2:TiO_2 nanostructure shown in Figure 16.1(a–c). All the diffraction peaks can be indexed as 2θ value positioned at 26.8, 29.7, 33.3, 37.5, 47.1, 51.3, 57.0, 62.0, and 68.3 corresponding to the (111), (112), (113), (002), (115), (117), (222), (134), (128) of Bragg reflection planes, respectively. SnO_2 nanostructure is good crystalline quality (JCPDS: 41-1445).[8] The tetragonal structure of SnO_2 nanostructure is confirmed by XRD analysis. The intensified sharp peaks confirm the final product that is highly crystalline in nature. The average grain sizes of the SnO_2, SnO_2:TiO_2 and Ag/SnO_2:TiO_2 samples were calculated and are 55, 50, and 56 nm, respectively.

16.3.2 STRUCTURAL ANALYSIS (FTIR)

The presence of functional group of nanostructures was characterized by Fourier transform infrared spectrometer (Perkin Elmer) range from 400 to 4000 cm^{-1}. Figure 16.2(a–c) shows FTIR spectra of SnO_2, SnO_2:TiO_2, and Ag/SnO_2:TiO_2 nanostructures. The broad absorption bands at 3455 and 1660 cm^{-1} are due to O–H stretching and bending vibrations of absorbed water in SnO_2 (a) surface. The sharp band absorption at 1384 cm^{-1} are due to C=C stretching and strong absorption band 613 cm^{-1} for Ag/SnO_2:TiO_2 (c). The appearance of large number of IR bending modes for particles indicating the homogeneous formations of particles Ag/SnO_2:TiO_2 nanocomposite.

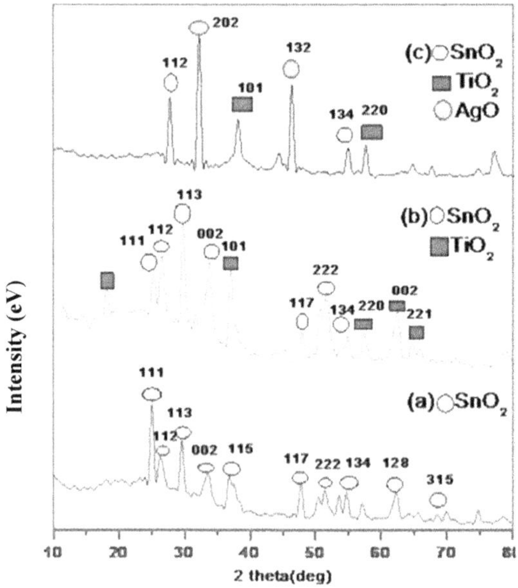

FIGURE 16.1 (1a–c) XRD patterns of (a) SnO_2 (b) SnO_2: TiO_2 (c) Ag/SnO_2: TiO_2 Nanostructures

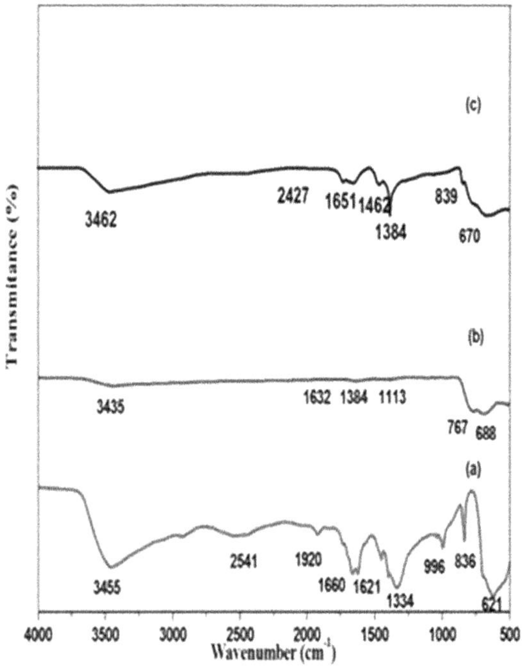

FIGURE 16.2 FTIR spectra of (a) SnO_2 (b) SnO_2: TiO_2 (c) Ag/SnO_2: TiO_2 Nanostructures

16.3.3 ABSORPTION STUDY (UV-VIS-SPECTROSCOPY)

The morphology of the synthesized product was inspected through a (JEOL-JSM – 5610 LV with INCA EDS) scanning electron microscope (SEM). Optical properties were studied by UV-vis spectroscopy (ShimadzuUV1700) as shown in Figure 16.3(a–c). The UV absorption was obtained for the sample Ag/Tio_2:SnO_2 having peak at ~287 nm which is less than that of SnO_2. The solvent used here plays an important role in adding the reduction path and nature of the product. The resulting emission characteristic exhibits the origination of one main strong emission peak around 287 nm. The energy band gap values of SnO_2, SnO_2:TiO_2, and Ag/SnO_2:TiO_2 nanostructures were 4.36, 4.33, 4.32eV, respectively.

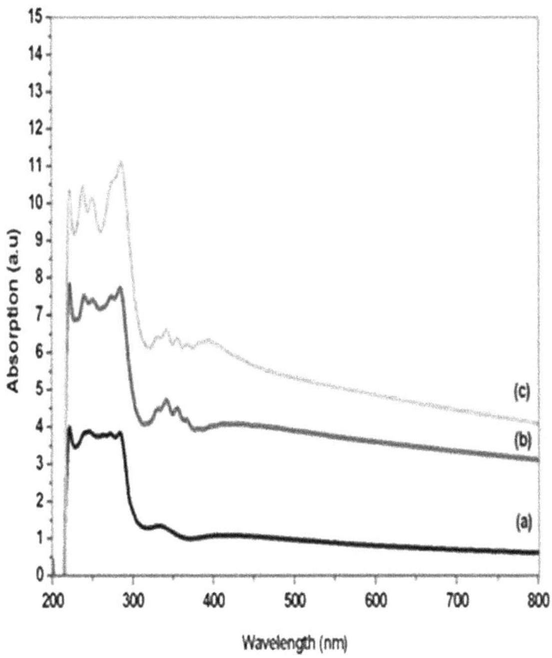

FIGURE 16.3 UV–Visible Spectra of (a) SnO_2 (b) SnO_2: TiO_2(c) Ag/SnO_2: TiO_2 Nanostructures

16.3.4 MORPHOLOGICAL ANALYSIS (SEM)

The surface morphology of SnO_2, SnO_2:TiO_2, and Ag/SnO_2:TiO_2 was shown in Figure 16.4(a–c). Figure 16.4a clearly shows that SnO_2 nanostructures are spherical in shape, and flake like structure was formed for TiO2:SnO_2

(Figure 16.4b). Ag/TiO_2:SnO_2 microstructure exhibits sphere like morphology (Figure 16.4c) with pores was visible on the surface of the nanostructure.

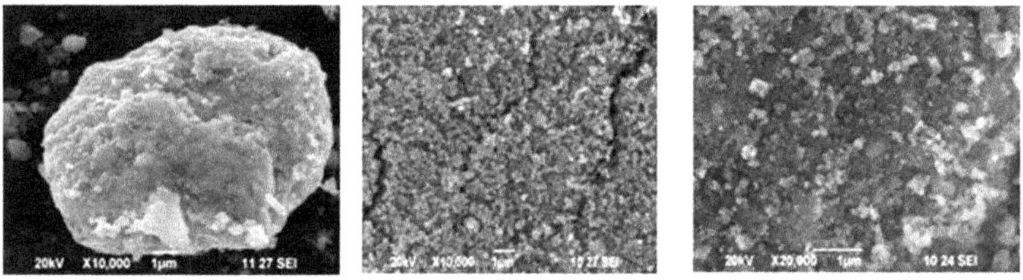

FIGURE 16.4 SEM of (a) SnO_2 (b) SnO_2:TiO_2(c) Ag/SnO_2:TiO_2 Nanostructures

16.3.5 XPS ANALYSIS

The survey spectrum of Ag/SnO_2:TiO_2 nanocomposite showed peaks of Sn, Ti, Ag, and O (Figure 16.5a). The Sn 3d level shows peaks at 483.85eV and 492.29eV which can be assigned to 3d5/2 and 3d3/2 of Sn, respectively (Figure 16.5b), confirming the +4 oxidation state in the tetragonal rutile structure of SnO_2.[9] Deconvolution of Ti 2p core level spectrum (Figure 16.5c) has two peaks at 455.8 and 461.6eV confirmed the presence of Ti element is in the state of +4 oxidation state. The result represents better coupling between TiO_2:SnO_2.[10] The deconvoluted peaks at 367.8eV and 373.8eV of Ag spectrum (Figure 16.5d) attributed to 3d5/2 and 3d3/2, respectively. The 6.0ev splitting within the Ag 3d shows formation of Ag on TiO_2.[11,12] The deconvolution of O 1s core level spectrum (Figure 16.5e, represented by a broad asymmetric peak 527.30eV could be attributed to the oxygen species such as O_2^- or O^- ions present within the matrix of Ag/TiO_2:SnO_2 nanocomposite.

16.4 FABRICATION AND MEASUREMENT SET-UP

The fabrication of gas sensor is as follows (1) The nanocomposite sensor material were mixed with PVP(.1 M) for binding in the ratio of 4:1 to form slurry (2) the slurry was coated on a glass substrate (3) Then the glass substrates was then dried in an oven at 100°C for 3 h. Gas sensor is placed inside the gas measurement set up chamber which consists of gas inlet and outlet path. The inlet gas path connected with gas mixer that controls the mass flow and ppm level of the gas. Silver leads are connected to gas sensing film. These leads are connected reaction chamber and Keithely source meter.

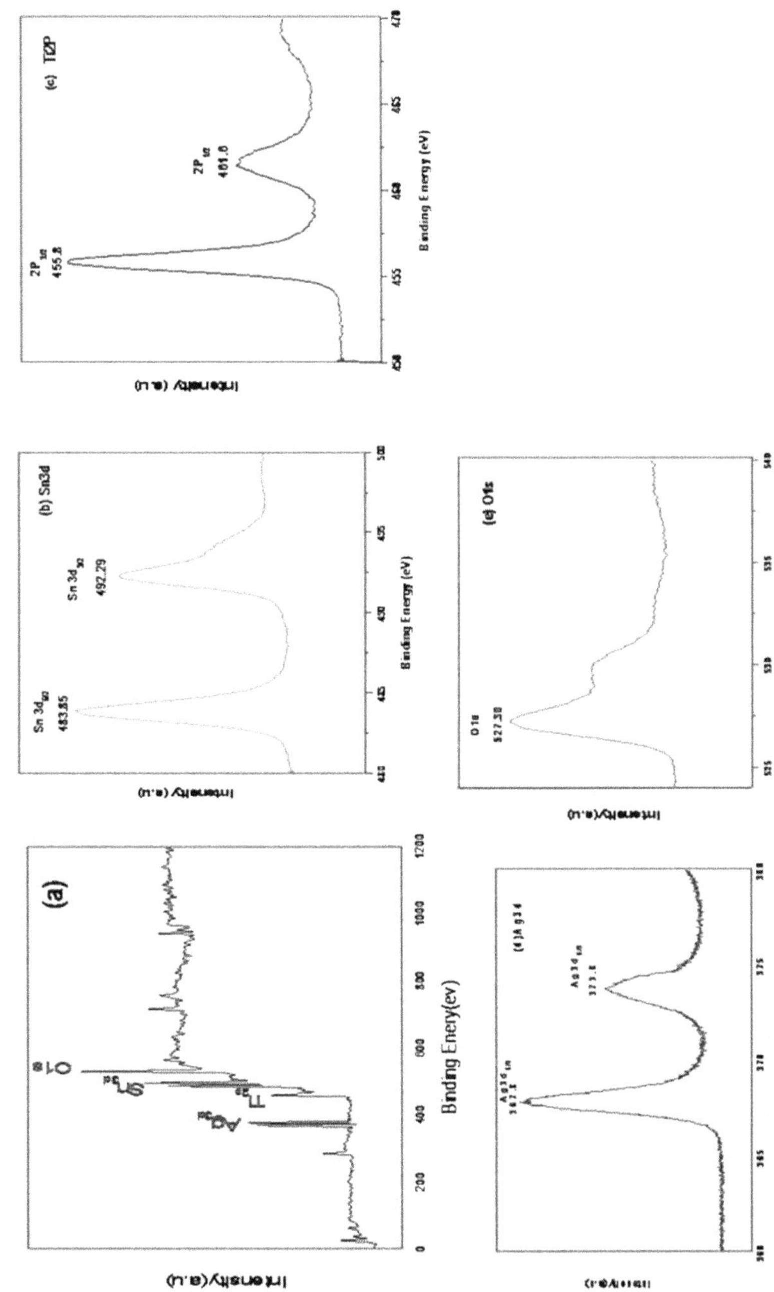

FIGURE 16.5 X-ray photoelectron spectrum of Ag/SnO_2:TiO_2 nanaocomposite where (a) survey spectrum, (b) Sn 3d region, (c) Ti 2p region and (d) Ag 3d region (e) O 1s region.

Reaction chamber is kept at room temperature and Keithely meter measures the variation in resistance of gas sensor.

16.5 SENSING MECHANISM

From the results, the resistance changing behaviors may be attributed to the adsorption and desorption of NH_3 molecules on the Ag-SnO_2:TiO_2 nanocomposite sensing film. The NH_3-sensing mechanisms of the Ag-SnO_2:TiO_2 nanocomposite gas sensor at room temperature may be explained based on the formation of p–n heterojunctions between p-type Ag and n-type metal–oxide–semiconductor (MOS) of SnO_2:TiO_2.[13] In air, oxygen species (O_2^-) chemisorbed on the surface of Ag-SnO_2:TiO_2 nanocomposite film will capture free electrons from the conduction band of SnO_2:TiO_2 and the Fermi level in band gap of Ag according to the surface reaction: (1) O_2 (ads) + $e^- \rightarrow O_2^-$ (ads), leads the formation of depletion layers on the interface of the Ag-SnO_2:TiO_2 sensing film and cause the high resistance of the sensor in air when exposure to NH_3 vapor, NH_3 molecules as a reducing gas will react with oxygen species according to the reaction: (2) $4NH_3$ (gas) + $5O_2^-$ (ads) $\rightarrow$ 4NO (gas) + $6H_2O$ (gas) + $5e^-$, lead the discharge of electrons into the conduction band of SnO_2:TiO_2 on the sensor surface.[14,15] This process results in a decrease of the thickness of the electron depletion layer leading to decrease in the resistance of the sensor.

The SnO_2:TiO_2 nanoparticles are randomly interwoven with the Ag nanostructure to form a nanocomposite network that can enhance the surface area of the sensing film to increases the ability for adsorption and desorption of gas molecules.[16] When the p–n heterojunctions are formed, the electrons in SnO_2:TiO_2 will transfer to Ag, whereas the holes will transfer from Ag to SnO_2:TiO_2 because the n-type SnO_2:TiO_2 has a higher Fermi level than the p-type Ag. Once the Ag-SnO_2:TiO_2 sensor is exposed to NH_3 vapor, the reaction between NH_3 molecules and oxygen species will occur on the n-type SnO_2:TiO_2 surface, so that the electrons are released from the ionized oxygen species back to the conduction band of n-type SnO_2:TiO_2 and combine with holes of p-type Ag, leading to reduction of the p–n heterojunction electron depletion layer width and decrease the potential barrier height. The resistance of the Ag-SnO_2:TiO_2 sensor thus decreases upon exposure to NH_3 vapor at room temperature for 25–100 ppm as shown in Figure 16.6. Figure 16.7 compared Ag/SnO_2:TiO_2-coated thick film shows good response to the Ammonia compare to other gases at room Temperature.

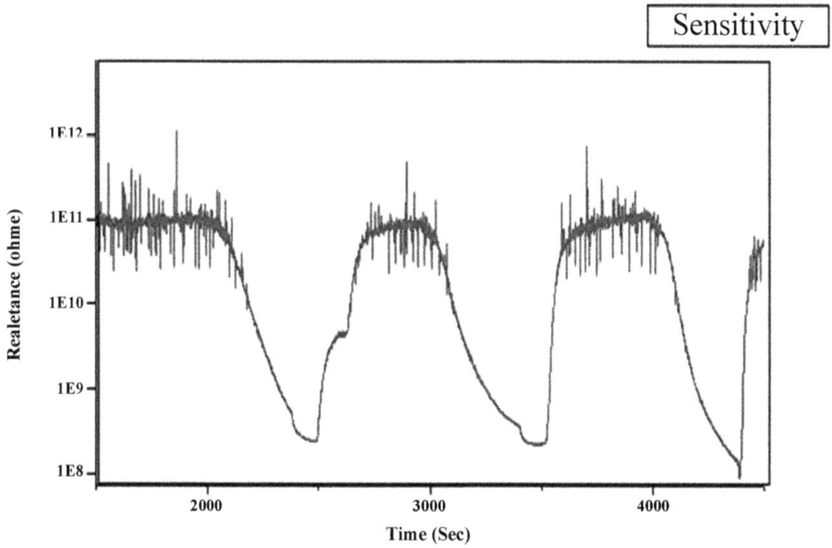

FIGURE 16.6 Sensitivity of NH3 (25 ppm to 100 ppm) at Room Temperature

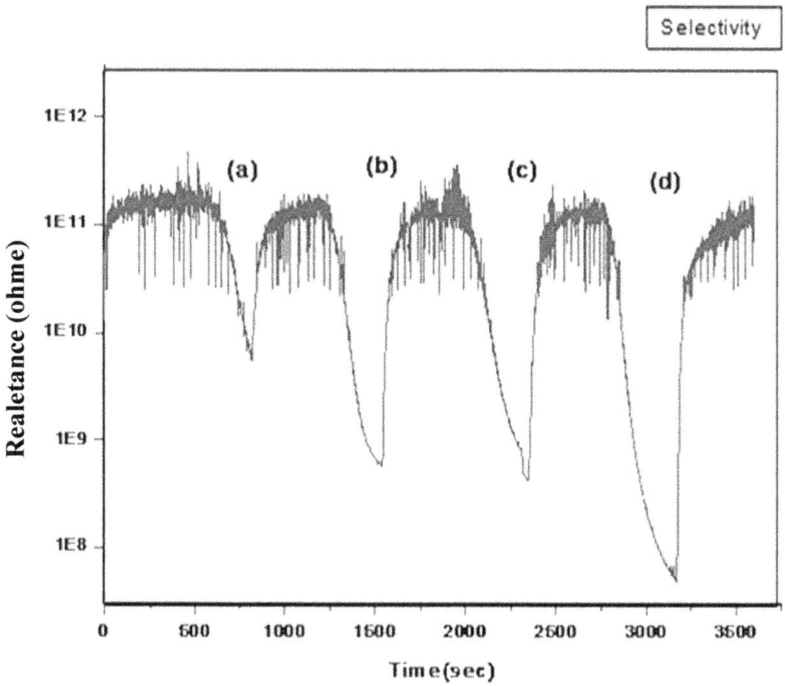

FIGURE 16.7 Selectivity of (a) Formaldehyde (b) Ethanol (c) Acetone (d) Ammonia at 100 ppm in RT

16.6 CONCLUSION

In summary, Ag-SnO_2:TiO_2 nanocomposites were successfully synthesized by the hydro thermal method associated with co-precipitation method and systematically characterized for NH_3 sensing at room temperature. The crystalline structure and particle size were characterized by XRD, functional groups were identified by FTIR, and a strong absorption peak was observed at 287 nm by UV-vis spectroscopy which are confirmed the presence of Ag on surface of SnO_2:TiO_2 nanocomposite. As seen from the gas-sensing results, Ag- SnO_2:TiO_2 gas sensor demonstrated a rapid decrease of resistance upon exposure to NH_3 and fully recovered to its baseline values in air with good repeatability. The Ag-SnO_2:TiO_2 gas sensor exhibited ultrahigh selectivity toward NH_3 against various volatile organic compounds (VOCs) and environmental gases at room temperature with no effect of humidity in the range 18% RH. The NH_3-sensing mechanism of the Ag-SnO_2:TiO_2 nanocomposite gas sensor has been proposed based on the formation of p–n heterojunctions of p-type Ag and n-type SnO_2:TiO_2 via the low-temperature oxidizing reaction process. The observed ultrahigh NH_3 selectivity and stability are proved that the Ag-SnO_2:TiO_2 nanocomposite can be considered as a promising material for practical application of NH_3 detection at room temperature.

ACKNOWLEDGMENT

We would like to thank Prof. Dr. K. Jeyadheepan, School of Electrical and Electronics Engineering, SASTRA Deemed to be University, Thanjavur 613401, for providing instrument to study gas sensing characterization.

KEYWORDS

- **gas sensors**
- **nanocomposites**
- **tansitional metal oxides**
- **hydrothermal method**

REFERENCES

1. Batzill, M. Surface Science Studies of Gas Sensing Materials: SnO_2. *Sensors* 2006, *6*, 1345–1366.
2. Wang, C.; Yin, L.; Zhang, L.; Xiang, D.; Gao, R. Metal Oxide Gas Sensors: Sensitivity and Influencing Factors. *Sensors* 2010, *10*, 2088–2106.
3. Williams, D. E.; Henshaw, G. S.; Prat, K. F. E.; Peat, R. Reaction–Diffusion Effects and Systematic Design of Gas-Sensitive Resistors Based on Semiconducting Oxides. *J. Chem. Soc. Faraday Trans.* **1995**, *91*, 4299–4307.
4. Fergus, J. W. *Sens. Actuators B Chem.* **2008,** *33*, 198–202.
5. Sheini, N. A.; Rohani, M. IOP Conference, International Conference on Materials and Application for Sensors and Transducers, Greece, 2015; *Mater. Sci. Eng.* **2016**, *2008*.
6. Sun, Y. F.; Liu, S. B.; Meng, F. L.; Liu, J. Y.; Jin, Z.; Kong, L. T.; Liu, J. H. Metal Oxide Nanostructures and Their Gas Sensing Properties: A Review. *Sensors* 2012, *12*, 2610–2631.
7. Selvan, T.; Spatz, J. P.; Klok, H. A.; Möller, M. Gold–Polypyrrole Core–Shell Particles in Diblock Copolymer Micelles. *Adv. Mater.* 1998, *10*, 132–134.
8. Pan, J.; Ganesan, R.; Shen, H.; Mathur, S. Plasma-Modified SnO_2 Nanowires for Enhanced Gas Sensing. *J. Phys. Chem.* **2010,** *114*, 8245–8250.
9. Ding, X.; Zeng, D.; Xie, C. *Sens. Actuators B* **2010,** *149*, 336–344.
10. Xue, X.; Chen, Z.; Ma, C.; Xing, L.; Chen, Y.; Wang, Y.; Wang, T. Controllable Modulation of the Electronic Structure of ZnO($10\bar{1}0$) Surface by Carboxylic Acids†. *J. Phys. Chem. C* **2010,** *114*, 3968–3972.
11. Liu, Y.; Wang, X.; Yang, F.; Yang, X. *Materials* **2008,** *114*, 431–439.
12. Wu, M.; Yang, B.; Lv, Y.; Fu, Z.; Xu, J. *Appl. Surf. Sci.* **2010,** *256*, 7125–7130.
13. Zhang, D.; Wu, Z.; Zong, X. *Sens. Actuators B* **2019,** *289*, 32–41.
14. Kim, J. H.; Mirzaei, A.; Kim, H. W.; Kim, S. S. *Sens. Actuators B* **2018,** *267*, 597–607.
15. Miller, D. R.; Akbar, S. A.; Morris, P. A. Nanoscale Metal Oxide-Based Heterojunctions for Gas Sensing: A Review. *Sens. Actuators B* **2014**, *204*, 250–272.
16. Sagadevan, S.; Podder, J. Optical and Electrical Properties of Nanocrystalline SnO_2 Thin Films Synthesized by Chemical Bath Deposition Method. *Soft Nanosci. Lett.* **2015,** *5* (4), 55–64.

CHAPTER 17

Struvite Fertilization in Horticulture as the Closed Phosphorus Cycle

ANNA JAMA-RODZEŃSKA[1], JÓZEF SOWIŃSKI[1],
KATARZYNA ADAMZCEWSKA-SOWIŃSKA[2], and ANDRZEJ DZIUBA[3]

[1]*Institute of Agroecology and Plant Production, Wroclaw University of Environmental and Life Sciences, Wroclaw, Poland*

[2]*Department of Horticulture, Wroclaw University of Environmental and Life Sciences, Wroclaw, Poland*

[3]*Krevox Europejskie Centrum Ekologiczne Sp. z o.o., Żurawia, Warszawa, Poland*

ABSTRACT

Struvite can constitute an efficient alternative to traditional phosphorous-based fertilizers in horticulture cultivation. Struvite as P fertilizer was used in the testing of butter lettuce (*Lactuca sativa* L.) under greenhouse conditions in the Research and Experimental Station of the University of Environmental and Life Sciences in Wroclaw (Psary). In a pot experiment (12 dm^3), the effect of struvite (STR) as a source of P for lettuce by comparing its effectiveness with that of superphosphate enriched (SUP) was conducted. The result of this experiment showed that STR can constitute a source of P as an attractive product for the fertilizer market, with agronomic and environmental benefits. Higher but no statistically significant value of mass of lettuce leaves, number of leaves, number of dried leaves, width of the rosette, and SPAD was observed in pots where struvite (STR) was applied in comparison with the control object and superphosphate-enriched fertilization (SUP). STR characterizes a high influence on increasing lettuce yield.

Bio-Based Polymers and Composites: Properties, Durability, and Applications. Amadou Belal Gueye, Sabu Thomas, Nandakumar Kalarikkal, & Modou Fall (Eds.)

17.1 INTRODUCTION

Great emphasis is placed on the so-called circular economy, that is, striving for reuse of materials with valuable properties.[1] The future direction of sewage sludge utilization that fits into the concept of a circular economy may be the production of fertilizers from it and the recovery of nutrients.[2,3] Nutrient recovery is practiced in a few European Union countries and in Poland to a small extent.[4] Sewage sludge is a valuable source of nutrients essential for proper plant growth and development. More than half of the sewage sludge solid fraction is organic matter, which has a significant impact on physical, chemical, and biological soil properties,[3] fertility, and microbiological balance. The presence of heavy metals, organic pollutants, and pathogenic microorganisms can be a limiting factor in their utilization.[5] Sustainable management of phosphorus is realized by using the model of Circular Economy, where the added value of sewage sludge should be kept as long as possible.[6–8] The sustainable utilization of phosphorus plays an important role in the closed cycle of this element to ensure food security over a long period as a fertilizer in horticulture.[3–5]

In the available literature, we can find a description of fertilizer production technologies based on various types of waste.[6–8] Fertilizers based on sewage sludge can largely contribute to solving the problem of increasing amount of waste and by-products by reducing the cost of their management.[11] Currently, the development of new technologies for disinfection, sterilization, and nutrient enrichment of sewage sludge allows the production of sustainable granular fertilizers.

Currently, the most commonly used technology of phosphorus recovery comprises phosphorus minerals precipitation from sludge, ash, or leachate. Phosphorus can be precipitated as struvite (magnesium ammonium phosphate hexahydrate, $MgNH_4PO_4 \cdot 6H_2O$).[9,10] Many techniques and technologies have been developed in the last decades in phosphorus recovery to receive struvite.[9–12]

The struvite recovery has been as follows: The leachate produced during the dewatering of digested sludge is fed into a continuous mixing reactor (Figure 17.1) where rapid mixing is carried out by means of a special mixing system called Turbomix. Struvite precipitation is initiated as the pH ranges from 7.6 to 8.3 by dosing magnesium salts. The integrated lamellar settling tank ensures the exact separation of the struvite lumps produced and the treated wastewater.[13–19]

FIGURE 17.1 Installation for phosphorus recovery. (Photo A. Dziuba)

Sewage sludge (waste) as a source, European Ecological Centrum named Krevox installation and Phosgreen phosphate fertilizer presented in Figure 17.2.

FIGURE 17.2 Close phosphorus cycle. Phosgreen production based on sewage sludge resources. (Photo A. Dziuba and A. Jama-Rodzeńska)

The evaluation of struvite as a phosphorus fertilizer requires further research. In order to examine the effectiveness, studies were carried out with the use of struvite (trade name Phosgreen) in lettuce cultivation. The issue of the research was to evaluate the value and potential properties of struvite fertilizer made from sewage sludge transformation as a source of phosphorus compared to triple superphosphate.

17.2 MATERIALS AND METHODS

In the pilot studies in a greenhouse, pot experiment on selected horticulture media was conducted with struvite (STR) fertilizer compared to triple superphosphate (SUP). The experiment with a test plant–butter lettuce (*Lactuca sativa* L.) of the effectiveness of struvite fertilizer was carried out in spring 2020 under greenhouse conditions in the Research and Didactic Station of the University of Life Sciences in Wroclaw (Psary) in 12 dm^3 pots.

The greenhouse pot experiment was set up with the following treatments with four replicates. The first-order factor was various horticulture media. We examined five types of that: (1) deacidified peat substrate (DP), (2) deacidified peat substrate with willow compost[1] (DP + WC), (3) mineral soil (MS), (4) mineral soil with the addition of soil improver (MS + SI), and (5) deacidified peat substrate with willow compost and soil improver (DP + WC + SI). The second-order factor was phosphorus fertilizer: control—no phosphorus (C), triple superphosphate (SUP), and struvite (STR). The mineral soil samples were collected before starting the experiment for chemical analysis from depth 0–25 cm. Deacidified peat substrate characterized standard nutrient concentration which was modified by adding mineral fertilizer for lettuce requirements. The following mineral fertilizers were used once: ammonium nitrate (AN), potassium sulfate (PS), granulated triple superphosphate (SUP), and granulated struvite (STR). Doses of fertilizers for peat substrates calculated in mg/l were as follows: AN- 294, PS– 400, SUP 300, STR 500, and for soil (MS): AN -294, PS – 300, SUP 150, STR 250. Lettuce seeds were sown directly into the pots filled with soil and horticultural media. After harvesting in the middle of July, biometric measurements of lettuce were performed included: weight of rosette, number of lettuce leaves, number of leaves over-dried, width of rosette, and SPAD.

Data from independent morphological measures were subjected to Anova/Manova statistical analysis in Statistica software (version 13.1 Stat-Soft. Poland). Two-factor variance analysis (ANOVA) was performed.

17.3 RESULTS

The results of the biometric measurements of main effects showed that mass of fresh rosette, number of leaves, number of dried leaves, width of rosette of leaves, and SPAD were significantly affected by the horticulture media, whereas P fertilization induced no significant response on lettuce (Table 17.1).

Plants fertilized with struvite (STR) in the experiment produced a greater mass of rosette compared to control (C) plants without any P application and triple superphosphate (SUP). Struvite (STR) treated lettuce had also greater number of leaves, width of rosette of leaves, and SPAD value, however,

TABLE 17.1 Influence of Selected Horticultural Media and Phosphorus Fertilizers on Morphological Traits of Lettuce.

Experiment factor	Selected morpohological characteristics and SPAD				
	Mass of rosette (g)	**Number of leaves**	**Number of dried leaves**	**Width of rosete (cm)**	**SPAD**
Horticultural media (A)					
DP	81.52	21.75	4.26	22.67	4.46
DP + WC	3.90	8.63	0.08	10.38	2.76
MS	57.87	17.13	0.97	25.72	3.66
MS + SI	65.82	16.02	1.30	25.13	3.58
MS + SI + SI	3.34	7.33	0.90	8.83	2.13
P value	$p < 0.000^{***}$	$p < 0.000^{***}$	$p < 0.000^{***}$	$p < 0.000^{***}$	$p < 0.000^{***}$
Phosphorus fertilizer (B)					
Control (C)	30.89	11.61	0.55	16.75	3.16
SUP	50.99	15.62	1.73	19.82	3.33
STR	51.10	16.14	2.16	20.63	3.67
P value	0.382	0.236	0.111	0.513	0.518

**significant $p < 0.01$

***significant $p < 0.001$

without significant differences. Unfavorable feature—number of dried leaves was also found in the pots with struvite (STR). The highest rosette weight, number of leaves, number of dried leaves, and SPAD were found on deacidified peat, therefore, it can be stated that this horticultural media is the most appropriate and recommended for future experiments.

17.4 DISCUSSION

The presented research data suggest that struvite may constitute a substitute of phosphorus fertilizer to be used as a P source for lettuce cultivation. The results obtained in conducted scientific research with struvite and its impact on the biomass of lettuce head confirm this statement. Ref. [20] confirmed that plants fertilized with struvite characterize with higher yield, phosphorus concentration, and phosphorus uptake compared to commercial phosphorus fertilizers. Ref. [21] stated that struvite (STR) fertilization resulted in higher yield compared to control and standard NPK fertilization treatment in grass and cereal cultivation. In turn, according to Ref. [22], similar yields of crops fertilized with struvite (STR) and triple superphosphate (SUP) were found. Ref. [23] also presented similar yields of wheat fertilized with struvite and ammonium phosphate. In our study, struvite (STR) fertilization had no significant effect on selected morphological features. The values of the examined features were comparable or even higher than obtained with the triple superphosphate (SUP) application.

17.5 CONCLUSIONS

This study investigated the potential of struvite as a P source for lettuce in five different horticulture media. Soil and deacidified peat substrate properties affected significantly the examined features. The greatest results of the examined features were obtained on deacidified peat. Due to the promising results of the first series with struvite and deacidified peat, further experiments will be carried out on the preferred media with different doses of calcium carbonate aiming at the changes of availability of phosphorus from struvite fertilization. STR fertilization indicated the potential value of P as a comparable source to triple superphosphate. The main indicator in favor of struvite use in lettuce cultivation was biomass and width rosette, respectively, 65.4% and 23.2% higher compared to control.

KEYWORDS

- **phosphorus**
- **circular economy**
- **lettuce**
- **struvite**
- **mass of leaves**
- **number of leaves**

REFERENCES

1. Smol, M.; Kulczycka, J.; Lelek, Ł.; Gorazda, K.; Wzorek, Z. Life Cycle Assessment (LCA) of the Integrated Technology for the Phosphorus Recovery from Sewage Sludge Ash (SSA) and Fertilizers Production. *Arch. Environ. Protect.* **2020,** *46* (2), 42–52.
2. Gorazda, K.; Tarko, B.; Wzorek, Z.; Kominko, H.; Nowak, A. K.; Kulczycka, J.; Henclik, A.; Smol, M. Fertilizers Production from Ashes After Sewage Sludge Combustion – A Strategy Towards Sustainable Development. *Environ. Resour.* **2017,** *154*, 171–180.
3. Kominko, H.; Gorazda, K.; Wzorek, Z. The Possibility of Organo-Mineral Fertilizer Production from Sewage Sludge. *Waste Biomass Valori.* **2017,** *8*, 1781–1791.
4. Jama-Rodzeńska, A.; Białowiec, A.; Koziel, J. A.; Sowiński, J. Waste to Phosphorus: A Transdisciplinary Solution to P Recovery From Wastewater Based on the TRIZ Approach. *J. Environ. Manag.* **2021,** *8*, 1781–1791.
5. Wierzbowska, J.; Kovačik, P.; Sienkiewicz, S.; Krzebietke, S.; Bowszys, T. Determination of Heavy Metals and Their Availability to Plants in Soil Fertilized with Different Waste Substances. *Environ. Monit. Assess.* **2018,** *190*, 567–579.
6. Mazeika, R.; Staugaitis, G.; Baltrusaitis, J. Engineered Pelletized Organo-Mineral Fertilizers (OMF) from Poultry Manure, Diammonium Phosphate and Potassium Chloride. *ACS Sustain. Chem. Eng.* **2016,** *4*, 2279–2285.
7. Rao, J. R.; Watabe, M.; Stewart, T. A.; Millar, B. C.; Moore, J. E. Pelleted Organo-Mineral Fertilisers from Composted Pig Slurry Solids, Animal Wastes and Spent Mushroom Compost for Amenity Grasslands. *Waste Manag.* **2017,** *27*, 1117–1128.
8. Deeks L. K.; Chaney, K.; Murray, C.; Sakrabani, R.; Gedara, S.; Le, M. S.; Tyrrel, S.; Pawlett, M.; Read, R.; Smith, G. H. A new Sludge-Derived Organo-Mineral Fertilizer Gives Similar Crop Yields as Conventional Fertilizers. *Agron. Sustain. Dev.* **2013,** *3*, 1708–1721.
9. Cornel, P.; Schaum C. Phosphorus Recovery from Wastewater: Needs, Technologies and Costs. *Water Sci. Technol.* **2009,** *59* (6), 1069–1076.
10. Jama-Rodzeńska, A.; Białowiec, A.; Kozieł, J.; Sowiński, J. Waste to Phosphorus: A Transdisciplinary Solution to P Recovery from Wastewater Based on the TRIZ Approach. *J. Environ. Manag.* **2021,** *287*, 1–16.

11. Childers, D. L.; Corman, J.; Edwards, M.; Elser, J. J. Sustainability Challenges of Phosphorus and Food: Solutions from Closing the Human Phosphorus Cycle. *Bioscience* **2011,** *61* (2), 117–124. https://doi.org/10.1525/bio.2011.61.2.6
12. Bindraban P. S.; Dimkpa C. O.; Pandey R. Exploring Phosphorus Fertilizers and Fertilization Strategies for Improved Human and Environmental Health. *Biol. Fertil. Soils* **2020,** *56*, 299–317.
13. Bradford-Hartke, Z.; Lane J.; Lant P.; Leslie G. Environmental Benefits and Burdens of Phosphorus Recovery from Municipal Wastewater. *Environ. Sci. Technol.* **2015,** *49* (14), 8611–8622.
14. Herzel, H.; Krüger, O.; Hermann, L.; Adam, C. Sewage Sludge Ash-A Promising Secondary Phosphorus Source for Fertilizer Production. *Sci. Total Environ.* **2016,** *542* (Pt B), 1136–1143. https://doi.org/10.1016/j.scitotenv.2015.08.059
15. Cieślik, B.; Konieczka, P. A Review of Phosphorus Recovery Methods at Various Steps of Wastewater Treatment and Sewage Sludge Management. The Concept of "No Solid Waste Generation" and Analytical Methods. *J. Clean. Prod.* **2017,** *142*, 1728–1740.
16. Egle, L.; Rechberger, H.; Krampe, J.; Zessner, M. Phosphorus Recovery from Municipal Wastewater: An Integrated Comparative Technological, Environmental and Economic Assessment of P Recovery Technologies. *Sci. Total Environ.* **2016,** *71*, 522–544.
17. Egle, L.; Rechberger, H.; Zessner, M. Overview and Description of Technologies for Recovering Phosphorus from Municipal Wastewater. *Resour. Conserv. Recycl.* **2015,** *105* (B), 325–346.
18. Günther, S.; Grunert, M.; Müller, S. Overview of Recent Advances in Phosphorus Recovery for Fertilizer Production. *Eng. Life Sci.* **2018,** *18* (7), 434–440.
19. Amann, A.; Zoboli, O.; Krampe, J.; Rechberger, H.; Zessner, M.; Egle, L. Environmental Impacts of Phosphorus Recovery from Municipal Wastewater. *Resour. Conserv. Recycl.* **2018,** *130*, 127–139. https://doi.org/10.1016/j.resconrec.2017.11.002
20. Hertzberger, A. J.; Cusick, R. D.; Margenot, A. J. A Review and Meta-Analysis of the Agricultural Potential of Struvite as a Phosphorus Fertilizer. *Soil Sci. Soc. Am. J.* **2020,** *84* (3), 653–671.
21. Borowik, M.; Mikos-Szymańska, M.; Wyzińska, M.; Zdunek, A.; Schab, S.; Sułek, A. Effect of Struvite Fertilizers on Yield and Structural Characteristics of Spring Wheat. *Przem. Chem.* **2018,** *3*, 463–466. https://doi.org/10.15199/62.2018.3.24
22. Talboys, P. J.; Heppell, J.; Roose, T.; Healey, J. R.; Jones, D. J.; Withers, P. J. A. Struvite: A Slow-Release Fertilizer for Sustainable Phosphorus Management? *Plant Soil* **2016,** *401*, 109–123. https://doi.org/10.1007/s11104-015-2747-3
23. Degryse, F.; Baird, R.; da Silva, R. C.; McLaughlin, M. J. Dissolution Rate and Agronomic Effectiveness of Struvite Fertilizers—Effect of Soil pH, Granulation and Base Excess. *Plant Soil* **2017,** *410*, 139–152. https://doi.org/10.1007/s11104-016-2990-2

Index

A

Absorption study (UV-Vis-spectroscopy), 291
Ag-doped CuO (CuAgxO) nanocrystals (NC), 225–226
  characterization details, 227–228
  elemental compositions present in Ag, 240
  experimental materials, 227
  high quality, 240
  nanowires, 226
  preparation of, 227
  result and discussion
    comparison, 231
    FT-IR Peaks and vibrational assignments of, 232
    luminescence study (PL), 236–238
    morphological study (SEM/EDS), 238–240
    optical analysis, 233–236
    spectral analysis (FTIR), 230–233
    structural investigation, 228–230
    surface analysis, 238–240
    UV-visible absorption spectrum, 233–236
α-Glucans
  alternans
    structure and function, 139
  dextran, 135–137
  mutans
    structure and function, 137
  pullulan
    structure and function, 138–139
  reuterans
    structure and function, 137–138
Anthracenemethanol (AC), 34, 35

B

Bagasse waste
  agrowaste materials, 164
  chlorofluorocarbon (CFC), 164
  experimentation, 168
  investigators, 164
  lambda meter, 165
  materials used, 166
  sample preparation, 166
  thermal insulation, 164
Bakelite, 84, 85
β-glucans
  structure and function, 140
Bio debris, 199
Biopolymer blend, 62–63
Biopolymeric films, 79
Biosynthesis, 132, 135
Biotechnological applications
  biomedicine and pharmaceutical industry, 91–92
  cosmetic sector, 93
  in food industry, 92
  nutraceutical industry, 94
  packaging and food packaging, 93–94
Butyl rubber, 86

C

Carbon nanosphere (CNS), 110, 111
  pectin
    characterization, 115–116
    diffusion coefficient, 121–123
    electro active surface area (ESA), 118–119
    electrode, fabrication, 114
    electron transfer rate constant, 121–123
    extraction, 112–113
    glucose oxidase (GOX), 119–120
    glucose sensor, 123
    interference tests, 124
    spectral characterization, 116–117
    surface electron transfer kinetics, 120–121
    surface excess concentration, 121–123
    synthesis, 113
Ceramic-based nanocomposites (CBN), 30
Ceramic-ceramic-based nanocomposites (CCBN), 30
Chemical crosslinking, 3

Chitosan
adsorbents, 61
biopolymer blend, 62–63
hydroxyapatite (HA) composites, 63
magnetic nanocomposites, 63
nanochitosan, 62
nanoclay composites, 63–64
steps and synthesis, 61
Chitosan (5-1,4-linked glucosamine), 89
Chlorofluorocarbon (CFC), 164
Cross-linked polymer, 2

D

Dental composite, 214
diametral tensile strength (DTS), 215
polymerization shrinkage, 214
statistical evaluation, 215
Dental composite paste (ORMO-48), 213–214
Diametral tensile strength (DTS), 215
Doped CuO nanocomposites, antimicrobial activity, 253
Drug delivery systems, 11–12
Dynamic light scattering (DLS), 214

E

Elastomer, 85
Electro active surface area (ESA), 118–119
Environmental pollution remediation, 33
limitations, 35
nanomaterials used
environmental pollutants, decontaminations, 38–39
PAHs remediation, 36–37
PAHs remediation
anthracenemethanol (AC), 34, 35
graphene oxide (GO), 34
green synthesis method, 34
iron oxide nanoparticles (IONPs), 34
pollutant remediation, 35
Enzymes
defined, 110
Equilibrium percentage swelling (EPS), 11
Exopolysaccharides
commercial potential and applications, 149, 150, 152–153
functional properties, 148–149
structure–function, 146, 148
Exopolysaccharides (EPS), 130

F

Fiber-reinforced polymer (FRP)
experimental
frequency domain spectroscopy, 182
materials, 179–180
morphology study, 181
PDC measurements, 181
specimen fabrication, 180
water absorption, 181–182
polarization and depolarization current (PDC), 178
result and discussion
absorption current measurements, 187
DC conductivity, 186–187
dielectric constant, 190–191
dielectric loss, 191–193
morphological properties, 182, 183
PDC measured, 183, 184, 185–186
water absorption, 187, 188–189
Fibers, 84
Food industries, 14
Fructans
inulin type
structure and function, 142–143
levan type
structure and function, 140–141

G

Gas sensors, 288
advantages, 288
fabrication of, 292–294
result and discussion
absorption study (UV-Vis-spectroscopy), 291
morphological analysis (SEM), 291–292
structural analysis (FTIR), 289–290
structural analysis (XRD), 289
XPS analysis, 292
sensing mechanism, 294–295
synthesis process, 288–289
Gelatin polymer, 88
Global environment
gas sensors, 288
Glucose oxidase (GOX), 119–120
Graphene oxide (GO), 34
Graphene-based Cu-S nanoparticles
syntheses, 275, 276, 282

H

Heavy metal ions
modified lignin-based adsorbents, 59–60
Heteropolysaccharides
structural evaluation, 143–146
structural pattern, 143
High-molecular-weight (HMW), 23
Homopolysaccharides, 131
biosynthesis, 132, 135
physicochemical properties and functions, 133–134
Hydroxy apatite (HAP), 212
Hydroxyapatite (HA) composites, 63
Hygienic products, 13

I

Indium sulfide nanostructures
environmental clean, 266–267
experimental methods
characterization, 268
materials, 267
preparation with different sulfur sources, 267–268
preparation, 265
result and discussions
optical properties, 271
photo catalytic experiment, 270–271
SEM analysis, 270
X-ray diffraction method, 268–270
synthesis of, 266
Insulation board
rice husk and bagasse waste
agrowaste materials, 164
chlorofluorocarbon (CFC), 164
experimentation, 168
investigators, 164
lambda meter, 165
materials used, 166
results and discussion, 169–171, 173
sample preparation, 166
thermal insulation, 164
Iron oxide nanoparticles (IONPs), 34

K

Kirby Bauer disk diffusion (KBDD) technique, 256

L

Lactic acid bacteria (LAB), 130
β-glucans
structure and function, 140
exopolysaccharides
commercial potential and applications, 149, 150, 152–153
functional properties, 148–149
structure–function, 146, 148
fructans
structure and function, 140–141, 142–143
α glucans
dextran, 135–137
mutans
structure and function, 137
pullulan
structure and function, 138–139
reuterans
structure and function, 137–138
structure and function, 139
heteropolysaccharides
structural evaluation, 143–146
structural pattern, 143
homopolysaccharides, 131
biosynthesis, 132, 135
physicochemical properties and functions, 133–134
Lignin, 53
adsorbents and properties, 54
covalent incorporation, 57
epoxy resins, 58
functionalization, 55
nitrogen containing functional groups, 55
oxygen containing functional groups, 55
phenol formaldehyde, 58
polymeric materials, 57
polyurethane (PU), 58
regeneration and reuse, 57
sulfur containing functional groups, 56–57
Little band hole, 198
Low-molecular-weight (LMW), 23
Luminescence study (PL), 236–238

M

Metal oxides
nanostructures of, 254–255
Metal-based nanocomposites (MBN), 30

Model organisms, 255
KBDD technique, preparation for antimicrobial study, 256
materials, 256
result and discussion, 256
antibacterial properties, 256–258
antifungal activity, 258–259
comparative study, 259–262
undoped and various transition metals-Doped CuO NPs, 261
sample selection criteria, 255–256
Morphological analysis (SEM), 291–292
Morphological study (SEM/EDS), 238–240

N

Nanolignin, 58–59
Natural polymers, 2, 87
alginate, 89
biotechnological applications
biomedicine and pharmaceutical industry, 91–92
cosmetic sector, 93
in food industry, 92
nutraceutical industry, 94
packaging and food packaging, 93–94
chitosan (5-1,4-linked glucosamine), 89
gelatin polymer, 88
keratin, 88
lignins, 89–90
pectin, 88
poly(3-hydroxybutyrate) (P3HB), 90
starch, 88–89
Natural rubber, 85
Non-polymer-based nanocomposites (NPBCs), 30
Nutrient recovery, 300
Nylon, 84–85

O

Optical absorption analysis (UV-Vis), 279–280
Optical analysis, 233–236

P

Pectin, 110
CNS composite
characterization, 115–116
diffusion coefficient, 121–123
electroactive surface area (ESA), 118–119
electrode, fabrication, 114
electron transfer rate constant, 121–123
extraction, 112–113
glucose oxidase (GOX), 119–120
glucose sensor, 123
interference tests, 124
spectral characterization, 116–117
surface electron transfer kinetics, 120–121
surface excess concentration, 121–123
synthesis, 113
experimental
apparatus, 112
reagents and chemicals, 111–112
Pharmaceutical, 11
Photocatalytic, 266
Photoluminescence (PL), 255–256
Plastics, 81–82
Polarization and depolarization current (PDC), 178
Poly(3-hydroxybutyrate) (P3HB), 90
Poly lactic acid (PLA)-stabilized
experimental
modified hummers' method, 277
RGO/CuS composite synthesis, 277
result and discussion, 277
optical absorption analysis (UV-Vis), 279–280
photo conductivity studies, 282
reaction time effect, 281–282
structural analysis (XRD), 277–279
surface morphological analysis (SEM/EDS), 280–2801
syntheses, 275, 276
Polybutadiene rubber (BR), 86
Polycyclic aromatic hydrocarbons (PAHs)
challenges, 22–23
nanocomposites
adsorption, 33
catalysis, 33
transformation, 33
nanomaterials and nanocomposites
central concept, 28–29
pollutants, 24–27
remediation
anthracenemethanol (AC), 34, 35

graphene oxide (GO), 34
green synthesis method, 34
iron oxide nanoparticles (IONPs), 34
pollutant remediation, 35
remediation technologies
bioremediation technology, 31–32
types, 29
ceramic-based nanocomposites (CBN), 30
ceramic-ceramic-based nanocomposites (CCBN), 30
metal-based nanocomposites (MBN), 30
nanocomposites, 30–31
non-polymer-based nanocomposites (NPBCs), 30
polymer-based nanocomposites (PBNC), 30
types and toxicity
high-molecular-weight (HMW), 23
low-molecular-weight (LMW), 23
Polyesters, 85
Polymer composites, 111
Polymer modifications, 2
Polymer-based nanocomposites (PBNC), 30
Polymers, 77
biopolymeric films, 79
natural polymers, 87
alginate, 89
chitosan (5-1,4-linked glucosamine), 89
gelatin polymer, 88
keratin, 88
lignins, 89–90
pectin, 88
poly(3-hydroxybutyrate) (P3HB), 90
starch, 88–89
plastic degradation, 79
synthetic polymers, 80
properties, 80–81
types
bakelite, 84, 85
butyl rubber, 86
elastomer, 85
fibers, 84
natural rubber, 85
nylon, 84–85
plastics, 81–82
polybutadiene rubber (BR), 86
polyesters, 85
polyvinyl alcohol (PVA), 82
rayon, 84
silicone rubber, 86
thermoplastic elastomer (TPE), 86–87
Polyurethane (PU), 58
Polyvinyl alcohol (PVA), 82

R

Rayon, 84
Rice husk and bagasse waste
agrowaste materials, 164
chlorofluorocarbon (CFC), 164
experimentation, 168
investigators, 164
lambda meter, 165
methodology
materials used, 166
sample preparation, 166
results and discussion, 169–171, 173
thermal insulation, 164

S

SABs
crosslinking in, 9–10
properties of, 10
Sewage sludge, 301
Silicone rubber, 86
Spectral analysis (FTIR), 230–233
Structural analysis (FTIR), 289–290
Structural analysis (XRD), 289
Structural investigation, 228–230
Structured analysis (XRD), 201–202
Struvite fertilization, 299, 300
close phosphorus cycle, 301
discussion, 304
horticultural media
influence of, 303
materials and methods, 301–302
phosphorus recovery
installation for, 301
recovery, 300
results, 302–304
sewage sludge, 301
Superabsorbents, 3
application
agriculture, 14
bio-medication, 12

cosmetics, 13
drug delivery systems, 11–12
food industries, 14
hygienic products, 13
pharmaceutical, 11
SABs, 13
tissue engineering, 12
water purification, 14
wound dressing, 12
classification
SABs, 3–5
preparation techniques, 5
chemical crosslinking, 6–7
copolymerization crosslinking, 7–9
physical crosslinking, 6
SABs
crosslinking in, 9–10
properties of, 10
swelling behavior
equilibrium percentage swelling (EPS), 11
Surface analysis (XPS), 204
Surface electron transfer kinetics, 120–121
Surface excess concentration, 121–123
Swelling behavior
equilibrium percentage swelling (EPS), 11

T

Thermoplastic elastomer (TPE), 86–87
Tissue engineering, 12
Titania-doped bio silica, 200
Transition metals (Ni/Co/Cu)
experimental detail
bio debris, 199
characterization, 201
materials, 199
silica gel, 199–200
synthesis, 200
titania-doped bio silica, 200
little band hole, 198
results and discussion
morphological analysis, 203–204
photocatalytic activity, 204–208
structured analysis (XRD), 201–202
surface analysis (XPS), 204
silica, 199
standards and guidelines, 198
waste water treatment, 198

U

UV-visible absorption spectrum, 233–236

V

Visible light cure (VLC), 212

W

Waste water treatment, 198
chemical modifications
combining chitosan with, 65
crosslinking, 64–65
grafting, 65
ION templating method, 65
Chitosan
adsorbents, 61
biopolymer blend, 62–63
hydroxyapatite (HA) composites, 63
magnetic nanocomposites, 63
nanochitosan, 62
nanoclay composites, 63–64
steps and synthesis, 61
dyes
decolorization of, 52
heavy metal ions
modified lignin-based adsorbents, 59–60
lignin, 53
adsorbents and properties, 54
covalent incorporation, 57
epoxy resins, 58
functionalization, 55
nitrogen containing functional groups, 55
oxygen containing functional groups, 55
phenol formaldehyde, 58
polymeric materials, 57
polyurethane (PU), 58
regeneration and reuse, 57
sulfur containing functional groups, 56–57
nanolignin, 58–59
Water purification, 14
Wound dressing, 12

X

X-ray diffraction method, 268–270

Z

Zinc oxide (ZnO), 212
  dental composite, 214
    diametral tensile strength (DTS), 215
    polymerization shrinkage, 214
    statistical evaluation, 215
  dynamic light scattering (DLS), 214
  materials and methods
    dental composite paste (ORMO-48), 213–214
    ORMO-48, resin mixture, 213
    zinc oxide nanostructure (ZNONPS), 213
  results and discussion
    diametral tensile strength, 218
    ORMO-48 paste, 215–216
Zinc oxide nanostructure (ZNONPS), 213
Zn-CoS/Graphene nanocomposite, 243–244
  experimental
    adsorption studies, sample preparation for, 246
    characterization details, 245
    electrode preparation, 245
    photocatalytic measurement, sample preparation for, 245–246
    synthesis of ZN-doped CoS, 244–245
  result and discussion
    electrochemical study, 249–250
    optical studies (UV-Vis spectroscopy), 246–247
    photocatalytic performance, 248–249
    SEM analysis, 248
    XRD analysis, 246

For Product Safety Concerns and Information please contact our EU
representative GPSR@taylorandfrancis.com
Taylor & Francis Verlag GmbH, Kaufingerstraße 24, 80331 München, Germany

www.ingramcontent.com/pod-product-compliance
Ingram Content Group UK Ltd.
Pitfield, Milton Keynes, MK11 3LW, UK
UKHW022209190926
475702UK00003B/277
* 9 7 8 1 7 7 4 9 1 5 3 2 5 *